Recent Advances in Analytical Techniques

(Volume 1)

Edited by

Atta-ur-Rahman, FRS
Kings College, University of Cambridge, Cambridge, UK

Sibel A. Ozkan
Department Of Basic Pharmaceutical Sciences, Faculty of Pharmacy, Ankara University, 06560 Yenimahalle/Ankara, Turkey

Rida Ahmed
TCM and Ethnomedicine Innovation & Development Laboratory, School of Pharmacy, Hunan University of Chinese Medicine, Changsha 410208, P.R. China

General:

1. Any dispute or claim arising out of or in connection with this License Agreement or the Work (including non-contractual disputes or claims) will be governed by and construed in accordance with the laws of the U.A.E. as applied in the Emirate of Dubai. Each party agrees that the courts of the Emirate of Dubai shall have exclusive jurisdiction to settle any dispute or claim arising out of or in connection with this License Agreement or the Work (including non-contractual disputes or claims).
2. Your rights under this License Agreement will automatically terminate without notice and without the need for a court order if at any point you breach any terms of this License Agreement. In no event will any delay or failure by Bentham Science Publishers in enforcing your compliance with this License Agreement constitute a waiver of any of its rights.
3. You acknowledge that you have read this License Agreement, and agree to be bound by its terms and conditions. To the extent that any other terms and conditions presented on any website of Bentham Science Publishers conflict with, or are inconsistent with, the terms and conditions set out in this License Agreement, you acknowledge that the terms and conditions set out in this License Agreement shall prevail.

Bentham Science Publishers Ltd.
Executive Suite Y - 2
PO Box 7917, Saif Zone
Sharjah, U.A.E.
Email: subscriptions@benthamscience.org

BENTHAM SCIENCE

CONTENTS

Xunbin Wei, Yuanzhen Suo, Pengfei Hai and *Xi Zhu*

PREFACE

Recent advances in analytical techniques have led to the development of effective strategies for rapid separation of complex mixtures with high resolution and their identification. The present 1st volume of the book series *"Recent Advances in Analytical Techniques"* is a compilation of six outstanding reviews, written by the leading researchers in the field. Each review is focused on an important aspect of analytical techniques including sample preparation, development of electronic sensors, *in vivo* monitoring of tumor cells, advances in glow discharge spectroscopy, characterization of volatiles using gas chromatography and application of nanomaterials in biomedical applications. Since many contaminants present in crude samples can interfere during the analysis, it is critically important to find new sample preparation strategies which are efficient and practical. Namera and Saito have discussed conventional and new techniques for sample preparation to analyze biological, food and environmental substances in chapter 1.

In chapter 2, Pereira *et al.* present the use of new materials used in the development of electrochemical sensors. The application of new types of electrochemical sensors in analytical chemistry has increased recently due to their numerous advantages such as high sensitivity, selectivity, stability, low cost and use in simple instrumentation. They can be applied specifically in the analysis of clinical and environmental samples to provide higher reactivity and selectivity. Promising organic and inorganic materials have been used as electrochemical sensors such as composites based on graphene and carbon nanotubes, Molecularly Imprinted Polymers (MIP) and/or Ionically Imprinted Polymers (IIP). Some inorganic compounds such as nanoparticles of noble metals and metal oxides have also been applied as modifiers of electrodes to enhance their electrochemical properties or to increase the surface area of electrodes.

In vivo flow cytometry is an emerging tool used to monitor circulating tumor cells (CTCs) *in vivo*. It is a non-invasive method, and allows monitoring of CTCs in their native biological environment. Wei *et al.* have discussed the basic principles of fluorescence based and photoacoustic based *in vivo* flow cytometry and reviewed a number of studies on cancer therapy using *in vivo* flow cytometry. Potential applications and disadvantages of *in vivo* flow cytometry in cancer therapy are also discussed.

Glow discharges (GDs) coupled to optical emission or to mass spectrometry have been widely investigated during the last three decades for a variety of direct solid analytical applications. The advances associated with GDs include low matrix effects, high sensitivity and resolution, and multi-elemental analysis. Lobo and Pereiro have discussed the basic principles of the use of GDs along with recent instrumental advances and applications for optical emission and mass spectrometry. GD time-of-flight mass spectrometry has been discussed in detail since it is promising in terms of high depth resolution, fast acquisition rates and time-gated detection. It has also shown a great potential in obtaining elemental and molecular information as well as in the characterization of advanced materials such as multilayers, thin film solar cells and polymers.

In the fifth chapter, Martin *et al.*, have reviewed recent technological advancements in gas chromatography (GC) for the detection and identification of volatile constituents of beer. The analysis of beer volatile fraction is challenging due to the presence of CO_2, and the diversity of chemical structures found in it with different polarities, volatilities, and a wide concentration range. It requires effective extraction techniques to recover the analytes of interest, while minimizing the production of artefacts during the extraction process. The advantages and disadvantages of different extraction techniques have been discussed along with significant improvements in the chromatographs (namely the multidimensional ones), detection systems, columns technology and algorithms that contribute to the reduction of analysis time, making the methods more expeditious and user-friendly.

Nanomaterials have been investigated for a number of biomedical applications such as drug delivery, biosensors, tissue engineering, and bio-imaging. During the past few decades, several life threatening diseases such as cancers and some common bacterial infections have been treated using photodynamic therapy (PDT). It is based on the photochemical reactions between light and tumour tissues through photosensitizing agents. Riaz *et al.* have reviewed the present applications and future prospects of the various materials developed as photodynamic therapeutic agents in the last chapter.

We are deeply grateful to all the authors for their excellent contributions which should be of wide interest to the readers. We are also grateful to Mr. Mahmood Alam (Director Publications) and his excellent team comprising Mr. Shehzad Naqvi (Senior Manager Publications) and Mr. Omer Shafi (Assistant Manager Publications) for their untiring efforts.

Atta-ur-Rahman, *FRS*
Kings College, University of Cambridge,
Cambridge,
UK

Sibel A. Ozkan
Department Of Basic Pharmaceutical Sciences,
Faculty of Pharmacy, Ankara University,
06560 Yenimahalle/Ankara,
Turkey

Rida Ahmed
TCM and Ethnomedicine Innovation & Development Laboratory, School of Pharmacy,
Hunan University of Chinese Medicine,
Changsha 410208, P.R. China

List of Contributors

Adelaide Almeida   Departamento de Biologia & CESAM, Universidade de Aveiro, Campus Universitário Santiago, 3810-193 Aveiro, Portugal

Akira Namera   Department of Forensic Medicine, Institute of Biomedical and Health Sciences, Hiroshima University, Hiroshima, Japan

Arnaldo C. Pereira   Departamento de Ciências Naturais, Universidade Federal de São João del Rei, 36301-160, São João del Rei, MG, Brasil

Cátia Martins   Departamento de Química & QOPNA, Universidade de Aveiro, Campus Universitário Santiago, 3810-193 Aveiro, Portugal
Departamento de Biologia & CESAM, Universidade de Aveiro, Campus Universitário Santiago, 3810-193 Aveiro, Portugal

Daniela N. Silva   Departamento de Ciências Naturais, Universidade Federal de São João del Rei, 36301-160, São João del Rei, MG, Brasil

Débora A.R. Moreira   Departamento de Ciências Naturais, Universidade Federal de São João del Rei, 36301-160, São João del Rei, MG, Brasil

Juliana F. Giarola   Instituto de Química, Universidade Estadual de Campinas, 13083-970, Campinas – SP, Brasil

Juraj Bujdak   Comenius University in Bratislava, Faculty of Natural Sciences, Department of Physical and Theoretical Chemistry, 842 15 Bratislava, Slovakia
Institute of Inorganic Chemistry Slovak Academy of Science, Bratislava, Slovakia

Lara Lobo   Department of Physical and Analytical Chemistry, Faculty of Chemistry, University of Oviedo, 33006 Oviedo, Spain

Pengfei Hai   Department of Biomedical Engineering, Washington University in St. Louis, St. Louis, Missouri, USA

Rosario Pereiro   Department of Physical and Analytical Chemistry, Faculty of Chemistry, University of Oviedo, 33006 Oviedo, Spain

Sílvia M. Rocha   Departamento de Química & QOPNA, Universidade de Aveiro, Campus Universitário Santiago, 3810-193 Aveiro, Portugal

S.M. Ashraf   Materials Research Laboratory, Department of Chemistry, Jamia Millia Islamia, New Delhi-110025, India

Takeshi Saito   Department of Emergency and Critical Care Medicine, Tokai University School of Medicine, Kanagawa, Japan

Ufana Riaz   Materials Research Laboratory, Department of Chemistry, Jamia Millia Islamia, New Delhi-110025, India

Xi Zhu   Med-X Research Institute and School of Biomedical Engineering, Shanghai Jiao Tong University, Shanghai, China

Xunbin Wei   Med-X Research Institute and School of Biomedical Engineering, Shanghai Jiao Tong University, Shanghai, China

Yuanzhen Suo   Med-X Research Institute and School of Biomedical Engineering, Shanghai Jiao Tong University, Shanghai, China

CHAPTER 1

Recent Advances in Unique Sample Preparation Techniques for Biological and Environmental Analysis

Akira Namera[1,*] and **Takeshi Saito**[2]

[1] *Deaprtment of Forensic Medicine, Graduate School of Biomedical and Health Sciences, Hiroshima University, Hiroshima, Japan*

[2] *Department of Emergency and Critical Care Medicine, Tokai University School of Medicine, Kanagawa, Japan*

Abstract: Remarkable techniques for the separation and detection of small quantities of analytes have been developed in recent years. However, it is still difficult to directly analyze species of interest in complex matrices. Although some methods have been reported for direct injection into an analytical instrument, removal of interfering substances during sample preparation is an important step in the analytical process. This procedure is usually tedious and time consuming. To reduce the tedium of this task and the time required for sample preparation, many unique extraction techniques have been introduced and applied to the analysis of substances in environmental, food, and biological samples. This chapter describes useful sample preparation techniques, including conventional and newly developed ones, for determining analytes of interest in biological, environmental, and food sources.

Keywords: Headspace extraction, Liquid-liquid extraction, Microextraction, Protein precipitation, Sample preparation, Solid-phase extraction.

INTRODUCTION

Separation and detection techniques that allow small amounts of analytes to be separated and detected, especially chromatography and mass spectrometry, have experienced remarkable development in last decade. A particular focus in metabolomics and proteomics is the identification and monitoring of low-molecular-weight metabolites and high-molecular-weight proteins, which may reflect toxic or medical conditions. High resolution mass spectrometry is needed to routinely accomplish the more challenging objectives in these fields. High

* **Corresponding author Akira Namera:** Department of Forensic Medicine, Institute of Biomedical and Health Sciences, Hiroshima University, Hiroshima, Japan; Tel: +81-82-257-5172; Fax: +81-82-257-5174; E-mail: namera@hiroshima-u.ac.jp

Atta-ur-Rahman, Sibel A. Ozkan & Rida Ahmed (Eds.)

resolution mass spectrometry is also finding an increasing application in environmental and food science, because of the risk posed by unexpected toxic substances in food and the environment. However, it is difficult to inject samples directly into these instruments for the identification of toxic components, because proteins, lipids, and other contaminants in biological, environmental, and food samples solidify and clog in the instrument, and hinder chromatographic separation and analytic ionization. Also, the concentrations of analytes are usually very low in comparison to that of interfering substances. Therefore, sample preparation remains a very important step in obtaining accurate results.

Analytical procedures usually consist of sampling, sample preparation, column separation, detection, and data analysis. Each step is crucial to success, but sample preparation requires more than 70% of the total analysis time and is the so-called "bottleneck" of the process [1 - 3]. Although new extraction techniques have been developed for extraction and enrichment of analytes from sample matrices, liquid–liquid extraction (LLE) is still widely used for sample preparation due to its high efficiency, ease of operation, and low cost. Solid-phase extraction (SPE) has been developed to overcome the limitations of LLE and is currently one of the most widespread extraction methods for pretreatment of environmental, food, and biological samples. SPE is simpler, more convenient, and easier to automate than LLE. However, LLE requires large amounts of hazardous solvents, and SPE requires a quantity of solvent two or three times that of the sorbent bed volume to ensure high recovery of analytes. Moreover, the extracted and eluted solutions must be concentrated by evaporation in both cases. New techniques have been developed to reduce the size of previously used devices and to enable the injection of all analytes into a single piece of equipment such as a gas chromatograph (GC) or liquid chromatograph (LC). These conventional LLE and SPE techniques, which avoid complicated, time-consuming steps and eliminate the use or hazardous solvents, fall under the umbrella of "green chemistry".

In this chapter, sample preparation techniques including the conventional and newly developed ones are introduced for the determination of analytes of interest in biological, environmental, and food samples as shown in Fig. (**1**).

HEADSPACE EXTRACTION

Headspace extraction is one of the most important sampling methods for volatile analytes, which mainly include aromas, odors, and solvents in water, food, and biological materials [4 - 6]. Because non-volatiles are not transferred to the gas phase (headspace), a clean extraction can be obtained from complex matrices. Static or dynamic headspace extraction is used for volatile analytes. In the static headspace method, which is also called the equilibrium method, the sample vial is heated so that the analyte in the vial reaches equilibrium with the gaseous phase (headspace). After the analyte reaches equilibrium, a portion of the headspace in the sample vial is introduced manually or automatically into a GC while equilibrium is maintained. In the dynamic headspace method, the headspace is introduced continuously into a GC by constantly passing a purge gas through the headspace or sample. Although the headspace is introduced directly into a GC in this method, the analyte occasionally must be concentrated by trapping on some adsorbents because the analyte exists in a large volume of the purge gas. The method that uses a trap adsorbent is called the purge and trap method. Unlike in the static method, the analyte does not necessarily need to reach equilibrium in the vial in the dynamic headspace method.

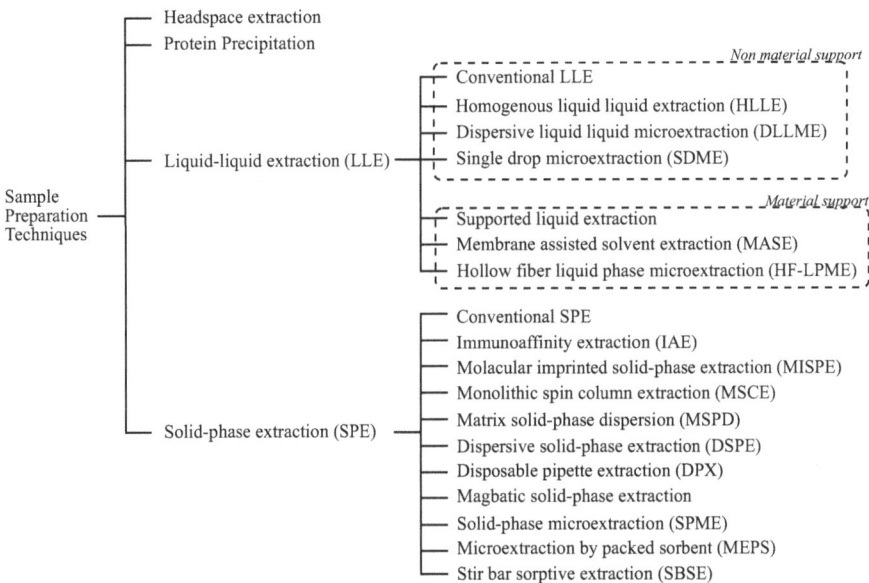

Fig. (1). Schematic classification of sample preparation techniques for extraction and enrichment of analyte.

The general headspace extraction procedure is schematically shown in Fig. (**2**). As seen, the sample is placed in a vial that is sealed with a septum. After equilibration by heating, the headspace is directly injected to a GC.

The analyte partitioned between the liquid sample and the headspace according to the following equation [4],

$$K = \frac{C_{spl}}{C_{gas}} \tag{1}$$

where K is the partition coefficient (or distribution ratio), and C_{spl} and C_{gas} are the analyte concentrations in the liquid phase and headspace, respectively. From mass balance considerations, the total (initial) amount of analyte in the vial is given by, where C_{int} is the initial analyte concentration in the sample, V_{spl} is the liquid phase volume, and V_{gas} is the headspace volume.

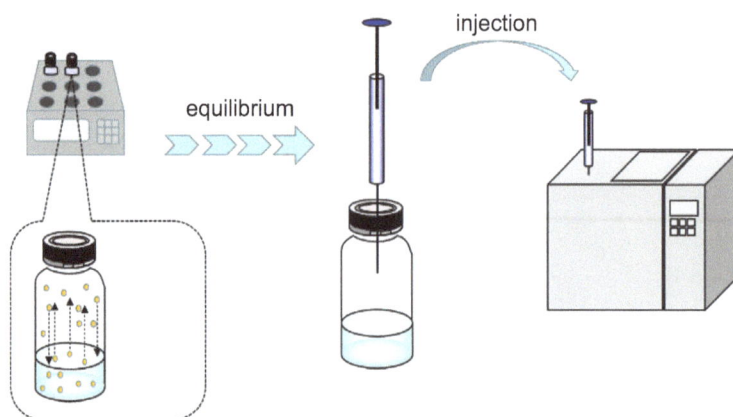

Fig. (2). Scheme of general extraction procedure in headspace sampling.

$$C_{int}V_{spl} = C_{gas}V_{gas} + C_{spl}V_{spl} \qquad (2)$$

Combination of equations (1) and (2) results in the following expression.

$$C_{gas} = \frac{C_{int}}{K + \dfrac{V_{gas}}{V_{spl}}} \qquad (3)$$

As illustrated by equation (3), it is possible to determine the concentration of an analyte in a sample by analyzing its concentration in the headspace after equilibrium, because C_{gas} is correlated with C_{int}. Theoretically, a highly volatile analyte, which has a small K value, will transfer more completely into the headspace giving a high headspace concentration. The value of K can be altered by changing the temperature at which the vial is equilibrated or by changing the composition of the sample matrix. In the case of ethanol, K decreases from 1355 to 216 as the temperature of the vial is increased from 40 to 80 °C [4]. As a result, lower concentrations can be detected by heating. However, water vapor can interfere with analyte separation and detection, if the temperature of the vial is increased too much.

Typical reconstructed selected ion monitoring chromatograms are shown for determination of volatile organic compounds in ground waters in Fig. (**3**) [7]. When the limit of detections (LOD) of the target analytes in water were calculated by a signal to noise ratio of 3, the LODs were from 1 to 100 ng/L. This technique is suitable for simultaneous trace determination of all target compounds that permit an environmental survey of both parent and degradation products.

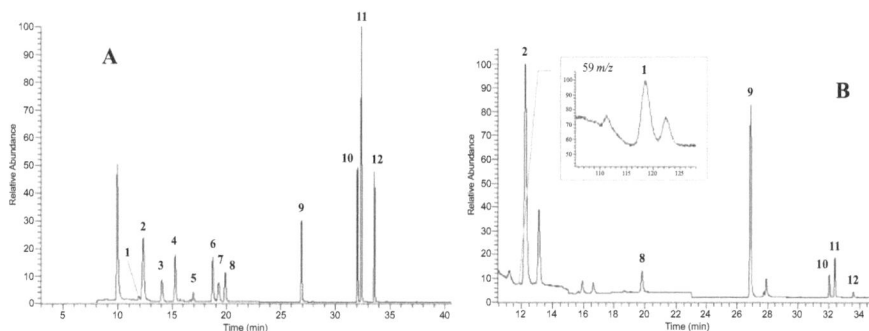

Fig. (3). Total ion chromatogram of volatile compounds in SIM mode for a 10 mg/L standard (**A**) and detail of groundwater sample (**B**) by an extraction of a static headspace method.
Peak identification: 1= *tert*-butyl alcohol, 2= *tert*-butyl ether-d_3 + *tert*-butyl ether, 3= di-isopropyl ether, 4= ethyl *tert*-butyl ether, 5= *tert*-butyl formate, 6=benzene, 7= *tert*-amyl methyl ether, 8=fluorobenzene, 9=toluene, 3 10=ethylbenzene, 11=*m*+*p*-xylene and 12=*o*-xylene. (From [7] with permission of Elsevier).

In Fig. (**4**), good chromatographic separation was achieved for all target compounds (acetaldehyde, acetone, methanol, ethanol, n-propanol (internal standard) and acetic acid) spiked into blood with retention times of 1.9, 2.5, 4.35, 5.62 and 6.10 min, respectively [8]. Ethanol and acetone was also detected from the plasma in the patient who received ethanol-containing medication.

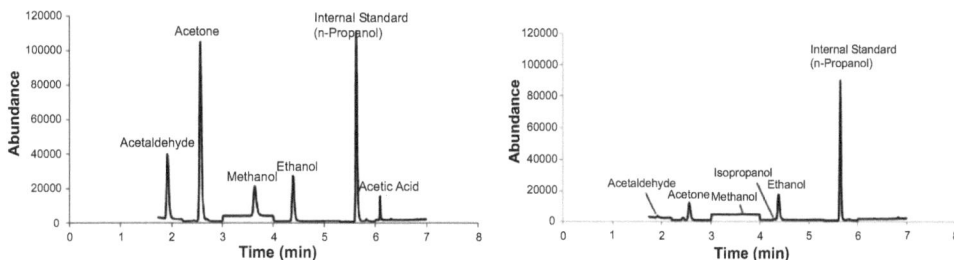

Fig. (4). GC-MS analysis of plasma from healthy volunteer spiked to give a final concentration of 100 mg/L, and from a neonatal patient before receiving any ethanol containing medication. (From [8] with permission of Springer).

Although headspace sampling is very simple for extraction of volatile compounds in sample, a disadvantage of static headspace sampling is that the limit of detection is relatively high. So, it is difficult to detect an analyte at low concentration by static headspace sampling. In samples containing analytes of different volatilities, the concentration of the more volatile analyte is enriched in the headspace. Dynamic headspace extraction can improve this shortcoming. However, a drawback of dynamic headspace sampling is the requirement of complex instrumentation including a purge gas device, purge gas, and a sample vial, sorbent, or cryogenic trapping unit equipped with a heating device.

PROTEIN PRECIPITATION

Protein precipitation is a simple pretreatment method that is used in proteomic and metabolomic studies. The mechanism is based on the decrease in aqueous solubility caused by changing the charge of the protein or addition of a precipitating reagent also named precipitants. The reagents used include acids, salts, metal ions, and organic solvents [9]. Suitable precipitants were evaluated by monitoring the amount of protein remaining in solution after precipitation [10]. In this study, trichloroacetic acid (10%, w/v), metaphosphoric acid (5%, w/v), zinc sulfate (10%, w/v), sodium chloride (0.5 M), acetonitrile, ethanol, methanol, and ammonium sulfate (saturated) were surveyed. Each precipitant was added to human plasma at ratios of 0.5:1 to 4:1. Solutions were vortexed and centrifuged at 3000 rpm. The absorbance of the resulting supernatant was measured at 280 nm. The plasma protein remaining in the supernatant was compared to that in non-precipitated plasma. Results are summarized in Table **1**. Precipitants effective in protein removal were zinc sulfate, acetonitrile, and trichloroacetic acid at precipitant to plasma volume ratios of 2:1 or greater.

Table 1. Comparison of protein precipitation efficiency of precipitants in different lots of human plasma [10].

Precipitants		% Protein precipitation efficiency						
		Ratio of precipitant to plasma						
		0.5:1	1:1	1.5:1	2:1	2.5:1	3:1	4:1
Acids	Trichloroacetic acid	91.4	91.8	91.5	91.0	91.2	91.3	91.4
	m-phosphoric acid	89.4	90.5	90.3	90.2	90.7	90.5	90.0
Metal ions	Zinc sulphate	89.2	96.8	96.8	99.0	99.0	99.0	>99.9
Organic	Acetonitrile	3.6	88.7	91.6	92.1	93.2	93.5	94.9
	Ethanol	0.1	78.2	87.2	88.1	89.8	91.8	92.0
	Methanol	13.4	63.8	88.2	89.7	90.0	91.1	91.5
Salts	Ammonium sulphate	24.8	50.1	94.0	84.2	90.4	90.4	89.0

% Protein precipitation efficiency=([total plasma protein - protein remaining in supernatant] / total plasma protein) X 100.

Although protein precipitation is simple and handy, many amounts of phospholipids are remained in the aliquot after protein precipitation as shown in Fig. (**5**) [11]. Moreover, the ionization of the analytes with electrospray ionization of LC-MS is sometimes affected by the remained phospholipids in the aliquot as shown in Fig. (**6**) [12]. In that case, more suitable preparation methods are required for the extraction of the analytes. The choice is important for obtaining accurate results, because the combination of precipitant and mobile phase used in LC-MS has a large influence on the matrix effect [10].

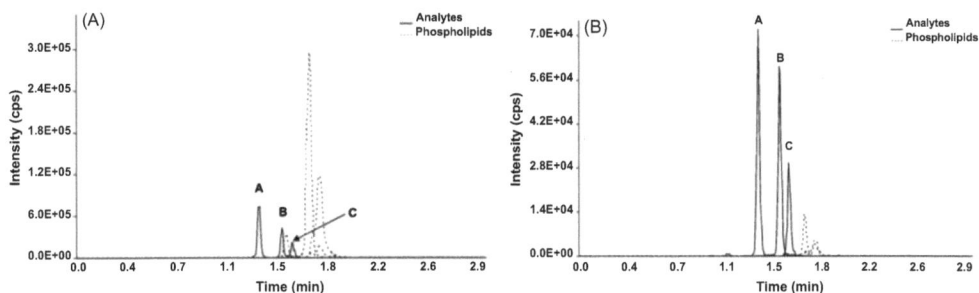

Fig. (5). Mass chromatograms of a spiked human plasma sample after PPT procedure (**A**) and the same human plasma sample spiked with the three model analytes at 5 ng/ml after HybridSPE-Precipitation (**B**). (From [11] with permission of Elsevier).

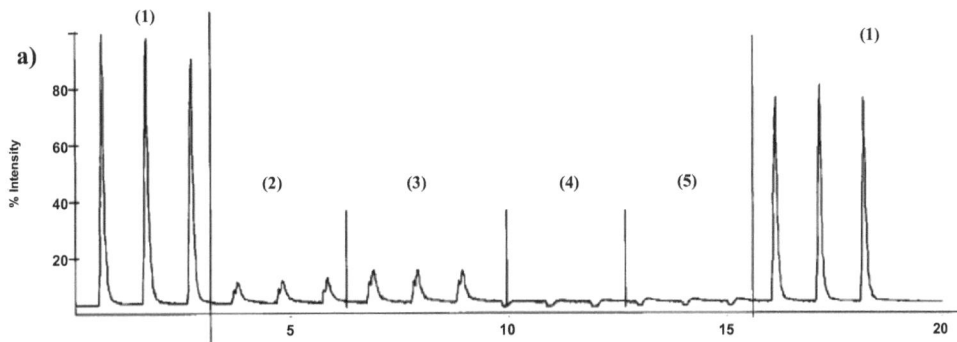

Fig. (6). Comparison of the effect of sample preparation method on the FIA response of phenacetin. SRM extracted ion chromatogram of triplicate FIA of phenacetin spiked into prepared plasma samples obtained by different sample preparation methods showing the response differences of the tested sample preparation methods compared to mobile phase. (From [12] with permission of John Wiley & Sons).
(1) Mobile phase spiked with phenacetin, (2) Oasis SPE plasma extract spiked with phenacetin, (3) Filtered Oasis SPE plasma extract spiked with phenacetin, (4) Plasma protein precipitation sample spiked with phenacetin, (5) Filtered plasma protein precipitation sample spiked with phenacetin Panel.

LIQUID BASED EXTRACTION

Liquid-Liquid Extraction

Liquid-liquid extraction (LLE) is one of the most popular sample preparation techniques for extraction and purification of analytes of interest in complex pharmaceutical, environmental, and food matrices. In LLE, analyte partitioning is driven by the difference in solubility between two immiscible phases; one is the aqueous phase, and the other is the organic phase (Fig. **7**).

Some analytes are classified as either ionic or non-ionic compounds. Non-ionic compounds are smoothly transferred to the organic layer depending on their distribution coefficients (*K*) or partition coefficients (*P*) and are easily extracted with an organic solvent. However, ionic compounds such as basic or acidic compounds exist in a mixture of ionic and non-ionic forms depending on the pH value of the solution. When extracting with an organic solvent, ionic analytes can be converted to the non-ionic form by changing the pH of the solution. The pH of the solution is usually higher/lower than the p*K*a of each analyte.

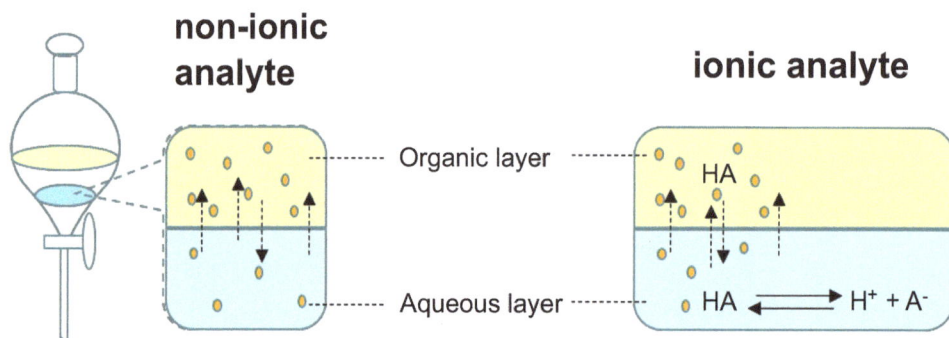

Fig. (7). Partitioning equilibria in liquid-liquid extraction.

At equilibrium, the analyte is distributed between the two phases according to the value of its distribution coefficient (*K*) or partition coefficient (*P*). The value usually is expressed logarithmically as *log K* or *log P*. The partition coefficient between water and 1-octanol, P_{ow}, typically is employed, because water (or a buffer solution) and 1-octanol are used in pharmaceutical and environmental sciences. *Log K* and *log P*$_{ow}$ also have been calculated by ChemDraw® and HSPiP® (Hansen Solubility Parameters in Practice) software.

The distribution coefficient, expressed here as *K*, is defined as the ratio of solute concentrations between two solvents as shown in the following equation [13],

$$K = \frac{C_{org}}{C_{aq}} = \frac{M_{org}V_{org}}{M_{aq}V_{aq}} = \frac{M_{org}V_{org}}{(M_{int}-M_{org})V_{aq}} \tag{4}$$

where C_{org} is the concentration of solute extracted into the organic phase at equilibrium, and C_{aq} is the concentration remaining in water. M_{int} is the amount of sample before extraction, M_{org} is the amount extracted into the organic phase at equilibrium, M_{aq} is the amount remaining in water, V_{org} is the volume of organic solvent, and V_{aq} is the volume of water. Equation (4) may be rearranged to give

$$M_{org} = M_{int} \left(\frac{KV_{org}}{V_{aq} + KV_{org}} \right) \tag{5}$$

Recovery following the first extraction is given by

$$R(\%) = \frac{M_{org}}{M_{int}} \times 100 = \left(\frac{KV_{org}}{V_{aq} + KV_{org}} \right) \times 100 \tag{6}$$

After the first extraction, the amount of analyte left in water is

$$M_{aq} = M_{int} - M_{org} = M_{int} - M_{int} \left(\frac{KV_{org}}{V_{aq} + KV_{org}} \right)$$
$$= M_{int} \left(\frac{V_{aq}}{V_{aq} + KV_{org}} \right) \tag{7}$$

After the second extraction, the amount remaining in water is equals

$$M_{aq(2)} = M_{aq} - M_{org(2)}$$
$$= M_{aq} - M_{aq} \left(\frac{KV_{org}}{V_{aq} + KV_{org}} \right)$$
$$= M_{int} \left(\frac{V_{aq}}{V_{aq} + KV_{org}} \right)^2 \tag{8}$$

Therefore, after n extractions, the amount of analyte remaining in water is,

$$M_{aq(n)} = M_{int} \left(\frac{V_{aq}}{V_{aq} + KV_{org}} \right)^n \tag{9}$$

and the recovery after n extractions is

$$R_{(n)}(\%) = \frac{M_{int} - M_{aq(n)}}{M_{int}} \times 100$$
$$= 1 - \left(\frac{V_{aq}}{V_{aq} + KV_{org}} \right)^n \times 100 \tag{10}$$

The following points are evident from the above equations: (1) the amount of analyte extracted depends on the value of K, (2) a larger volume of extracting solvent is more effective in a single extraction, (3) many portions of the extracting solvent with smaller volumes are more effective than one portion with large volume, and (4) recovery does not depend on the concentration of analyte in the sample.

When it is unclear which solvent to use for an extraction, a solvent miscibility chart (Table **2**) is helpful in selecting an immiscible solvent combination [13, 14]. Hexane, ethyl acetate, and dichloromethane are commonly used as extraction solvents. An experimental study has described the relationship between analyte recovery and *log P*. An analyte for which *log P* is greater than 3 can be extracted to greater than 70% by one hexane extraction [15]. An analyte for which *log P* is greater than 1 can be extracted to greater than 70% by one dichloromethane extraction. Methyl *t*-butyl ether is suitable for extraction of hydrophilic analytes. However, its selectivity is high, which makes it unsuitable for simultaneous extractions in screening analyses [16].

Table 2. Properties of solvents.

Solvent	Snyder Polarity Index	BP (°C)	Solubility in water (% w/w)
n-Hexane	0	68.9	0.001
Cyclohexane	0	80.7	0.01
Carbon tetrachloride	1.7	76.5	0.08
i-Propyl ether	2.2	68.3	
Toluene	2.3	101.6	0.051
Diethyl ether	2.8	35	6.89
Methyl t-butyl ether (MTBE)	2.9	55.2	4.8
Benzene	3	80.1	0.18
Chloroform	3.4-4.4	61.2	0.815
Dichloromethane	3.4	40	1.6
Tetrahydrofuran	4.2	66	100
Ethylene dichloride	3.7	83.5	0.81
Methyl ethyl ketone (MEK)	4.5	80	24
Acetone	5.4	56.3	100
Acetonitrile	6.2	81.6	100
Ethyl acetate	4.3	77.1	8.7
Dimethyl sulfoxide	6.5	189	100
2-Propanol	4.3	82.4-117.7	100
Ethanol	5.2	78.3	100
Methanol	6.6	64.7	100
Acetic acid	6.2	117.9	100
Water	9	100	

☐ : immiscible with water

Emulsification, which is a combination of small organic and water droplets from two immiscible phases, is a bottleneck in LLE. An emulsion may form upon vigorous shaking of the combined phases. It is hard to break an emulsion, and a long time is needed to establish phase separation. Some approaches taken to break emulsions include addition of sodium chloride to the aqueous phase, centrifugation, cooling, and filtration of both phases. An effective general approach for breaking emulsions has not been found.

Clean extraction can be achieved by LLE. However, lipids and other endogenous substances, which affect chromatographic separation and mass spectrometric detection, cannot be completely removed by LLE. To remove these contaminants, an acetonitrile–hexane partition is commonly used to exclude lipids from sample extracts. Analytes are partitioned into the acetonitrile phase, and lipids are partitioned into the hexane phase. However, analytes of low polarity also transfer into the hexane phase, and recovery is decreased [17]. These approaches also shorten the lifetime of columns and chromatographic systems making further clean-up of extracts mandatory [18 - 20].

Supported Liquid Extraction

Conventional LLE uses large volumes of solvents that often are hazardous and can pollute the environment. Instances of emulsification also make the LLE process tedious and time consuming. Supported liquid extraction, which also is known as solid-supported liquid extraction or supported liquid-liquid extraction, is an established method that can replace conventional LLE for analyte extraction. The method is similar to solid phase extraction, because a solid material packed into a column or cartridge is used in the extraction. However, the principle of analyte extraction is same as LLE. In general, a powder or granule of high porous diatomaceous earth is used as the support material. Commercially available examples include Extrelut® (Merck), Chem Elute (Agilent Technologies), InertSep® K-solute (GL Sciences), and ISOLUTE® SLE+ (Biotage). Many examples of analyte extraction and clean-up have been reported in forensic analysis [21 - 24]. An aqueous sample is adsorbed on the solid material. After 10 to 15 min, a thin aqueous layer forms on the surface of the solid. The extracting solvent is then added and allowed to percolate by gravity through the column or cartridge. As the solvent contacts the thin aqueous layer on the solid surface, the analyte is transferred from the aqueous to organic phase by the same principle as LLE. Because no vigorous shaking is required, emulsions do not form. Thus, surfactants and fatty materials can be studied by this technique. The extraction of drugs in whole blood also can be achieved without formation of emulsions. Analyte recovery is greater than in conventional LLE, because of more effective contact between the aqueous layer and organic solvent.

SMALL SCALE LIQUID BASED EXTRACTION

Conventional LLE utilizes large volumes of solvent, and the process can be tedious and time consuming. The organic solvent used for extraction also must be evaporated to concentrate the analyte. To avoid these steps, some special techniques have been developed to reduce the time involved and volume of solvent required.

Homogeneous Liquid–Liquid Extraction (HLLE)

HLLE is a simple and powerful preconcentration technique that is based on the high solubility of an organic solvent in water at high temperature. A uniform state of solution characteristically forms in the process. After this homogeneous solution is cooled and centrifuged, a small water–immiscible sediment phase is obtained and separated without vigorous mechanical shaking (Fig. **8**). HLLE reduces the extraction time, process cost, and consumption of and exposure to organic solvent [25].

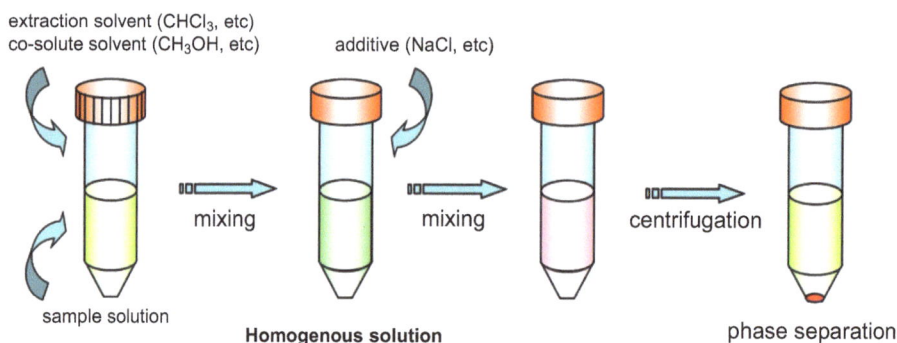

Fig. (8). Scheme of general extraction procedures in homogeneous liquid-liquid extraction.

Although conventional solvent extraction employs two immiscible solvents, HLLE extracts the solute from a homogeneous solution into a very small volume of sediment formed by phase separation. The theory of HLLE is similar to that of LLE. In HLLE, there is no true interface between water and the extracting solvent. In other words, the surface area between the aqueous and organic phases is infinitely large. Therefore, transfer of analytes from the aqueous to organic phase is fast, equilibrium is established quickly, and extraction time is short. The procedure is simple and requires only a change of temperature. A ternary component solvent system or a perfluorinated surfactant system are two common modes of homogeneous liquid–liquid extraction. Recently, two-phase separation has been accomplished by addition of salt (salting out effect), a change of pH, and a change of temperature. Homogeneous liquid–liquid extraction has been successfully applied to the extraction of organic and inorganic analytes [26 - 30].

The following factors should be considered in obtaining optimal extraction conditions: extraction solvent (type and amount), co-solute solvent (type and amount), additives (buffer for pH control), ionic strength, and extraction time. The mixing ratio of extracting solvents is important for optimizing analyte recovery. Chloroform typically is used as the extracting solvent, because it is only slightly soluble in water and has greater density than aqueous solutions. Moreover, it readily forms sediments at the bottom of the conical tube. The co-solute solvent is selected on the basis of its miscibility with the organic and aqueous phases. Acetonitrile and methanol have been examined as co-solute solvents for dissolving chloroform in aqueous solution. A homogeneous solution of chloroform in water in the presence of methanol is created by solvation of chloroform by methanol molecules. The ability of methanol to do so is decreased in the presence of NaCl. Hence, chloroform is separated as an immiscible phase from the aqueous solution. The NaCl concentration has been varied from 1 to 25% to study effect of salt concentration on extraction efficiency.

Dispersive Liquid–Liquid Microextraction (DLLME)

DLLME, which is based on a ternary component solvent system as in HLLE and cloud point extraction, was proposed by Assadi and co-workers in 2006 [31]. In this method, an appropriate mixture of extracting and dispersing solvents is forcefully injected into an aqueous sample by syringe causing a cloudy solution to form. The analyte is extracted into fine droplets of the extracting solvent. After extraction, phase separation is accomplished by centrifugation, and the analyte enriched in the sediment phase is quantitated by an appropriate instrumental method (Fig. **9**). The advantages of DLLME include ease of operation, low sample volume, low cost, and high recovery. Recently, many patterns of DLLME have been proposed, and they have been compared using low-density solvents or ideal containers [32]. However, it is difficult to find a suitable combination of extracting and dispersing solvents to obtain accurate results. The principles of the technique and its application in the separation, pre-concentration, and determination of organic and inorganic compounds in biological samples have been reviewed recently [33 - 47].

In DLLME, the enrichment factor (EF) is used to evaluate mass transfer of the analyte. The analyte concentrations in the initial and sedimented phases equal C_0 and C_{sed}, respectively. EF is defined by the following formula

$$EF = \frac{C_{sed}}{C_o} \qquad (11)$$

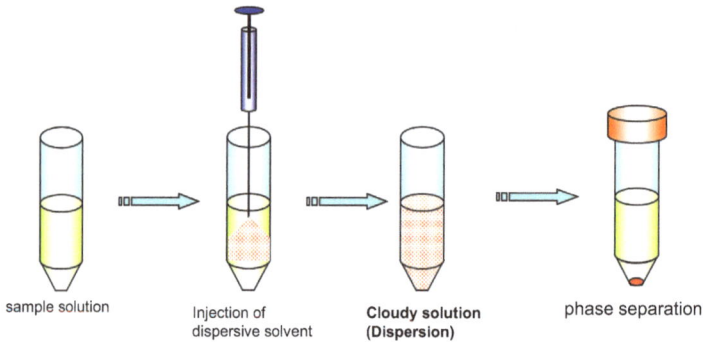

sample solution   Injection of   **Cloudy solution**   phase separation
                        dispersive solvent   **(Dispersion)**

Fig. (9). Scheme of general extraction procedures in dispersive liquid–liquid microextraction.

The C_{sed} of polyaromatic hydrocarbons (PAHs) ranges from 0.5 to 2.5 mg/L in C_2Cl_4.

Extraction recovery (ER) is defined as the percentage of total analyte (n_0) extracted into the sedimented phase (n_{sed}),

$$ER(\%) = \frac{n_{sed}}{n_0} \times 100 = \frac{C_{sed} \times V_{sed}}{C_0 \times V_{aq}} \times 100$$
$$= \frac{V_{sed}}{V_{aq}} \times EF \times 100 \tag{12}$$

where V_{sed} and V_{aq} are the sedimented phase and sample solution volumes, respectively.

The identity and proportions of extracting and dispersing solvents are important in DLLME. Extracting solvents should have a density greater than water, be capable of extracting target analytes, and exhibit good chromatographic behavior. Carbon disulfide, carbon tetrachloride, and tetrachloroethylene, which have densities of 1.26, 1.59, and 1.62 mg/L, respectively, have been used as extracting solvents. Miscibility of the dispersing solvent with organic and aqueous phases also is a consideration. In the original article [31], several extracting and dispersing solvents were examined in variable proportions. Thypical chromatograms of PAHs and BTEX in water are shown in Fig. (**10**).

Extraction time, which was examined over a 0–60 min range under constant experimental conditions, had no influence on extraction efficiency. The surface area between the extracting solvent and aqueous phase was infinitely large. Transfer of analytes from the aqueous to extracting phase was fast, and equilibrium was achieved rapidly resulting in short extraction times.

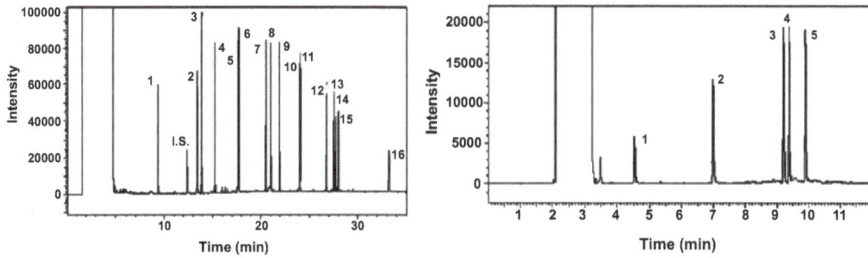

Fig. (10). DLLME–GC-FID analysis of PAHs and BTEX in water sample.
Peak identification: Left chromatogram; 1 = naphthalene, 2 = acenaphthylene, 3 = acenaphthene, 4 = fluorene, 5 = phenanthrene, 6 =anthracene, 7 = fluoranthene, 8 = pyrene, 9 = benzofluorene, 10 = benzo[a]anthracene, 11 = chrysene, 12 = benzo[e]acephenanthylene, 13 = benzo[e]pyrene, 14 = benzo[a]pyrene, 15 = perylene, 16 = benzo[ghi]perylene, I.S. = biphenyl. Right chromatogram; 1 = benzene, 2 = toluene, 3 = ethyl benzene, 4 = m- or p-xylene, 5 = o-xylene. (From [31] with permission of Elsevier).

Single Drop Microextraction (SDME)

To further reduce extraction volume, SDME was introduced independently by Liu and Dasgupta [48] and Jeannot and Cantwell in 1996 [49]. In SDME, a microsyringe is inserted into an aqueous sample with a small drop of water-immiscible organic solvent held at the tip of the needle (Fig. **11**).

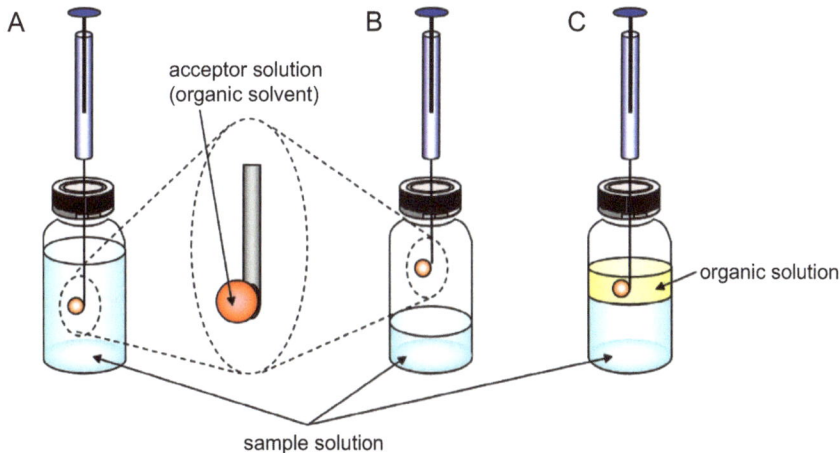

Fig. (11). Different modes of single drop microextraction.

After exposure for a prescribed time, the microdrop is retracted into the microsyringe, and the contents are injected into a gas chromatograph. As in LLE, the distribution coefficient is expressed as

$$K = \frac{C_{org}}{C_{aq}} \tag{13}$$

where C_{org} and C_{aq} are the equilibrium concentrations of analyte in organic solvent and water, respectively. From mass balance, the total analyte in the sample vial is given by,

$$C_i V_{aq} = C_{org} V_{org} + C_{aq} V_{aq} \qquad (14)$$

where C_i is the initial concentration of analyte in the sample, V_{aq} is the volume of the aqueous phase, and V_{org} is the volume of organic solvent.

By combining equations (13) and (14), the following equation is obtained.

$$C_{org} = C_i \frac{K}{1 + K \frac{V_{org}}{V_{aq}}} \qquad (15)$$

According to equation (15), it is possible to determine the analyte concentration in the organic phase at equilibrium, because C_{org} is correlated with C_i. The following factors should be considered in obtaining optimal extraction conditions: solvent (type and volume), additives (buffer for pH control), agitation, extraction time, and sample volume. Although maximum sensitivity and precision are obtained by stirring until equilibrium is reached, it is not necessary to for this condition to extract the analyte provided that stirring conditions and time are reproduced.

In this method, analytes can be extracted from a sample by one of two modes: direct SDME (Fig. **11A**) or headspace SDME (HS-SDME) (Fig. **11B**). In direct SDME, a single microdrop of organic solvent on the tip of a syringe is immersed in an aqueous sample. HS-SDME is similar to direct SDME, except that a microdrop of high-boiling extraction solvent is exposed to the headspace of the sample. Three-phase SDME (Fig. **11C**), in which an aqueous microdrop is used as an acceptor phase, has also been developed. The analyte in the aqueous sample is first extracted into the organic phase and then back-extracted into the aqueous receiving phase. Therefore, cleaner extraction can be obtained.

Although SDME is cheap and a reduced volume of toxic organic solvent is required for extraction, a major problem of the technique is that the microdrop is easily dislodged from the microsyringe while stirring the aqueous sample. Furthermore, the technique is not suitable for dirty samples, because particulate impurities make the drop unstable and are potentially harmful to the analytical instrument.

HS-SDME is based on an equilibrium between three phases. Slow mass transfer in the aqueous phase and analyte diffusion into the microdrop are limiting steps in the overall process. Stirring the solution improves mass transfer in the aqueous phase and induces convection in the headspace. Therefore, equilibrium between the aqueous and vapor phases is achieved more rapidly, and analysis time is reduced. High-temperature HS-SDME is performed by evaporating the organic solvent during extraction. Several reviews have reported application of this technique to the determination of drugs and medicines in biological materials [50 - 60].

Membrane Assisted Solvent Extraction (MASE)

MASE, which is based on small-scale LLE, was introduced by Hauser and Popp in 2001 [61]. MASE is based on the transport of analytes through a low-density polyethylene membrane into a small amount of organic solvent (Fig. **12**). The organic phase is then injected into a GC. Because water is undesirable for large-volume injections, the non-porous membrane excludes water from the organic extract. A membrane bag allows the transport of analytes from all sides of the bag in MASE, in contrast to a flat-sheet membrane (Fig. **12A**) [62]. The bag is commercially available from Gerstel® (Mulheim, Germany), which also supplies a fully automated system (Fig. **12B**).

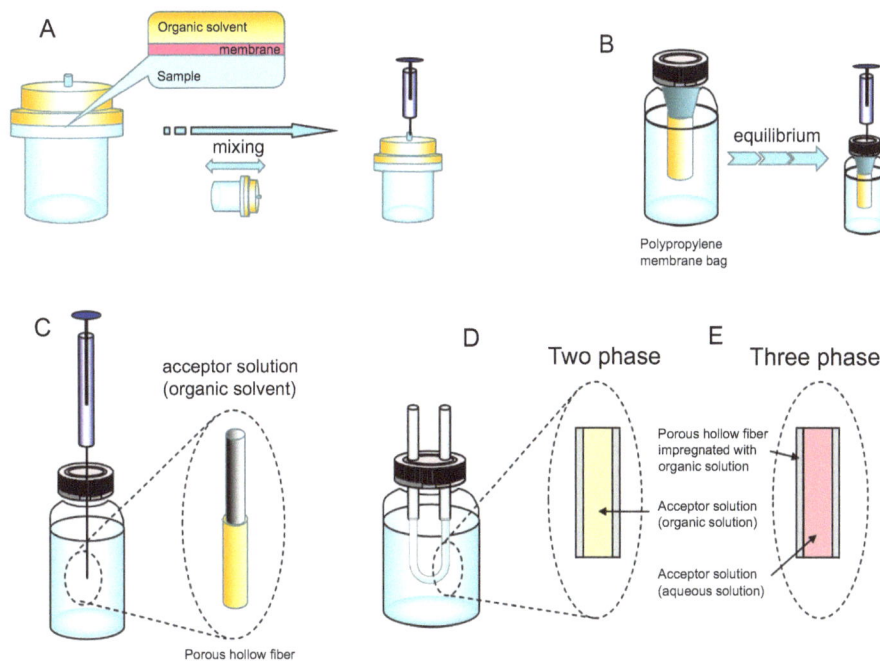

Fig. (12). Different modes of membrane supported microextraction.

The following factors should be considered in obtaining an optimal extraction: extraction solvent (type and amount), additives (buffer for pH control), ionic strength, agitation, and extraction time. The mixing ratio of extracting solvents is important for optimizing analyte recovery.

The main advantage of MASE is the circumvention of phase separation by use of a non-porous membrane. However, analyte diffusion and mass transfer are slow, and a long time is required to achieve high recovery. Therefore, it is preferable to perform MASE at a high temperature. Another drawback is that initial extraction with a new bag cannot be performed unless the bag is thoroughly preconditioned. This fact is important, because preconditioning requires large quantities of organic solvent to remove interfering compounds from the membrane.

The next generation of LPME, hollow-fiber LPME (HF-LPME), was introduced by Rasmussen and co-workers in 1999 [63] as a preconcentration technique for capillary zone electrophoresis (CZE). In HF-LPME, a porous hollow fiber filled with microliter quantities of acceptor solution is immersed in the sample solution. The analyte is extracted into the acceptor solution through pores in the hollow fiber, which are filled with a solvent chosen on the basis of the analyte partition coefficient. The preconcentration factor (PF) is used to indicate recovery,

$$PF = \frac{C_{a,final}}{C_{s,initial}} \tag{16}$$

where $C_{a,final}$ is the final analyte concentration in the sample extract (acceptor phase), and $C_{s,initial}$ is the initial concentration of analyte in the sample.

The extraction efficiency (EE) is defined as the percentage of total analyte present in the original sample, $n_{s,initial}$, that has been transferred to the acceptor phase, $n_{a,final}$,

$$EE(\%) = \frac{n_{a,final}}{n_{s,initial}} \times 100 = \left(\frac{V_a}{V_s}\right) \times PF \times 100 \tag{17}$$

where V_a and V_s are the volumes of the acceptor and sample solutions, respectively.

Two modes of LPME, two-phase (Fig. **12D**) and three-phase (Fig. **12E**), were used originally. In two-phase HF-LPME, the acceptor solution can be the same organic solvent that is immobilized in the pores of the hollow fiber and injected into a chromatograph. In three-phase HF-LPME, the acceptor solution can be aqueous. Analytes are extracted from the sample solution into the organic membrane and subsequently back extracted into the aqueous acceptor solution.

The following factors should be examined to obtain an optimal extraction: extraction solvent (type and amount), additives (buffer for pH control), ionic strength, agitation, and extraction time. Recovery depends on the time required to reach equilibrium as mass transfer from sample to acceptor through the hollow fiber is slow.

The chromatograms of ephedrine in (a) non-spiked urine sample, (b) direct injection at concentration of 500 ng mL−1, and (c) a urine sample related to a volunteer who treated with ephedrine using HF-LPME are shown in Fig. (**13**) [64]. This technique demonstrated several advantages over the other extraction methods especially high sample clean-up. Under the optimized conditions, preconcentration factors of 35 for urine and 8 for human plasma were obtained using HF-LPME, respectively. The calibration curves showed good linearity for urine and plasma samples by both methods with the coefficient of estimations higher than 0.98. The limits of detection were obtained 60 and 200 ng/mL by HF-LPME for urine and plasma samples.

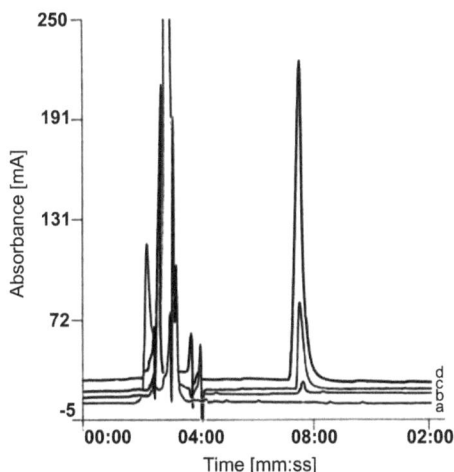

Fig. (13). The chromatograms corresponding to non-spiked urine sample (a); (b) direct injection of ephedrine at concentration of 500 ng/mL (b); extraction of a urine sample related to a volunteer who treated with ephedrine, using HF-LPME (c), respectively. (From [64] with permission of Elsevier).

Some disadvantages of LPME using hollow fiber membranes are the following. (1) The membrane barrier between the aqueous and organic phases reduces the extraction rate and increases extraction time. (2) In two-phase LPME, a large amount of solvent is needed to elute analytes from the lumen and pores of the fiber, and the process is time consuming. (3) Creation of air bubbles on the surface of the hollow fiber reduces the transport rate and decreases the reproducibility of extraction. (4) In real samples such as blood plasma, urine, and wastewater, adsorption of hydrophobic substances on the fiber surface may block the pores. Several reviews have reported application of this technique to the determination of drugs and medicines in biological materials [65 - 74].

A new approach, electromembrane extraction, has been introduced to decrease equilibration time [75 - 77]. In this method, electrodes are placed in the sample solution and the acceptor solution that is inside the hollow fiber. The extraction vial is filled with sample, and the pH is adjusted to produce a net ionic charge on the analyte(s). A voltage is applied to the electrodes, and charged analytes in the sample are drawn across the membrane to the acceptor solution. For cations, the cathode is located in the acceptor solution, and the anode is located in the sample; the polarity is reversed for anions. To complete the procedure, the voltage is turned off, and the acceptor solution is collected for analysis by a convenient technique. High voltages may lead to electrolysis, bubble formation, and analyte decomposition due to redox reactions at the electrodes.

SOLID BASED EXTRACTION

Solid-Phase Extraction (SPE)

SPE is the process of separating analytes of interest based on differences in affinity for a sorbent (solid phase) and a solvent (liquid phase). These differences depend on physical and chemical properties. The following reviews are helpful for method development in environmental, food, and biological analysis [78 - 81]. Usually the analyte of interest and impurities are adsorbed on the surface of a sorbent by passing a sample solution through a column. Impurities are selectively removed by washing with a solvent that has a low affinity for the analyte. The analyte is then eluted with a high affinity solvent. The extraction mechanism is similar to LLE, but detailed understanding is difficult to achieve, because factors such as hydrophobicity, solubility, hydrogen bonding, and $\pi-\pi$ interactions influence the affinity of the sorbent for the analyte. SPE can be classified by three separation modes: normal phase, reversed phase, and ion-exchange extraction. These procedures are typically used for clean-up and concentration of analytes in samples. Classification as normal or reversed phase is based on the polarity of the sorbent and eluting solvent. In normal phase, silica gel is used as the sorbent, and a non-polar solvent such as hexane is used for elution. In reversed phase, silica gel coated with an octyl or octadecyl surface functionality is used as the sorbent, and a polar solvent such as methanol or chloroform is used for elution. As in LLE, there is a relationship between recovery and the P_{ow} of the analyte, although this term depends on the sorbent used. In experimental studies, analytes with *log P* of 2 to 3 can be extracted completely by C_{18} surfaces [82], analytes with *log P* greater than 1 can be extracted completely by a polymer sorbent (PS-1, InertSep PLS-2) [79, 82, 83], and analytes with *log P* is less than 1 (water preferring) can be extracted completely by activated carbon [84].

Immunoaffinity Extraction (IAE)

The conventional sorbents are silica and polymer sorbents with chemically bonded *n*-alkyl functional groups such as octyl or octadecyl groups on the surface. The extraction mechanism is based on hydrophilic interactions; therefore, the selectivity is often low in the extraction of a trace analyte in a complex matrix. Immunoaffinity extraction (IAE), which is one of the immuno-based sample preparation techniques, is used to improve the selectivity [85 - 87]. In this extraction method, antibodies for the target analytes are immobilized on a sorbent; this is called an immunosorbent. The sorbent recognizes the target molecule through antigen-antibody interaction. Usually, a column packed with the immunosorbent is used for the extraction of the analytes in a sample solution. Although IAE had been used as a purification process in biochemistry, its utility in microanalysis recently began to be recognized after it was applied to the extraction of a hormone in biological materials. In IAE, the sample is applied to a sorbent bonded with the antibodies. The targets are tightly bonded to the antibody immobilized on the sorbent and substances with no affinity, including endogenous substances, can be washed out of the column by passing a buffer through it. The analyte is finally eluted from the column by (1) changing the pH, (2) changing the polarity by adding methanol or acetone, (3) adding a protein denaturation reagent (guanidine, urea, or chaotropic salts), and (4) adding a cross-reactive antigen. Usually, the analyte is eluted by denaturing the immunosorbent; therefore, the immunosorbent column is a single-use column. When the analyte is eluted with mild conditions such as a buffer, however, the column can be used for repeated extractions [88 - 90]. The most significant difference between IAE and conventional SPE is that the immunosorbent must be stored under wet conditions, usually in phosphate-buffered saline.

Typical chromatograms of the elution fractions resulting from the use of C18 silica and of the immunosorbent are shown in Fig. (**14**) [91]. A similar extraction recovery close to 100% was found, but the chromatogram corresponding to the immunosorbent (Fig. **14C**) shows that fewer interfering compounds were co-extracted compared to the extraction by the conventional C18 support (Fig. **14B**).

Recently, aptamers (single-stranded oligonucleotides or peptides) that recognize and bind to a specific target molecule have been focused on as potential antibodies [92 - 94]. Aptamers offer advantages over antibodies, as they can be engineered completely in a test tube, are readily produced through chemical synthesis, possess desirable storage properties, and elicit little or no immunogenicity in therapeutic applications. The typical chromatogram of ochratoxin extracted by oligosorbent (aptamer) is shown in Fig. (**14A**). As expected, an extraction recovery close to 100% and a very clean baseline were observed that confirms the performances of the oligosorbent extraction. However, the comparison with the chromatogram corresponding to the immunosorbent shows that the selectivity is similar around the elution peak of ochratoxin, allowing an easy quantification in both cases. These first results are very promising for a future use of oligosorbents dedicated for the determination of OTA from various complex samples.

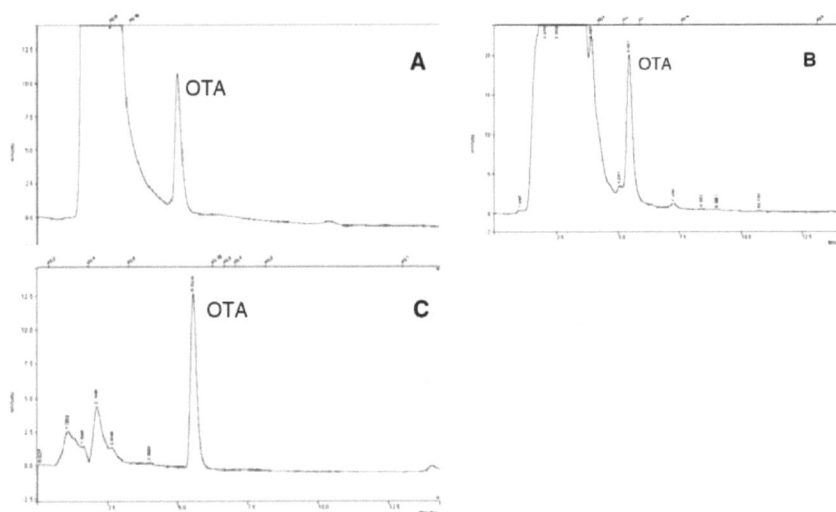

Fig. (14). HPLC chromatograms of ochratoxin of a red wine sample spiked at 2 µg/L on the oligosorbent based on a covalent immobilization (**a**), on the conventional C18 silica cartridge (**b**), and on the immunosorbent (**c**). (From [91] with permission of Springer).

Molecularly Imprinted Solid-Phase Extraction (MISPE)

Molecular imprint technology has recently been applied to the modification of sorbent surfaces in SPE. A template of the analyte of interest is formed on the surface of the sorbent to adsorb a single analyte or a class of structurally related analytes with high selectivity. Selectivity is greatly influenced by the kind and amount of cross-linking agent used in the synthesis of the imprinted polymer, as is the case in the synthesis of antibodies [95 - 98]. Selectivity also is determined by covalent and non-covalent interactions between the target molecule and monomer functional groups [97, 99]. MISPE has many applications in environmental [100] and food [101] analysis. Its use in the analysis of drugs, metabolites, and biomarkers is area of recent interest, although commercially available MISPE cartridges are limited.

Laboratory preparation of a molecularly imprinted sorbent is easy, but it is more difficult to make an MIP having the necessary characteristics. The combination of monomer unit, molecular template, and diluting solvents must be investigated by trial and error, and considerable time and effort are required for optimization. Additional limitations encountered in routine applications include difficulties in using rare chemicals as templates, removing template molecules within an MIP from the sorbent, and preparing an MIP of water soluble chemicals.

The typical chromatograms obtained by injecting the elution fraction from MIP and NIP are shown in Fig. (**15**) [102]. expected, only a low amount of alfuzosin (2%) was observed using the non-imprinted polymer. A recovery of 60% was obtained for the MIP. In addition, co-medications were not retained on the MIP. Those results clearly highlight the selectivity of the optimized extraction procedure associated to the MIP.

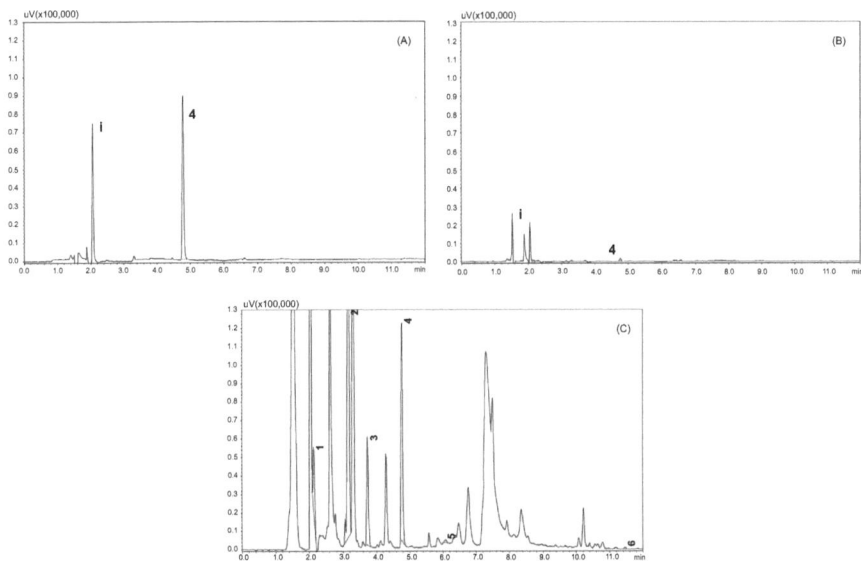

Fig. (15). Chromatograms of alfuzosin and co-medications in serum spiked with 50 ng after the percolation on MIP 1 (**A**) and NIP 1 (**B**) and without treatment (**C**).
Peak identification: 1 = nicotine, 2 = acetaminophen, 3 = caffeine, 4 = alfuzosin, 5 = aspirin, 6 = ibuprofen, i = non-identified compound. (From [102] with permission of Elsevier).

Monolithic Spin Column Extraction (MSCE)

When a sorbent particle is packed into a column or syringe barrel (cartridge), extraction typically is performed by manual handling or by use of a specific manifold. The manifold allows multiple samples to pass simultaneously. Pressure in the cartridge is increased by manual extraction, but is reduced with manifold extraction. However, it is difficult to control pressure and maintain the constant flow needed for reproducibility within the cartridge. Recently, a monolithic silica spin column, which uses gravitational force to carry out extraction, has been developed. The device consists of monolithic silica disks (4.3-mm i.d., 1.5-mm thickness) packed into a spin column [103, 104]. The column is commercially available from GL Sciences (Tokyo, Japan) under the trade name MonoSpin. Monolithic silica disks, which have a surface area greater than that of silica particles, are attached to the support body.

Sample loading, washing, and elution of target drugs are carried out simply by centrifugation (Fig. **16**). Many samples can be processed simultaneously. The method has advantages of simple operation, requirement of a low elution volume, and no need of solvent evaporation. First-time users should pay attention to flow rate, preferably by use of a manifold, and avoid drying the SPE column to obtain good accuracy. It is difficult for beginners to use the manifold. Centrifugation is necessary only for spin-column extraction, and the time invested here can be used to prepare for the next step. The extraction ratio is affected by the pH of the sample solution and the rate of rotation. Optimal extraction conditions should be chosen based on these two factors.

In most applications, the surface of the monolith is bonded with a C_{18} moiety. Multifunctional moieties have been attached to the silica surface to meet more challenging objectives. Several types of spin columns have been developed: C_{18}-SCX, C_{18}-TiO, and C_{18}-C. Simultaneous extraction of acidic and basic drugs can be performed with high recovery using C_{18}-SCX without interference from endogenous substances [105 - 108]. The sorbent is suitable for drug screening in bioanalysis.

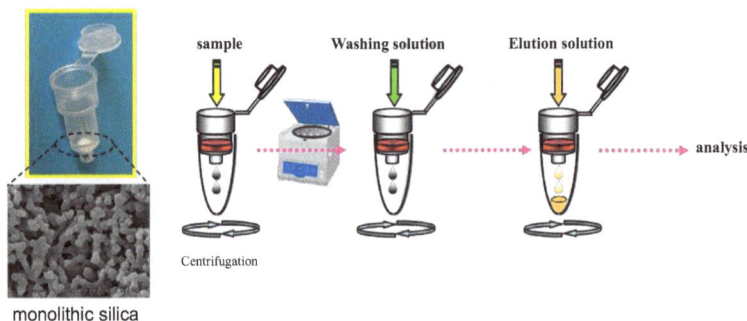

Fig. (16). Scheme of general extraction procedures in monolithic silica monospin extraction.

The typical chromatograms obtained for the urine sample of a drug abuse patient are shown in Fig. (**17**) [104]. When the peak purity of amphetamines was checked by monitoring the UV spectrum pattern from 200 to 350 nm, no peak was overlapped to the tested amphetamines, and high selectivity of this method was verified. The curves exhibited linearity in the concentration range of 0.2–20 µg/mL, the minimum detectable levels in the urine were 0.1 µg/mL. The recovery of amphetamines in urine was from 96 to 111%, and the value of RSD for intra- and inter-day variations in the urine samples with amphetamine concentrations of 1.0 and 10µg/mL were 2.8 and 10.4%, respectively.

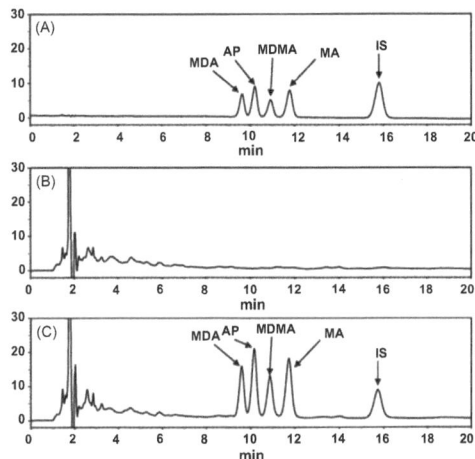

Fig. (17). Chromatograms of amphetamines extracted from normal and drug containing urine samples by using the spin column: (**A**) authentic solution, (**B**) normal urine, and (**C**) drug-containing urine (5µg/mL). Peak identification: AP = amphetamine, MA = methamphetamine, MDA = methylenedioxyamphetamine, MDMA = methylenedioxymethamphetamine. (From [104] with permission of Elsevier).

Matrix Solid-Phase Dispersion (MSPD)

Matrix solid-phase dispersion (MSPD) was first introduced for extraction of drug residues from bovine muscle (Fig. **18**) [109]. MSPD offers the possibility of simultaneous extraction and removal of co-extracted substances, which simplifies the overall process [110 - 113]. In this procedure, the sample architecture is disrupted by mechanical blending with a solid bonded-phase support (dispersant) such as silica, alumina, Florisil, or diatomaceous earth. Analytes tend to be less strongly bonded to the support phase after mechanical blending. In addition, sample pretreatment is performed under mild conditions (atmospheric pressure and room temperature), which minimize degradation. MSPD is simple and cheap, because specific equipment is not required, and the method consumes less solvent due to the low affinity of analyte for the dispersant. Many applications of the procedure have been reported for extracting target molecules from highly viscous, solid, or semisolid samples.

The following factors have been examined for their effect on extraction: (1) particle size of the dispersant, (2) dispersant characteristics, (3) dispersant-sample ratio, (4) chemical modification of the matrix or matrix/solid-support blend, and

(5) selection of eluents and the sequence of their application to a column. Particle sizes in the range of 40–100 μm are recommended for this procedure, because smaller particles (3–10 μm) lead to extended elution times and the need for excessive pressure or a vacuum to obtain adequate flow. Dramatic effects of polarity may be observed using lipophilic bonded phases such as C_{18} and C_8 or hydrophilic phases such as underivatized silica. Factors (1) through (4) are most commonly considered, but their impact varies from application to application.

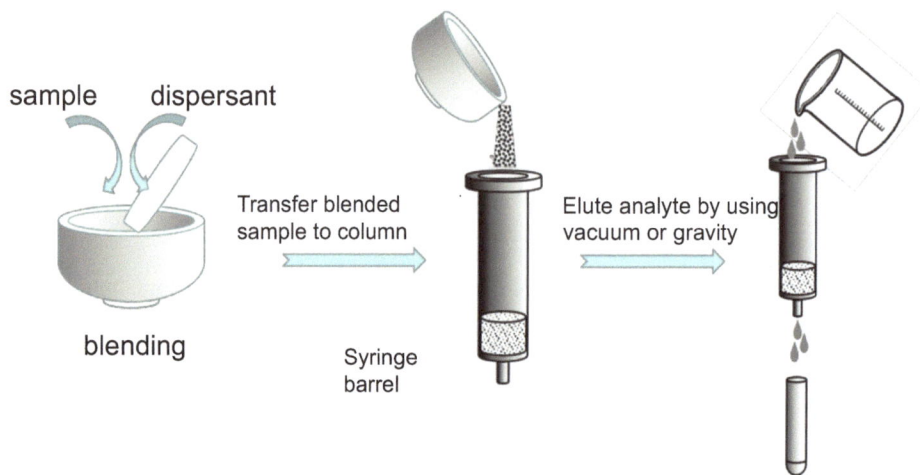

Fig. (18). Scheme of general extraction procedures in matrix solid-phase dispersion.

In a typical procedure, a 0.5 g of sample of liver or fruit is placed in a glass or agate mortar containing 2 g of a dispersant such as octadecylsilyl-derivatized silica (ODS-C_{18}). The solid support and sample are manually blended for about 30 s with a glass or agate pestle. Internal standards or spikes may be added prior to this step. The blended material is then packed into the syringe barrel, and the analyte is eluted with a solvent such as dichloromethane, ethyl acetate, or hexane.

Dispersive Solid-Phase Extraction (DSPE)

Dispersive SPE (DSPE) was introduced as part of a method for the simultaneous extraction and clean-up of pesticide residues in fruits and vegetables in 2003 by Anastassiades [114]. In this technique, sorbent is added directly to the sample solution without preconditioning. Consequently, analytes are extracted from the bulk solution without the requirement of a column, cartridge, or disk (Fig. **19**). Extraction time and simplicity are favorable, because the need to pass the sample solution slowly through a column is avoided. However, few workers have used this direct extraction procedure. To increase selectivity, different adsorbents such as graphene [115], multiwalled carbon nanotubes [116], and sol–gel organic–inorganic hybrids [117] have been developed and applied to the extraction of analytes from various samples.

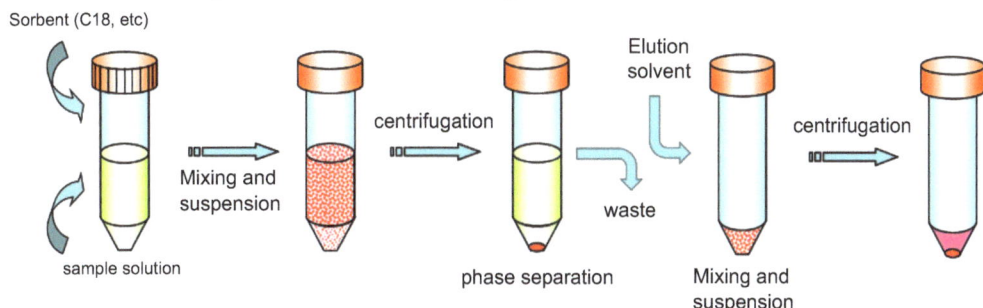

Fig. (19). Scheme of general extraction procedures in dispersive solid-phase extraction.

The original method, which involves salting out by LLE using acetonitrile and purification by two-step dispersive solid-phase extraction, is identified as QuEChERS for Quick, Easy, Cheap, Effective, Rugged, and Safe [114]. Modified procedures have been introduced by the EU and the AOAC. In recent years, the method has been applied to the extraction of small levels of contaminants in non-vegetable foods such as meat and fish. The number of applications has increased dramatically [118 - 121], and the technique has begun to be applied to the extraction of drugs from biological samples [122 - 124].

In the general procedure, acetonitrile is added to the homogenate or crushed sample. Sodium chloride and magnesium sulfate are added to the mixture for the purposes of salting out and dehydration. Citrate is sometimes added to simultaneously transfer basic, acidic, and neutral analytes into acetonitrile. The supernatant obtained after centrifugation is treated by dispersive solid-phase extraction. In DSPE, substances interfering with the analysis are removed by admixture with a highly porous material such as a C_{18}-coated particle or graphitic carbon. However, pigments, polar matrices, and fatty substances, which affect chromatographic separation and ionization suppression or enhancement, are not effectively eliminated by conventional methods. To improve this situation, combination of SPE with a material that can selectively remove lipids and pigments provides better clean-up and removal of matrix materials.

The typical chromatograms of sulfonylurea herbicides in soil obtained with/without clean-up by DSPE were shown in Fig. (**20**) [125]. The results indicate that DSPE with C18 could give a good clean-up effect. The signal was linear over the concentration range from 5.0 to 200 ng/g for the four sulfonylurea herbicides in soil, with the correlation coefficients between 0.9967 and 0.9987, the LODs between 0.5 and 1.2 ng/g, and LOQs between 1.0 and 2.4 ng/g, respectively.

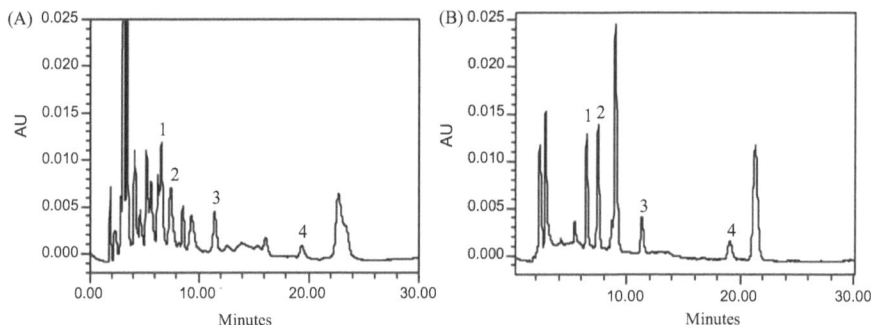

Fig. (20). The chromatograms of soil sample without clean-up by DSPE (**A**) and with clean-up by DSPE (C18) (**B**). The soil sample was spiked with each sulfonylurea at 30 ng/g.
Peak identification: 1 = metsulfuron-methyl, 2 = chlorsulfuron, 3 = bensulfuron-methyl, 4 = chlorimuron-ethyl. (From [125] with permission of Elsevier).

Disposable Pipette Extraction (DPX)

DPX was introduced by Brewer in 2003 as a sample preparation based on conventional SPE that allows on-line coupling with GC or LC [126]. A solid-phase sorbent is loosely packed into a disposable pipette tip, which is fritted at the top and bottom (Fig. **21**). After drawing sample into the tip, the analyte is adsorbed and concentrated on the sorbent by mixing the sorbent suspension with the sample by air bubbling. Dynamic mixing uses less sorbent and provides faster extraction than classical SPE. The analyte is eluted into a vial with an organic solvent followed by washing the sorbent, if needed. Commercially available reversed phase and cation exchange stationary phases are used for chromatographic separation. The sorbent particles in the DPX tips are small (10–20 μm in diameter) and have a high surface area. The adsorption mechanism involves hydrophobic and π–π interactions with the styrene–divinylbenzene sorbent. The extraction tip is sold commercially by Gerstel (Mühlheim a/d Ruhr, Germany). Combinations of these approaches provide selective enrichment of most pesticides and removal of polar matrix interferences. Analytes of medium polarity, whose *log P* values were 2 or less, had recoveries below 60%. Recoveries of 90–100% were obtained for pesticides with *log P* equal to 2.86 or greater. Pesticide retention in DPX-RP is related to *log P*, which enables estimation of pesticide recovery based on hydrophobicity [127].

The DPX method is rapid and takes just a few minutes to perform without solvent evaporation. However, complete elution of the analyte from the sorbent typically requires 0.5 mL of solvent; thus, preconcentration of the analyte is low. Large-volume injections have to be performed to increase sensitivity.

The following factors should be considered in optimizing the extraction: sorbent (type and amount), additives (buffer), ionic strength, and eluting solvent and volume. The draw speed of the plunger that creates the turbulent air bubble is important in increasing the recovery of the analyte. To reduce the time required for sample preparation, a PTV interface or large-volume injection can be used to introduce the eluate directly without evaporation of solvent. Several reviews have reported application of this technique to the determination of drugs and medicines in biological materials [128 - 133].

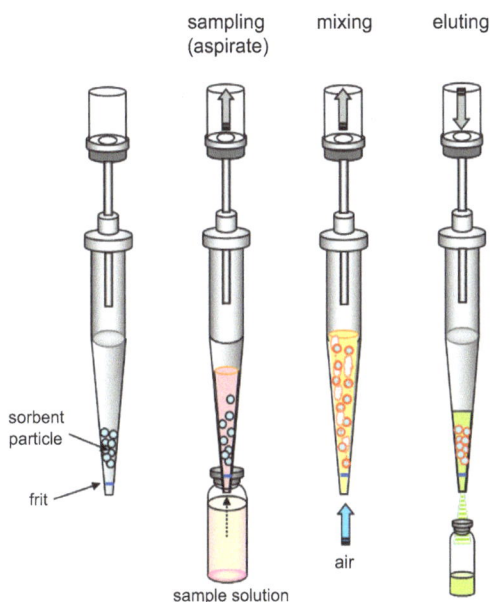

Fig. (21). Scheme of general extraction procedures in disposable-pipette extraction.

Magnetic Solid-Phase Extraction

Magnetic solid-phase extraction is a new procedure based on the use of magnetic or magnetizable sorbent particles [134]. In this procedure, a magnetic sorbent is added to the sample solution, and the suspension is stirred to adsorb the analyte (Fig. **22**). The sorbent containing adsorbed analyte is separated from the solution by a magnet placed outside the extraction vessel; centrifugation or filtration of the sample is unnecessary. The analyte is eluted with an appropriate solvent and analyzed. Magnetite was the original core material, but other magnetic materials such as metal oxide, silica-coated nanoparticles, carbon nanotubes, and molecular imprinted polymers have been used in this technique [135 - 138].

The pH of the sample solution, amount of sorbent, salting out effect, type and volume of desorbing solvent, and desorption and extraction time influence the extraction efficiency. Extraction equilibrium usually requires more than 10 min, and a typical extraction time is 15 to 20 min [139, 140]. Extraction equilibrium is reached quickly by using nanoparticle and microsphere in the sorbent, 3 min is a typical extraction time. Experiments with desorption times of 0.5 to 6 min were carried out with best results achieved after 2 min [141, 142]. There have been many reports of analyte extraction from environmental, food, and biological materials with high recovery.

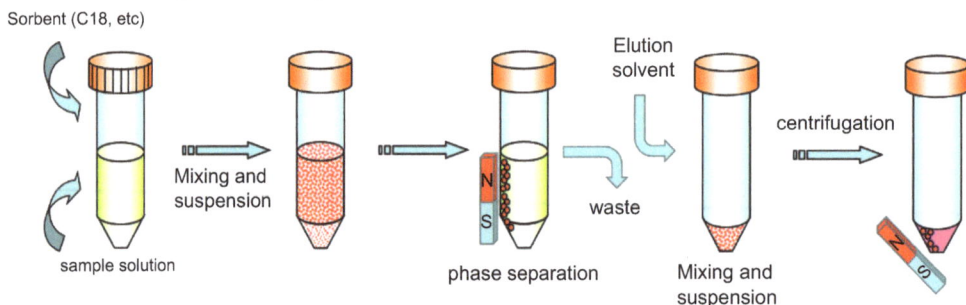

Fig. (22). Scheme of general extraction procedures in magnetic SPE.

SMALL SCALE SOLID BASED EXTRACTION

As is the case with conventional LLE, SPE processes are sometimes tedious and time consuming, and evaporation cannot avoid concentrating the analyte. To circumvent these problems, special techniques have been developed to reduce the time and solvent volume of extraction.

Solid-Phase Microextraction (SPME)

SPME was introduced by Pawliszyn as a solvent-free technique for extraction and direct injection into GC [143, 144]. The SPME device looks like a modified micro-syringe consisting of a fiber coated with a thin layer of extracting phase (volume: 0.5 to 2 μL) inside the needle (Fig. **23A**). Mass transfer is governed by diffusion of the analyte from the sample matrix through the boundary layer to the extracting phase with the goal of establishing equilibrium between the two phases. The amount of analyte extracted at equilibrium is described by equation (18) [145],

$$n = \frac{K_{fs}V_f V_s C_0}{K_{fs}V_f + V_s} \qquad (18)$$

where n is the amount of analyte extracted, K_{fs} is the distribution coefficient between the extraction phase and sample matrix, V_f is the extraction phase volume, V_s is the sample matrix volume, and C_0 is the initial concentration of analyte in the sample matrix. When the sample volume is very large ($K_{fs}V_f << V_s$), equation (18) simplifies to:

$$n = K_{fs}V_fC_0 \tag{19}$$

As seen from equation (19), the amount of analyte extracted depends on its concentration in the sample matrix as in LLE theory. Therefore, a wide dynamic range is available for quantitating analytes.

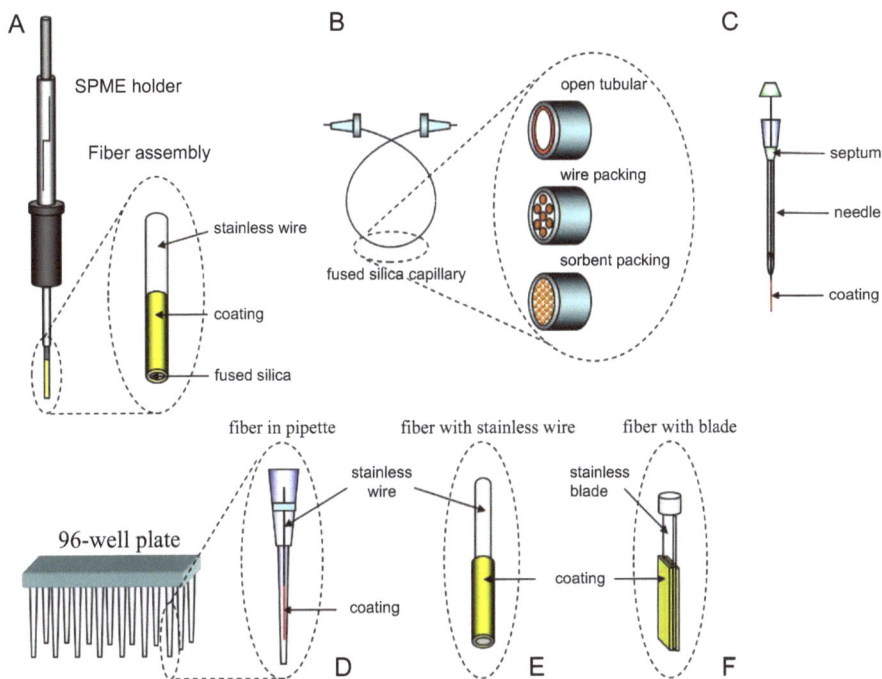

Fig. (23). Different modes of solid-phase microextraction.

Headspace and direct immersion sampling modes can be used to extract analytes. Because only the analyte volatilized from the sample is adsorbed in the headspace mode, the resulting chromatogram is simpler. Also, the lifetime of the coating fiber is extended in the headspace mode, because the sample matrix is not in close contact with the coating. Typical chromatograms of alcohols and esters in beer by SPME and static headspace (SHS) method are shown in Fig. (**24**) [146]. For ethyl acetate the limit of detection determined with the SHS method was lower than that found with the SPME method; for propanol the limits of detection were the same for both methods, whereas for the remaining compounds the limits of detection determined with the SPME method were lower than those determined with the SHS method, usually 2-5 times lower. And Alcohol and ester contents determined by using the SPME method were somewhat higher than those determined by SHS.

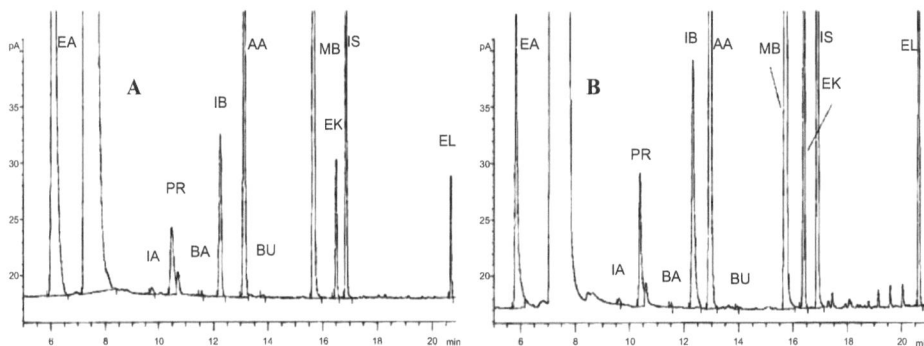

Fig. (24). Chromatograms of beer alcohols and esters determined by SHS (**A**) and SPME (**B**) methods. Peak identification: IS = internal standard, EA = ethyl acetate, IA = isobutyl acetate, PR = propanol, EB = ethyl butyrate, BA = butyl acetate, IB = isobutanol, AA = isoamyl acetate, BU = butanol, MB = methyl-1-butanol, EK = ethyl caproate, EL = ethyl caprylate. (from [146] with permission of American Chemical Society).

The selection of the coating material and its thickness are important factors in obtaining accurate results, because the amount of analyte adsorbed is affected by the coating and its thickness, and recovery and sensitivity depend on the coating used. Liquid-like polymeric materials (*e.g.*, commercial polydimethylsiloxane (PDMS) and polyacrylate) and solid porous materials (*e.g.*, divinylbenzene (DVB)/PDMS and Carboxen/PDMS) are commercially available coating substrates for extraction of analytes from various matrices. Recently, nanomaterials such as carbon nanotubes, graphene, and monoliths have been developed as coating materials [147 - 149]. PDMS is a first choice for screening purposes, but lacks selectivity compared to that of solid coatings. A suitable coating should be selected for each study to achieve high specificity and a long lifetime. Fibers should be carefully handled due to their fragility and tendency to rupture.

The following factors should be considered to obtain optimal extraction conditions: sampling mode, coating (type and thickness), additives (buffer), ionic strength, extraction temperature, extraction time, and agitation. pH control, salting-out, and agitation are crucial factors in increasing analyte recovery and improving sensitivity. Extraction time and temperature also affect extraction efficiency. Elevated temperature may create a less favorable coating–headspace partition, although heat may increase the vapor pressure of volatile analytes in the headspace mode based on the temperature dependence of the Henry's law constant. On-line sample preparation increases productivity and reduces human error during the extraction process. The fiber SPME is easily automated by fitting to a commercial autosampler such as the Combi PAL (CTC Analytics, Zwingen, Switzerland). After the initial report of SPME, applications, modifications, and improvements in methodology of the technique have increased dramatically [150 - 158].

A fully automated sample preparation technique, in-tube SPME, was introduced by the Pawliszyn's group in 1997 [159]. In contrast to conventional fiber SPME, an open or sorbent-packed glass tube is inserted in the sample loop of an auto-sampler (Fig. **23B**). The method resembles a column switching technique. The sample is transferred to the extraction layer or sorbent in the glass tube by repeated drawing and dispensing of the sample. After extraction, the adsorbed analyte is loaded into an analytical system for quantitation. Although the technique can be automated for direct introduction of extracted analytes into the apparatus, it is necessary to pretreat samples by filtration and/or dilution to prevent blockage of the narrow capillary column. To ensure optimal extraction conditions, the same factors mentioned for conventional SPME should be considered (*i.e.*, sampling mode, coating, additives, and extraction time). The number of sample draws and dispersals are crucial for increasing extraction recovery. Several reviews have summarized application of this technique for the determination of drugs and medicines in biological materials [160 - 164]. Recently, *in vivo* SPME has been developed for direct extraction of analytes from humans and animals (Fig. **23C**) [165 - 167]. The technique is promising for live sampling of biological materials in clinical trials to study the effect of drugs administered to patients.

SPME extraction usually is non-exhaustive, and absolute recovery is low. Although low recovery is a drawback, all extracted analyte can be introduced into an analytical instrument. Therefore, the overall amount recovered is greater than with the conventional technique. An additional drawback of SPME is the lengthy extraction time for achieving high recovery, because of low diffusion rates and slow mass transfer. To overcome this problem, in-house multi-fiber devices such as in-tip fiber SPME (Fig. **23D**) [168 - 170], 96-well wire (Fig. **23E**) [171, 172] and 96-blade SPME systems (Fig. **23F**) [173 - 175] have been developed to extract analytes simultaneously and reduce extraction time.

Microextraction by Packed Sorbent (MEPS)

MEPS is an integrated approach combining SPE and a syringe that was introduced by Abdel-Rehim in 2004. The miniaturized SPE device allows on-line coupling with gas and liquid chromatography and capillary electrophoresis [176, 177]. The mechanism of operation is similar to that of conventional SPE. Mass transfer in MEPS occurs by hydrophobic adsorption on the sorbent surface and partitioning into bonded alkyl chains. The adsorption capacity depends on the physical properties and amount of sorbent. In MEPS, a small amount of sorbent (1–4 mg) is packed at the bottom of the needle (Fig. **25**). The apparatus is available commercially from SGE Analytical Science (Ringwood, Australia). Commercially available silica, alkyl-bonded silica (C_2, C_8, C_{18}), ion-exchange resin, or a combined sorbent may be used. The main difference between this technique and conventional SPE is that the sorbent material is directly integrated into the needle. When the plunger of the syringe (100–250 µL) is pulled with the needle inserted in the sample, analytes are extracted by adsorption onto the packed bed as the sample passes through the solid phase. The solid phase is washed once with water or buffer to remove interfering substances, and the analytes are eluted with 20–100 µL of organic solvent. MEPS can be used as a sample preparation technique prior to GC or LC. MEPS has a shorter sample preparation time than LLE and SPE, because, although 20–100 µL of solvent is used for elution of the analyte, MEPS can be fully automated by coupling to programmed temperature vaporizing (PTV) interfaces or large-volume injections. With these injection techniques, it is possible to directly introduce the eluate without evaporating the eluent. In addition, extraction recovery (60–90%) is much greater than that achieved with SPME (1–10%).

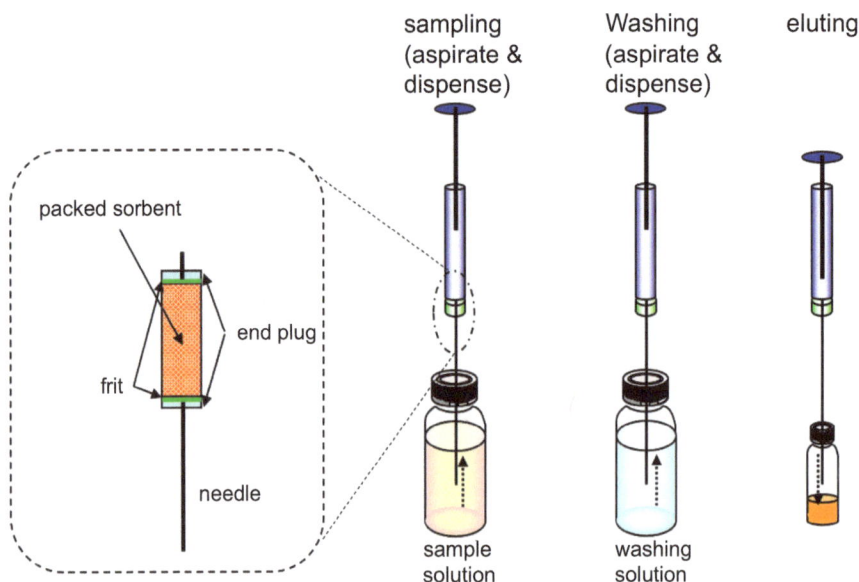

Fig. (25). Scheme of general extraction procedures in microextraction by packed sorbents.

The following factors should be considered for obtaining optimal extractions: sorbent (type and amount), additives (buffer), ionic strength, and elution solvent and volume. The injector type and injection mode are critical, if the method is performed on-line. Both headspace and direct immersion sampling modes can be used. Several reviews have reported applications of this technique to the determination of drugs and medicines in biological materials [177 - 182].

Stir Bar Sorptive Extraction (SBSE)

SBSE, which was developed by Sandra and coworkers in 1999 [183], is based on the same principle as SPME. In SBSE, an extraction coating is deposited on the surface of a magnetic stirring bar (Fig. **26**). The method differs spatially from SPME, where the extraction coating is deposited on the tip of a stainless steel wire. The amount of sorbent used in a commercial SBSE is about 125 μL, which is greater than the quantity used in SPME. The stir bar is sold by Gerstel (Mühlheim a/d Ruhr, Germany) under the trade name Twister, but coatings are limited to commercially available PDMS. To extract analytes from aqueous samples, the SBSE bar is immersed in the sample and stirred for a length of time depending on the matrix. After extraction, the bar is removed from the sample solution and wiped with a soft tissue. Analytes are then thermally desorbed in the injection port of a GC. The headspace mode can be used to extract analytes from the sample matrix as in SPME.

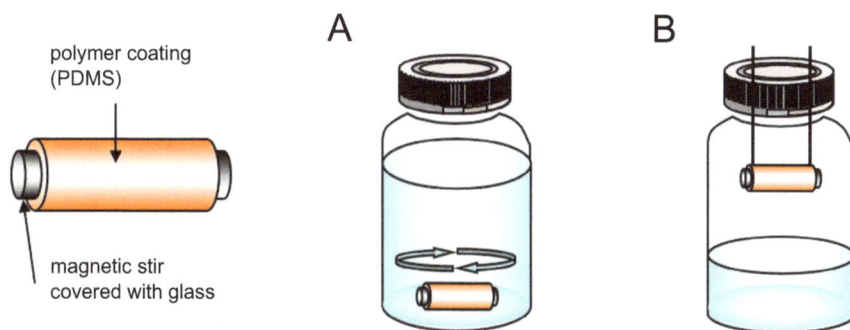

Fig. (26). Different modes of stir bar sorptive extraction.

The theory of SBSE is similar to that of SPME. The amount extracted at equilibrium is expressed by equation (20),

$$K = \frac{C_{SBSE}}{C_W} = \frac{m_{SBSE}}{m_W} \times \frac{V_W}{V_{SBSE}} \tag{20}$$

where m_{SBSE} and m_W equal the mass of analyte in SBSE and water, and V_{SBSE} and V_W are the volumes of SBSE and water, respectively. The phase ratio, β, which equals V_W / V_{SBSE}, is given by equation (21).

$$\beta = \frac{V_W}{V_{SBSE}} \tag{21}$$

Equation 20 can thus be rewritten as

$$K_{O/W} = \frac{m_{SBSE}}{m_W} \times \beta \tag{22}$$

where $K_{o/w}$ is the octanol-water partition coefficient of the analyte [183].

According to mass balance, the total amount of analyte in the sample is given by,

$$m_0 = m_{SBSE} + m_W \tag{23}$$

$$K = \frac{m_{SBSE}}{m_0 - m_{SBSE}} \times \beta \tag{24}$$

where m_0 is the initial amount of analyte in the sample matrix.

$$m_{SBSE} = m_0 \times \left(\frac{\frac{K_{O/W}}{\beta}}{1 + \frac{K_{O/W}}{\beta}} \right) \tag{25}$$

As is evident from equation (25), the amount of analyte extracted depends on its initial concentration in the sample matrix. Although the method is applicable to a wide range of sample contents, recovery is approximately proportional to $K_{o/w}/\beta$. In contrast to SPME, large amounts of sorbent are used in SBSE to increase recovery from the sample solution. The desorption process is limited by slow mass transfer, because a thicker coating is used. SBSE cannot extract strongly polar compounds due to the non-polar nature of the coating material (*i.e.*, PDMS). To overcome this limitation, other strategies including the use of a dual-phase stir bar, higher polarity coating materials, molecularly imprinted polymers, and monolithic SBSE coating materials have been proposed [184]. A drawback of the SBSE technique is that operations such as removal of the stir bar from the sample, its rinsing and drying, and extraction must be performed manually. Specialized equipment (a thermal desorption unit) is needed to thermally desorb the analytes. The most significant drawback of the method is that the sorbent can break or be dislodged during the extraction process.

The typical total ion chromatograms are shown in Fig. (**27**): wine samples were extracted by either SBSE or pentane/diethyl ether liquid/liquid extraction, then analysed by GC/MS in order to compare the three extraction techniques [185]. Greater differences in selectivity are seen for the liquid/liquid extract where more polar hydrophilic components such as phenyl ethanol (7) can be detected. With SBSE the same compounds could be extracted as by SPME, although the recovery of SBSE was much higher than that of SPME because the amount of sorbent of SBSE is much more than that of SPME.

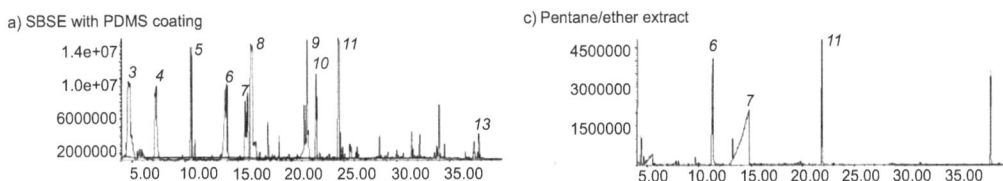

Fig. (27). Comparison of wine components extracted by SBSE and liquid/liquid extraction with GC/MS techniques.
Peak identification: 1 = ethanol, 2 = ethyl acetate, 3 = isoamyl alcohol, 4 = isoamyl acetate, 5 = ethyl hexanoate, 6 = phenylethyl alcohol, 7 = ethyl succinate, 8 = ethyl octanoate, 9 = ethyl decanoate, 10 = ethyl isoamyl succinate, 11 = butyl hydroxytoluene (IS), 12 = phenylethyl octanoate, 13 = ethyl-9-octadecenoate. (From [185] with permission of Springer).

To optimize extraction, the effects of extraction and desorption times, the number of steps needed to complete desorption into the solvent, and carryover after cleaning were examined. A disadvantage is that the procedure is not fully automated; thus, extraction times are longer than with SPME. Several reviews have reported application of this technique to the determination of drugs and medicines in biological materials [186 - 196].

CONCLUDING REMARKS

Several unique sample preparation techniques exist for the extraction and clean-up of analytes from complex matrices. Extraction devices have continued to be miniaturized, after the introduction of SPME by Pawliszyn. Initially, devices were not readily available, because they were made in-house in each laboratory. However, devices have become commercially available over the last two decades, and many applications have been reported. New devices and sorbents are currently being developed for the extraction and purification of analytes of interest in biological, environmental, and food materials. Recently, coupled techniques have been demonstrated for the extraction of analytes, and the sample size has been sufficiently decreased to allow miniaturization of the extraction procedure. The researchers expect that more convenient and practical techniques will be used for sample preparation in the future.

ABBREVIATIONS

DLLME	Dispersive liquid-liquid microextraction
DPX	Disposable pipette extraction
DSPE	Dispersive solid-phase extraction
GC	Gas chromatography
LC	Liquid chromatography
HF-LPME	Hollow-fiber liquid-phase microextraction
HLLE	Homogeneous liquid-liquid extraction
HS	Headspace
IAE	Immunoaffinity extraction
LLE	Liquid-liquid extraction
MASE	Membrane-assisted solvent extraction
MEPS	Microextraction by packed sorbent
MISPE	Molecularly imprinted solid-phase extraction
MS	Mass spectrometry
MSCE	Monolithic spin column extraction
MSPD	Matrix solid-phase dispersion
SBSE	Stir bar sorptive extraction
SDME	Single drop microextraction
SPE	Solid-phase extraction
SPME	Solid-phase microextraction

CONFLICT OF INTEREST

The authors declare no conflict of interest, financial or otherwise.

ACKNOWLEDGEMENTS

Declared none.

REFERENCES

[1]   Pavlović, D.M.; Babić, S.; Horvat, A.J.; Kaštelan-Macan, M. Sample preparation in analysis of pharmaceuticals. *Trends Analyt. Chem.,* **2007**, *26*(11), 1062-1075.
[http://dx.doi.org/10.1016/j.trac.2007.09.010]

[2]   Majors, R.E. Sample Preparation Perspectives: Trends in Sample Preparation and Automation—What the Experts Are Saying. *LC GC,* **1995**, *13*(9), 742-749.

[3]   Alpendurada, M.F. Solid-phase microextraction: a promising technique for sample preparation in environmental analysis. *J. Chromatogr. A,* **2000**, *889*(1-2), 3-14.
[http://dx.doi.org/10.1016/S0021-9673(00)00453-2] [PMID: 10985530]

[4]   Tipler, A. *An Introduction to Headspace Sampling in Gas Chromatography, Fundamentals and Theory.* Technical Note, PerkinElmer, Inc.: Waltham, MA, **2013-2014**.

[5]   Snow, N.H.; Bullock, G.P. Novel techniques for enhancing sensitivity in static headspace extraction-gas chromatography. *J. Chromatogr. A,* **2010**, *1217*(16), 2726-2735.
[http://dx.doi.org/10.1016/j.chroma.2010.01.005] [PMID: 20116067]

[6]   Soria, A.C.; García-Sarrió, M.J.; Sanz, M.L. Volatile sampling by headspace techniques. *Trends Analyt. Chem.,* **2015**, *71*, 85-99.
[http://dx.doi.org/10.1016/j.trac.2015.04.015]

[7]   Rosell, M.; Lacorte, S.; Ginebreda, A.; Barceló, D. Simultaneous determination of methyl tert.-butyl ether and its degradation products, other gasoline oxygenates and benzene, toluene, ethylbenzene and xylenes in Catalonian groundwater by purge-and-trap-gas chromatography-mass spectrometry. *J. Chromatogr. A,* **2003**, *995*(1-2), 171-184.
[http://dx.doi.org/10.1016/S0021-9673(03)00500-4] [PMID: 12800934]

[8]   Cordell, R.L.; Pandya, H.; Hubbard, M.; Turner, M.A.; Monks, P.S. GC-MS analysis of ethanol and other volatile compounds in micro-volume blood samples quantifying neonatal exposure. *Anal. Bioanal. Chem.,* **2013**, *405*(12), 4139-4147.
[http://dx.doi.org/10.1007/s00216-013-6809-1] [PMID: 23420137]

[9]   McDowall, R.D. Sample preparation for biomedical analysis. *J. Chromatogr. A,* **1989**, *492*, 3-58.
[http://dx.doi.org/10.1016/S0378-4347(00)84463-1] [PMID: 2670995]

[10]   Polson, C.; Sarkar, P.; Incledon, B.; Raguvaran, V.; Grant, R. Optimization of protein precipitation based upon effectiveness of protein removal and ionization effect in liquid chromatography-tandem mass spectrometry. *J. Chromatogr. B Analyt. Technol. Biomed. Life Sci.,* **2003**, *785*(2), 263-275.

[11]   Pucci, V.; Di Palma, S.; Alfieri, A.; Bonelli, F.; Monteagudo, E. A novel strategy for reducing phospholipids-based matrix effect in LC-ESI-MS bioanalysis by means of HybridSPE. *J. Pharm. Biomed. Anal.,* **2009**, *50*(5), 867-871.
[http://dx.doi.org/10.1016/j.jpba.2009.05.037] [PMID: 19553055]

[12]   Bonfiglio, R.; King, R.C.; Olah, T.V.; Merkle, K. The effects of sample preparation methods on the variability of the electrospray ionization response for model drug compounds. *Rapid Commun. Mass Spectrom.,* **1999**, *13*(12), 1175-1185.
[http://dx.doi.org/10.1002/(SICI)1097-0231(19990630)13:12<1175::AID-RCM639>3.0.CO;2-0] [PMID: 10407294]

[13]   Wells, M.J. Principles of Extraction and the Extraction of Semivolatile Organics from Liquids. In: *Sample Preparation Techniques in Analytical Chemistry*; Mitra, S., Ed.; John Wiley & Sons: New York, **2003**; Vol. 162, pp. 37-138.
[http://dx.doi.org/10.1002/0471457817.ch2]

[14]   Solvent miscibility table. [Online], 14th June **2016**. https://www.erowid.org/archive/rhodium/pdf/solvent.miscibility.pdf [Accessed: 14th June 2016].

[15]   Okumura, T.; Imamura, K. Simultaneous determination of pesticides by capillary gas chromatography/mass spectrometry. *Jap. J. Water Pollut. Res.,* **1991**, *14*(2), 109-122.
[http://dx.doi.org/10.2965/jswe1978.14.109]

[16]   Jemal, M.; Ouyang, Z.; Xia, Y.Q. Systematic LC-MS/MS bioanalytical method development that incorporates plasma phospholipids risk avoidance, usage of incurred sample and well thought-out chromatography. *Biomed. Chromatogr.,* **2010**, *24*(1), 2-19.
[http://dx.doi.org/10.1002/bmc.1373] [PMID: 20017121]

[17]   Gilbert-López, B.; García-Reyes, J.F.; Molina-Díaz, A. Sample treatment and determination of pesticide residues in fatty vegetable matrices: a review. *Talanta,* **2009**, *79*(2), 109-128.
[http://dx.doi.org/10.1016/j.talanta.2009.04.022] [PMID: 19559852]

[18]   Lo, D.S.; Chao, T.C.; Ng-Ong, S.E.; Yao, Y.J.; Koh, T.H. Acidic and neutral drugs screen in blood

with quantitation using microbore high-performance liquid chromatography-diode array detection and capillary gas chromatography-flame ionization detection. *Forensic Sci. Int.,* **1997**, *90*(3), 205-214.
[http://dx.doi.org/10.1016/S0379-0738(97)00170-9] [PMID: 9493336]

[19] Takeda, A.; Tanaka, H.; Shinohara, T.; Ohtake, I. Systematic analysis of acid, neutral and basic drugs in horse plasma by combination of solid-phase extraction, non-aqueous partitioning and gas chromatography-mass spectrometry. *J. Chromatogr. B Biomed. Sci. Appl.,* **2001**, *758*(2), 235-248.
[http://dx.doi.org/10.1016/S0378-4347(01)00189-X] [PMID: 11486834]

[20] García-Reyes, J.F.; Ferrer, C.; José Gómez-Ramos, M.; Fernández-Alba, A.R.; García-Reyes, J.F.; Molina-Díaz, A. Determination of pesticide residues in olive oil and olives. *Trends Analyt. Chem.,* **2007**, *26*(3), 239-251.
[http://dx.doi.org/10.1016/j.trac.2007.01.004]

[21] Logan, B.K.; Stafford, D.T. Liquid/solid extraction on diatomaceous earth for drug analysis in postmortem blood. *J. Forensic Sci.,* **1989**, *34*(3), 553-564.
[http://dx.doi.org/10.1520/JFS12676J] [PMID: 2738560]

[22] Venisse, N.; Marquet, P.; Duchoslav, E.; Dupuy, J.L.; Lachâtre, G. A general unknown screening procedure for drugs and toxic compounds in serum using liquid chromatography-electrospray-single quadrupole mass spectrometry. *J. Anal. Toxicol.,* **2003**, *27*(1), 7-14.
[http://dx.doi.org/10.1093/jat/27.1.7] [PMID: 12587676]

[23] Silvestro, L.; Savu, S.R. An update on solid phase-supported liquid extraction. *Bioanalysis,* **2015**, *7*(17), 2177-2186.
[http://dx.doi.org/10.4155/bio.15.144] [PMID: 26378936]

[24] Subedi, B.; Aguilar, L.; Robinson, E.M.; Hageman, K.J.; Björklund, E.; Sheesley, R.J.; Usenko, S. Selective pressurized liquid extraction as a sample-preparation technique for persistent organic pollutants and contaminants of emerging concern. *Trends Analyt. Chem.,* **2015**, *68*, 119-132.
[http://dx.doi.org/10.1016/j.trac.2015.02.011]

[25] Murata, K.; Ikeda, S. Homogeneous liquid-liquid extraction method. *Bunseki Kagaku,* **1969**, *18*(9), 1137.
[http://dx.doi.org/10.2116/bunsekikagaku.18.1137]

[26] Farajzadeh, M.A.; Bahram, M.; Zorita, S.; Mehr, B.G. Optimization and application of homogeneous liquid-liquid extraction in preconcentration of copper (II) in a ternary solvent system. *J. Hazard. Mater.,* **2009**, *161*(2-3), 1535-1543.
[http://dx.doi.org/10.1016/j.jhazmat.2008.05.041] [PMID: 18586388]

[27] Myasein, F.; Kim, E.; Zhang, J.; Wu, H.; El-Shourbagy, T.A. Rapid, simultaneous determination of lopinavir and ritonavir in human plasma by stacking protein precipitations and salting-out assisted liquid/liquid extraction, and ultrafast LC-MS/MS. *Anal. Chim. Acta,* **2009**, *651*(1), 112-116.
[http://dx.doi.org/10.1016/j.aca.2009.08.010] [PMID: 19733744]

[28] Zhang, J.; Rodila, R.; Gage, E.; Hautman, M.; Fan, L.; King, L.L.; Wu, H.; El-Shourbagy, T.A. High-throughput salting-out assisted liquid/liquid extraction with acetonitrile for the simultaneous determination of simvastatin and simvastatin acid in human plasma with liquid chromatography. *Anal. Chim. Acta,* **2010**, *661*(2), 167-172.
[http://dx.doi.org/10.1016/j.aca.2009.12.023] [PMID: 20113731]

[29] Shamsipur, M.; Hassan, J. A novel miniaturized homogenous liquid-liquid solvent extraction-high performance liquid chromatographic-fluorescence method for determination of ultra traces of polycyclic aromatic hydrocarbons in sediment samples. *J. Chromatogr. A,* **2010**, *1217*(30), 4877-4882.
[http://dx.doi.org/10.1016/j.chroma.2010.05.038] [PMID: 20566200]

[30] Zhao, F.J.; Tang, H.; Zhang, Q.H.; Yang, J.; Davey, A.K.; Wang, J.P. Salting-out homogeneous liquid-liquid extraction approach applied in sample pre-processing for the quantitative determination of entecavir in human plasma by LC-MS. *J. Chromatogr. B Analyt. Technol. Biomed. Life Sci.,* **2012**, *881-882*, 119-125.

[http://dx.doi.org/10.1016/j.jchromb.2011.12.003] [PMID: 22197609]

[31]   Rezaee, M.; Assadi, Y.; Milani Hosseini, M.R.; Aghaee, E.; Ahmadi, F.; Berijani, S. Determination of organic compounds in water using dispersive liquid-liquid microextraction. *J. Chromatogr. A,* **2006,** *1116*(1-2), 1-9.
[http://dx.doi.org/10.1016/j.chroma.2006.03.007] [PMID: 16574135]

[32]   Kocúrová, L.; Balogh, I.S.; Šandrejová, J.; Andruch, V. Recent advances in dispersive liquid–liquid microextraction using organic solvents lighter than water. A review. *Microchem. J.,* **2012,** *102*, 11-17.
[http://dx.doi.org/10.1016/j.microc.2011.12.002]

[33]   Xiong, C.; Ruan, J.; Cai, Y.; Tang, Y. Extraction and determination of some psychotropic drugs in urine samples using dispersive liquid-liquid microextraction followed by high-performance liquid chromatography. *J. Pharm. Biomed. Anal.,* **2009,** *49*(2), 572-578.
[http://dx.doi.org/10.1016/j.jpba.2008.11.036] [PMID: 19135820]

[34]   Xu, H.; Song, D.; Cui, Y.; Hu, S.; Yu, Q.; Feng, Y. Analysis of hexanal and heptanal in human blood by simultaneous derivatization and dispersive liquid-liquid microextraction then LC-APCI-MS-MS. *Chromatographia,* **2009,** *70*(5-6), 775-781.
[http://dx.doi.org/10.1365/s10337-009-1208-7]

[35]   Cruz-Vera, M.; Lucena, R.; Cárdenas, S.; Valcárcel, M. One-step in-syringe ionic liquid-based dispersive liquid-liquid microextraction. *J. Chromatogr. A,* **2009,** *1216*(37), 6459-6465.
[http://dx.doi.org/10.1016/j.chroma.2009.07.040] [PMID: 19674753]

[36]   Rezaee, M.; Yamini, Y.; Faraji, M. Evolution of dispersive liquid-liquid microextraction method. *J. Chromatogr. A,* **2010,** *1217*(16), 2342-2357.
[http://dx.doi.org/10.1016/j.chroma.2009.11.088] [PMID: 20005521]

[37]   Lili, L.; Xu, H.; Song, D.; Cui, Y.; Hu, S.; Zhang, G. Analysis of volatile aldehyde biomarkers in human blood by derivatization and dispersive liquid-liquid microextraction based on solidification of floating organic droplet method by high performance liquid chromatography. *J. Chromatogr. A,* **2010,** *1217*(16), 2365-2370.
[http://dx.doi.org/10.1016/j.chroma.2010.01.081] [PMID: 20181347]

[38]   Zarei, A.R.; Gholamian, F. Development of a dispersive liquid-liquid microextraction method for spectrophotometric determination of barbituric acid in pharmaceutical formulation and biological samples. *Anal. Biochem.,* **2011,** *412*(2), 224-228.
[http://dx.doi.org/10.1016/j.ab.2011.02.004] [PMID: 21303651]

[39]   Zgoła-Grześkowiak, A.; Kaczorek, E. Isolation, preconcentration and determination of rhamnolipids in aqueous samples by dispersive liquid-liquid microextraction and liquid chromatography with tandem mass spectrometry. *Talanta,* **2011,** *83*(3), 744-750.
[http://dx.doi.org/10.1016/j.talanta.2010.10.037] [PMID: 21147315]

[40]   Li, J.; Lu, W.; Ma, J.; Chen, L. Determination of mercury(II) in water samples using dispersive liquid-liquid microextraction and back extraction along with capillary zone electrophoresis. *Mikrochim. Acta,* **2011,** *175*(3), 301-308.
[http://dx.doi.org/10.1007/s00604-011-0679-z]

[41]   Soltani, S.; Ramezani, A.M.; Soltani, N.; Jouyban, A. Analysis of losartan and carvedilol in urine and plasma samples using a dispersive liquid-liquid microextraction isocratic HPLC-UV method. *Bioanalysis,* **2012,** *4*(23), 2805-2821.
[http://dx.doi.org/10.4155/bio.12.261] [PMID: 23216121]

[42]   Ranjbari, E.; Golbabanezhad-Azizi, A.A.; Hadjmohammadi, M.R. Preconcentration of trace amounts of methadone in human urine, plasma, saliva and sweat samples using dispersive liquid-liquid microextraction followed by high performance liquid chromatography. *Talanta,* **2012,** *94*, 116-122.
[http://dx.doi.org/10.1016/j.talanta.2012.03.004] [PMID: 22608423]

[43]   Ghambari, H.; Hadjmohammadi, M. Low-density solvent-based dispersive liquid-liquid microextraction followed by high performance liquid chromatography for determination of warfarin in

human plasma. *J. Chromatogr. B Analyt. Technol. Biomed. Life Sci.,* **2012**, *899*, 66-71.
[http://dx.doi.org/10.1016/j.jchromb.2012.04.035] [PMID: 22622064]

[44]   Zhang, J.; Li, M.; Li, L.; Li, Y.; Peng, B.; Zhang, S.; Gao, H.; Zhou, W. Investigation of the ultrasound effect and target analyte selectivity of dispersive liquid-liquid microextraction and its application to a quinocetone pharmacokinetic study. *J. Chromatogr. A,* **2012**, *1268*(1268), 1-8.
[PMID: 23122995]

[45]   Ma, J.; Lu, W.; Chen, L. Recent advances in dispersive liquid - liquid microextraction for organic compounds analysis in environmental water: A review. *Curr. Anal. Chem.,* **2012**, *8*(1), 78-90.
[http://dx.doi.org/10.2174/157341112798472170]

[46]   Kohler, I.; Schappler, J.; Sierro, T.; Rudaz, S. Dispersive liquid-liquid microextraction combined with capillary electrophoresis and time-of-flight mass spectrometry for urine analysis. *J. Pharm. Biomed. Anal.,* **2013**, *73*, 82-89.
[http://dx.doi.org/10.1016/j.jpba.2012.03.036] [PMID: 22494520]

[47]   Jain, R.; Singh, R. Applications of dispersive liquid–liquid micro-extraction in forensic toxicology. *Trends Analyt. Chem.,* **2016**, *75*, 227-237.
[http://dx.doi.org/10.1016/j.trac.2015.07.007]

[48]   Liu, H.; Dasgupta, P.K. Analytical chemistry in a drop. Solvent extraction in a microdrop. *Anal. Chem.,* **1996**, *68*(11), 1817-1821.
[http://dx.doi.org/10.1021/ac960145h] [PMID: 21619093]

[49]   Jeannot, M.A.; Cantwell, F.F. Solvent microextraction into a single drop. *Anal. Chem.,* **1996**, *68*(13), 2236-2240.
[http://dx.doi.org/10.1021/ac960042z] [PMID: 21619310]

[50]   He, Y.; Lee, H.K. Liquid-Phase Microextraction in a Single Drop of Organic Solvent by Using a Conventional Microsyringe. *Anal. Chem.,* **1997**, *69*(22), 4634-4640.
[http://dx.doi.org/10.1021/ac970242q]

[51]   Xiao, Q.; Hu, B.; He, M. Speciation of butyltin compounds in environmental and biological samples using headspace single drop microextraction coupled with gas chromatography-inductively coupled plasma mass spectrometry. *J. Chromatogr. A,* **2008**, *1211*(1-2), 135-141.
[http://dx.doi.org/10.1016/j.chroma.2008.09.089] [PMID: 18922539]

[52]   Lucena, R.; Cruz-Vera, M.; Cárdenas, S.; Valcárcel, M. Liquid-phase microextraction in bioanalytical sample preparation. *Bioanalysis,* **2009**, *1*(1), 135-149.
[http://dx.doi.org/10.4155/bio.09.16] [PMID: 21083193]

[53]   Jeannot, M.A.; Przyjazny, A.; Kokosa, J.M. Single drop microextraction development, applications and future trends. *J. Chromatogr. A,* **2010**, *1217*(16), 2326-2336.
[http://dx.doi.org/10.1016/j.chroma.2009.10.089] [PMID: 19932482]

[54]   Jain, A.; Verma, K.K. Recent advances in applications of single-drop microextraction: a review. *Anal. Chim. Acta,* **2011**, *706*(1), 37-65.
[http://dx.doi.org/10.1016/j.aca.2011.08.022] [PMID: 21995911]

[55]   Choi, K.; Kim, J.; Chung, D.S. Single-drop microextraction in bioanalysis. *Bioanalysis,* **2011**, *3*(7), 799-815.
[http://dx.doi.org/10.4155/bio.11.3] [PMID: 21452996]

[56]   Nuhu, A.A.; Basheer, C.; Saad, B. Liquid-phase and dispersive liquid-liquid microextraction techniques with derivatization: recent applications in bioanalysis. *J. Chromatogr. B Analyt. Technol. Biomed. Life Sci.,* **2011**, *879*(17-18), 1180-1188.
[http://dx.doi.org/10.1016/j.jchromb.2011.02.009] [PMID: 21376675]

[57]   Wu, H.F.; Kailasa, S.K.; Lin, C.H. Single drop microextraction coupled with matrix-assisted laser desorption/ionization mass spectrometry for rapid and direct analysis of hydrophobic peptides from biological samples in high salt solution. *Rapid Commun. Mass Spectrom.,* **2011**, *25*(2), 307-315.

[http://dx.doi.org/10.1002/rcm.4843] [PMID: 21192026]

[58] AlOthman, Z.A.; Dawod, M.; Kim, J.; Chung, D.S. Single-drop microextraction as a powerful pretreatment tool for capillary electrophoresis: A review. *Anal. Chim. Acta,* **2012,** *739,* 14-24.
[http://dx.doi.org/10.1016/j.aca.2012.06.005] [PMID: 22819045]

[59] Saraji, M.; Khaje, N. Recent advances in liquid microextraction techniques coupled with MS for determination of small-molecule drugs in biological samples. *Bioanalysis,* **2012,** *4*(6), 725-739.
[http://dx.doi.org/10.4155/bio.12.26] [PMID: 22452263]

[60] Han, D.; Row, K.H. Trends in liquid-phase microextraction, and its application to environmental and biological samples. *Mikrochim. Acta,* **2012,** *176*(1-2), 1-22.
[http://dx.doi.org/10.1007/s00604-011-0678-0]

[61] Hauser, B.; Popp, P. Membrane assisted solvent extraction of organochlorine compounds in combination with large volume injection / gas chromatography electron capture detection. *J. Sep. Sci.,* **2001,** *24*(7), 551-560.
[http://dx.doi.org/10.1002/1615-9314(20010801)24:7<551::AID-JSSC551>3.0.CO;2-2]

[62] Hause, B.; Popp, P.; Kleine-Benne, E. Membrane-assisted solvent extraction of triazines and other semi-volatile contaminants directly coupled to large-volume injection-gas chromatography-mass spectrometric detection. *J. Chromatogr. A,* **2002,** *963*(1-2), 27-36.
[http://dx.doi.org/10.1016/S0021-9673(02)00135-8] [PMID: 12187980]

[63] Pedersen-Bjergaard, S.; Rasmussen, K.E. Liquid-liquid-liquid microextraction for sample preparation of biological fluids prior to capillary electrophoresis. *Anal. Chem.,* **1999,** *71*(14), 2650-2656.
[http://dx.doi.org/10.1021/ac990055n] [PMID: 10424162]

[64] Fotouhi, L.; Yamini, Y.; Molaei, S.; Seidi, S. Comparison of conventional hollow fiber based liquid phase microextraction and electromembrane extraction efficiencies for the extraction of ephedrine from biological fluids. *J. Chromatogr. A,* **2011,** *1218*(48), 8581-8586.
[http://dx.doi.org/10.1016/j.chroma.2011.09.078] [PMID: 22024341]

[65] Rasmussen, K.E.; Pedersen-Bjergaard, S.; Krogh, M.; Ugland, H.G.; Grønhaug, T. Development of a simple in-vial liquid-phase microextraction device for drug analysis compatible with capillary gas chromatography, capillary electrophoresis and high-performance liquid chromatography. *J. Chromatogr. A,* **2000,** *873*(1), 3-11.
[http://dx.doi.org/10.1016/S0021-9673(99)01163-2] [PMID: 10757280]

[66] Pedersen-Bjergaard, S.; Rasmussen, K.E. Liquid-phase microextraction with porous hollow fibers, a miniaturized and highly flexible format for liquid-liquid extraction. *J. Chromatogr. A,* **2008,** *1184*(1-2), 132-142.
[http://dx.doi.org/10.1016/j.chroma.2007.08.088] [PMID: 17889886]

[67] Lee, J.; Lee, H.K.; Rasmussen, K.E.; Pedersen-Bjergaard, S. Environmental and bioanalytical applications of hollow fiber membrane liquid-phase microextraction: a review. *Anal. Chim. Acta,* **2008,** *624*(2), 253-268.
[http://dx.doi.org/10.1016/j.aca.2008.06.050] [PMID: 18706332]

[68] Saraji, M.; Khalili Boroujeni, M.; Hajialiakbari Bidgoli, A.A. Comparison of dispersive liquid-liquid microextraction and hollow fiber liquid-liquid-liquid microextraction for the determination of fentanyl, alfentanil, and sufentanil in water and biological fluids by high-performance liquid chromatography. *Anal. Bioanal. Chem.,* **2011,** *400*(7), 2149-2158.
[http://dx.doi.org/10.1007/s00216-011-4874-x] [PMID: 21442368]

[69] Ghambarian, M.; Yamini, Y.; Esrafili, A. Developments in hollow fiber based liquid-phase microextraction: principles and applications. *Mikrochim. Acta,* **2012,** *177*(3-4), 271-294.
[http://dx.doi.org/10.1007/s00604-012-0773-x]

[70] Guo, X.; He, M.; Chen, B.; Hu, B. Phase transfer hollow fiber liquid phase microextraction combined with electrothermal vaporization inductively coupled plasma mass spectrometry for the determination of trace heavy metals in environmental and biological samples. *Talanta,* **2012,** *101,* 516-523.

[http://dx.doi.org/10.1016/j.talanta.2012.10.017] [PMID: 23158357]

[71]   Bello-López, M.Á.; Ramos-Payán, M.; Ocaña-González, J.A.; Fernández-Torres, R.; Callejón-Mochón, M. Analytical applications of hollow fiber liquid phase microextraction (HF-LPME): A Review. *Anal. Lett.,* **2012**, *45*(8), 804-830.
[http://dx.doi.org/10.1080/00032719.2012.655676]

[72]   Farajzadeh, M.A.; Sorouraddin, S.M.; Mogaddam, M.R. Liquid phase microextraction of pesticides: a review on current methods. *Mikrochim. Acta,* **2014**, *181*(9), 829-851.
[http://dx.doi.org/10.1007/s00604-013-1157-6]

[73]   Carasek, E.; Merib, J. Membrane-based microextraction techniques in analytical chemistry: A review. *Anal. Chim. Acta,* **2015**, *880*, 8-25.
[http://dx.doi.org/10.1016/j.aca.2015.02.049] [PMID: 26092333]

[74]   Sharifi, V.; Nosrati, A. Application of hollow fiber liquid phase microextraction and dispersive liquid–liquid microextraction techniques in analytical toxicology. *J. Food Drug Anal.,* **2016**, *24*(2), 264-276.
[http://dx.doi.org/10.1016/j.jfda.2015.10.004]

[75]   Kjelsen, I.J.; Gjelstad, A.; Rasmussen, K.E.; Pedersen-Bjergaard, S. Low-voltage electromembrane extraction of basic drugs from biological samples. *J. Chromatogr. A,* **2008**, *1180*(1-2), 1-9.
[http://dx.doi.org/10.1016/j.chroma.2007.12.006] [PMID: 18164716]

[76]   Basheer, C.; Lee, J.; Pedersen-Bjergaard, S.; Rasmussen, K.E.; Lee, H.K. Simultaneous extraction of acidic and basic drugs at neutral sample pH: a novel electro-mediated microextraction approach. *J. Chromatogr. A,* **2010**, *1217*(43), 6661-6667.
[http://dx.doi.org/10.1016/j.chroma.2010.04.066] [PMID: 20488447]

[77]   Petersen, N.J.; Rasmussen, K.E.; Pedersen-Bjergaard, S.; Gjelstad, A. Electromembrane extraction from biological fluids. *Anal. Sci.,* **2011**, *27*(10), 965-972.
[http://dx.doi.org/10.2116/analsci.27.965] [PMID: 21985919]

[78]   Thurman, E.M.; Mills, M.S. *Solid-phase extraction: principles and practice*; John Wiley & Sons: New York, **1998**.

[79]   Hennion, M.C. Solid-phase extraction: method development, sorbents, and coupling with liquid chromatography. *J. Chromatogr. A,* **1999**, *856*(1-2), 3-54.
[http://dx.doi.org/10.1016/S0021-9673(99)00832-8] [PMID: 10526783]

[80]   Poole, C.F.; Gunatilleka, A.D.; Sethuraman, R. Contributions of theory to method development in solid-phase extraction. *J. Chromatogr. A,* **2000**, *885*(1-2), 17-39.
[http://dx.doi.org/10.1016/S0021-9673(00)00224-7] [PMID: 10941665]

[81]   Pichon, V. Solid-phase extraction for multiresidue analysis of organic contaminants in water. *J. Chromatogr. A,* **2000**, *885*(1-2), 195-215.
[http://dx.doi.org/10.1016/S0021-9673(00)00456-8] [PMID: 10941673]

[82]   Nakamura, M.; Nakamura, M.; Yamada, S. Conditions for solid-phase extraction of agricultural chemicals in waters by using n-octanol-water partition coefficients. *Analyst (Lond.),* **1996**, *121*, 469-475.
[http://dx.doi.org/10.1039/AN9962100469]

[83]   Tachon, R.; Pichon, V.; Barbe Le Borgne, M.; Minet, J.J. Comparison of solid-phase extraction sorbents for sample clean-up in the analysis of organic explosives. *J. Chromatogr. A,* **2008**, *1185*(1), 1-8.
[http://dx.doi.org/10.1016/j.chroma.2008.01.026] [PMID: 18272163]

[84]   Hennion, M.C. Graphitized carbons for solid-phase extraction. *J. Chromatogr. A,* **2000**, *885*(1-2), 73-95.
[http://dx.doi.org/10.1016/S0021-9673(00)00085-6] [PMID: 10941668]

[85]   Hennion, M.C.; Pichon, V. Immuno-based sample preparation for trace analysis. *J. Chromatogr. A,*

2003, *1000*(1-2), 29-52.
[http://dx.doi.org/10.1016/S0021-9673(03)00529-6] [PMID: 12877165]

[86]   Senyuva, H.Z.; Gilbert, J. Immunoaffinity column clean-up techniques in food analysis: A review. *J. Chromatogr. B Analyt. Technol. Biomed. Life Sci.*, **2010**, *878*(2), 115-132.
[http://dx.doi.org/10.1016/j.jchromb.2009.05.042] [PMID: 19525155]

[87]   Cruz-Vera, M.; Lucena, R.; Cárdenas, S.; Valcárcel, M. Highly selective and non-conventional sorbents for the determination of biomarkers in urine by liquid chromatography. *Anal. Bioanal. Chem.*, **2010**, *397*(3), 1029-1038.
[http://dx.doi.org/10.1007/s00216-010-3476-3] [PMID: 20127317]

[88]   Gude, T.; Preiss, A.; Rubach, K. Determination of chloramphenicol in muscle, liver, kidney and urine of pigs by means of immunoaffinity chromatography and gas chromatography with electron-capture detection. *J. Chromatogr. B Biomed. Appl.*, **1995**, *673*(2), 197-204.
[http://dx.doi.org/10.1016/0378-4347(95)00280-5] [PMID: 8611953]

[89]   Rejeb, S.B.; Cléroux, C.; Lawrence, J.F.; Geay, P.Y.; Wu, S.G.; Stavinski, S. Development and characterization of immunoaffinity columns for the selective extraction of a new developmental pesticide: Thifluzamide, from peanuts. *Anal. Chim. Acta*, **2001**, *432*(2), 193-200.
[http://dx.doi.org/10.1016/S0003-2670(00)01376-3]

[90]   Klinglmayr, C.; Nöbauer, K.; Razzazi-Fazeli, E.; Cichna-Markl, M. Determination of deoxynivalenol in organic and conventional food and feed by sol-gel immunoaffinity chromatography and HPLC-UV detection. *J. Chromatogr. B Analyt. Technol. Biomed. Life Sci.*, **2010**, *878*(2), 187-193.
[http://dx.doi.org/10.1016/j.jchromb.2009.08.016] [PMID: 19736050]

[91]   Chapuis-Hugon, F.; du Boisbaudry, A.; Madru, B.; Pichon, V. New extraction sorbent based on aptamers for the determination of ochratoxin A in red wine. *Anal. Bioanal. Chem.*, **2011**, *400*(5), 1199-1207.
[http://dx.doi.org/10.1007/s00216-010-4574-y] [PMID: 21221554]

[92]   Gulbakan, B.; Yasun, E.; Shukoor, M.I.; Zhu, Z.; You, M.; Tan, X.; Sanchez, H.; Powell, D.H.; Dai, H.; Tan, W. A dual platform for selective analyte enrichment and ionization in mass spectrometry using aptamer-conjugated graphene oxide. *J. Am. Chem. Soc.*, **2010**, *132*(49), 17408-17410.
[http://dx.doi.org/10.1021/ja109042w] [PMID: 21090719]

[93]   Dua, F.; Guoa, L.; Qina, Q.; Zhenga, X.; Ruana, G.; Lia, J.; Lib, G. Recent advances in aptamer-functionalized materials in sample preparation. *Trends Analyt. Chem.*, **2015**, *67*, 134-146.
[http://dx.doi.org/10.1016/j.trac.2015.01.007]

[94]   Płotka-Wasylka, J.; Szczepańska, N.; de la Guardia, M.; Namieśnik, J. Modern trends in solid phase extraction: New sorbent media. *Trends Analyt. Chem.*, **2016**, *77*, 23-43.
[http://dx.doi.org/10.1016/j.trac.2015.10.010]

[95]   Martín-Esteban, A. Molecularly-imprinted polymers as a versatile, highly selective tool in sample preparation. *Trends Analyt. Chem.*, **2013**, *45*, 169-181.
[http://dx.doi.org/10.1016/j.trac.2012.09.023]

[96]   Wen, Y.; Chen, L.; Li, J.; Liu, D.; Chen, L. Recent advances in solid-phase sorbents for sample preparation prior to chromatographic analysis. *Trends Analyt. Chem.*, **2014**, *59*, 26-41.
[http://dx.doi.org/10.1016/j.trac.2014.03.011]

[97]   Huang, D.L.; Wang, R.Z.; Liu, Y.G.; Zeng, G.M.; Lai, C.; Xu, P.; Lu, B.A.; Xu, J.J.; Wang, C.; Huang, C. Application of molecularly imprinted polymers in wastewater treatment: a review. *Environ. Sci. Pollut. Res. Int.*, **2015**, *22*(2), 963-977.
[http://dx.doi.org/10.1007/s11356-014-3599-8] [PMID: 25280502]

[98]   Chen, L.; Wang, X.; Lu, W.; Wu, X.; Li, J. Molecular imprinting: perspectives and applications. *Chem. Soc. Rev.*, **2016**, *45*(8), 2137-2211.
[http://dx.doi.org/10.1039/C6CS00061D] [PMID: 26936282]

[99]   Cheong, W.J.; Yang, S.H.; Ali, F. Molecular imprinted polymers for separation science: a review of reviews. *J. Sep. Sci.,* **2013**, *36*(3), 609-628.
[http://dx.doi.org/10.1002/jssc.201200784] [PMID: 23281278]

[100]   He, J.; Lv, R.; Zhan, H.; Wang, H.; Cheng, J.; Lu, K.; Wang, F. Preparation and evaluation of molecularly imprinted solid-phase micro-extraction fibers for selective extraction of phthalates in an aqueous sample. *Anal. Chim. Acta,* **2010**, *674*(1), 53-58.
[http://dx.doi.org/10.1016/j.aca.2010.06.018] [PMID: 20638499]

[101]   Mohajeri, S.A.; Hosseinzadeh, H.; Keyhanfar, F.; Aghamohammadian, J. Extraction of crocin from saffron (*Crocus sativus*) using molecularly imprinted polymer solid-phase extraction. *J. Sep. Sci.,* **2010**, *33*(15), 2302-2309.
[http://dx.doi.org/10.1002/jssc.201000183] [PMID: 20589782]

[102]   Hugon-Chapuis, F.; Mullot, J.U.; Tuffal, G.; Hennion, M.C.; Pichon, V. Selective and automated sample pretreatment by molecularly imprinted polymer for the analysis of the basic drug alfuzosin from plasma. *J. Chromatogr. A,* **2008**, *1196-1197*, 73-80.
[http://dx.doi.org/10.1016/j.chroma.2008.04.038] [PMID: 18466912]

[103]   Namera, A.; Nakamoto, A.; Nishida, M.; Saito, T.; Kishiyama, I.; Miyazaki, S.; Yahata, M.; Yashiki, M.; Nagao, M. Extraction of amphetamines and methylenedioxyamphetamines from urine using a monolithic silica disk-packed spin column and high-performance liquid chromatography-diode array detection. *J. Chromatogr. A,* **2008**, *1208*(1-2), 71-75.
[http://dx.doi.org/10.1016/j.chroma.2008.08.091] [PMID: 18790482]

[104]   Saito, T.; Yamamoto, R.; Inoue, S.; Kishiyama, I.; Miyazaki, S.; Nakamoto, A.; Nishida, M.; Namera, A.; Inokuchi, S. Simultaneous determination of amitraz and its metabolite in human serum by monolithic silica spin column extraction and liquid chromatography-mass spectrometry. *J. Chromatogr. B Analyt. Technol. Biomed. Life Sci.,* **2008**, *867*(1), 99-104.
[http://dx.doi.org/10.1016/j.jchromb.2008.03.018] [PMID: 18417428]

[105]   Namera, A.; Nakamoto, A.; Saito, T.; Miyazaki, S. Monolith as a new sample preparation material: recent devices and applications. *J. Sep. Sci.,* **2011**, *34*(8), 901-924.
[http://dx.doi.org/10.1002/jssc.201000795] [PMID: 21394910]

[106]   Namera, A.; Miyazaki, S.; Saito, T.; Nakamoto, A. Monolithic silica with HPLC separation and solid phase extraction materials for determination of drugs in biological materials. *Anal. Methods,* **2011**, *3*, 2189-2200.
[http://dx.doi.org/10.1039/c1ay05243h]

[107]   Namera, A.; Saito, T. Advances in monolithic materials for sample preparation in drug and pharmaceutical analysis. *Trends Analyt. Chem.,* **2013**, *45*, 182-196.
[http://dx.doi.org/10.1016/j.trac.2012.10.017]

[108]   Namera, A.; Saito, T. Spin column extraction as a new sample preparation method in bioanalysis. *Bioanalysis,* **2015**, *7*(17), 2171-2176.
[http://dx.doi.org/10.4155/bio.15.146] [PMID: 26340717]

[109]   Barker, S.A.; Long, A.R.; Short, C.R. Isolation of drug residues from tissues by solid phase dispersion. *J. Chromatogr. A,* **1989**, *475*, 353-361.
[http://dx.doi.org/10.1016/S0021-9673(01)89689-8] [PMID: 2777960]

[110]   Barker, S.A. Matrix solid-phase dispersion. *J. Chromatogr. A,* **2000**, *885*(1-2), 115-127.
[http://dx.doi.org/10.1016/S0021-9673(00)00249-1] [PMID: 10941670]

[111]   Barker, S.A. Matrix solid phase dispersion (MSPD). *J. Biochem. Biophys. Methods,* **2007**, *70*(2), 151-162.
[http://dx.doi.org/10.1016/j.jbbm.2006.06.005] [PMID: 17107714]

[112]   Kristenson, E.M.; Ramos, L.; Brinkman, U.A. Recent advances in matrix solid-phase dispersion. *Trends Analyt. Chem.,* **2006**, *2*(2), 96-111.

[http://dx.doi.org/10.1016/j.trac.2005.05.011]

[113]   Capriotti, A.L.; Cavaliere, C.; Foglia, P.; Samperi, R.; Stampachiacchiere, S.; Ventura, S.; Laganà, A. Recent advances and developments in matrix solid-phase dispersion. *Trends Analyt. Chem.*, **2015**, *71*, 186-193.
[http://dx.doi.org/10.1016/j.trac.2015.03.012]

[114]   Anastassiades, M.; Lehotay, S.J.; Štajnbaher, D.; Schenck, F.J. Fast and easy multiresidue method employing acetonitrile extraction/partitioning and dispersive solid-phase extraction for the determination of pesticide residues in produce. *J. AOAC Int.*, **2003**, *86*(2), 412-431.
[PMID: 12723926]

[115]   Wu, X.; Hong, H.; Liu, X.; Guan, W.; Meng, L.; Ye, Y.; Ma, Y. Graphene-dispersive solid-phase extraction of phthalate acid esters from environmental water. *Sci. Total Environ.*, **2013**, *444*, 224-230.
[http://dx.doi.org/10.1016/j.scitotenv.2012.11.060] [PMID: 23274241]

[116]   González-Curbelo, M.Á.; Herrera-Herrera, A.V.; Hernández-Borges, J.; Rodríguez-Delgado, M.Á. Analysis of pesticides residues in environmental water samples using multiwalled carbon nanotubes dispersive solid-phase extraction. *J. Sep. Sci.*, **2013**, *36*(3), 556-563.
[http://dx.doi.org/10.1002/jssc.201200782] [PMID: 23303564]

[117]   Omar, M.M.; Wan Ibrahim, W.A.; Elbashir, A.A. Sol-gel hybrid methyltrimethoxysilane-tetraethoxysilane as a new dispersive solid-phase extraction material for acrylamide determination in food with direct gas chromatography-mass spectrometry analysis. *Food Chem.*, **2014**, *158*, 302-309.
[http://dx.doi.org/10.1016/j.foodchem.2014.02.045] [PMID: 24731346]

[118]   Fernandes, V.C.; Domingues, V.F.; Mateus, N.; Delerue-Matos, C. Determination of pesticides in fruit and fruit juices by chromatographic methods. An overview. *J. Chromatogr. Sci.*, **2011**, *49*(9), 715-730.
[http://dx.doi.org/10.1093/chrsci/49.9.715] [PMID: 22586249]

[119]   Tadeo, J.L.; Pérez, R.A.; Albero, B.; García-Valcárcel, A.I.; Sánchez-Brunete, C. Review of sample preparation techniques for the analysis of pesticide residues in soil. *J. AOAC Int.*, **2012**, *95*(5), 1258-1271.
[http://dx.doi.org/10.5740/jaoacint.SGE_Tadeo] [PMID: 23175957]

[120]   Bruzzoniti, M.C.; Checchini, L.; De Carlo, R.M.; Orlandini, S.; Rivoira, L.; Del Bubba, M. QuEChERS sample preparation for the determination of pesticides and other organic residues in environmental matrices: a critical review. *Anal. Bioanal. Chem.*, **2014**, *406*(17), 4089-4116.
[http://dx.doi.org/10.1007/s00216-014-7798-4] [PMID: 24770804]

[121]   González-Curbelo, M.Á.; Socas-Rodríguez, B.; Herrera-Herrera, A.V.; González-Sálamo, J.; Hernández-Borges, J.; Rodríguez-Delgado, M.Á. Evolution and applications of the QuEChERS method. *Trends Analyt. Chem.*, **2015**, *71*, 169-185.
[http://dx.doi.org/10.1016/j.trac.2015.04.012]

[122]   Plössl, F.; Giera, M.; Bracher, F. Multiresidue analytical method using dispersive solid-phase extraction and gas chromatography/ion trap mass spectrometry to determine pharmaceuticals in whole blood. *J. Chromatogr. A*, **2006**, *1135*(1), 19-26.
[http://dx.doi.org/10.1016/j.chroma.2006.09.033] [PMID: 17049535]

[123]   Usui, K.; Hayashizaki, Y.; Hashiyada, M.; Funayama, M. Rapid drug extraction from human whole blood using a modified QuEChERS extraction method. *Leg Med (Tokyo)*, **2012**, *14*(6), 286-296.
[http://dx.doi.org/10.1016/j.legalmed.2012.04.008] [PMID: 22682428]

[124]   Anzillotti, L.; Odoardi, S.; Strano-Rossi, S. Cleaning up blood samples using a modified QuEChERS procedure for the determination of drugs of abuse and benzodiazepines by UPLC-MSMS(□). *Forensic Sci. Int.*, **2014**, *243*, 99-106.
[http://dx.doi.org/10.1016/j.forsciint.2014.05.005] [PMID: 24907511]

[125]   Wu, Q.; Wang, C.; Liu, Z.; Wu, C.; Zeng, X.; Wen, J.; Wang, Z. Dispersive solid-phase extraction followed by dispersive liquid-liquid microextraction for the determination of some sulfonylurea herbicides in soil by high-performance liquid chromatography. *J. Chromatogr. A*, **2009**, *1216*(29),

5504-5510.
[http://dx.doi.org/10.1016/j.chroma.2009.05.062] [PMID: 19523645]

[126]   Brewer, W.E. Disposable pipette extraction. U.S. Patent 6,566,145, 2003.

[127]   Guan, H.; Brewer, W.E.; Garris, S.T.; Morgan, S.L. Disposable pipette extraction for the analysis of pesticides in fruit and vegetables using gas chromatography/mass spectrometry. *J. Chromatogr. A,* **2010**, *1217*(12), 1867-1874.
[http://dx.doi.org/10.1016/j.chroma.2010.01.047] [PMID: 20144461]

[128]   Schroeder, J.L.; Marinetti, L.J.; Smith, R.K.; Brewer, W.E.; Clelland, B.L.; Morgan, S.L. The analysis of delta9-tetrahydrocannabinol and metabolite in whole blood and 11-nor-delt-9-tetrahydrocannabinol-9-carboxylic acid in urine using disposable pipette extraction with confirmation and quantification by gas chromatography-mass spectrometry. *J. Anal. Toxicol.,* **2008**, *32*(8), 659-666.
[http://dx.doi.org/10.1093/jat/32.8.659] [PMID: 19007518]

[129]   Ellison, S.T.; Brewer, W.E.; Morgan, S.L. Comprehensive analysis of drugs of abuse in urine using disposable pipette extraction. *J. Anal. Toxicol.,* **2009**, *33*(7), 356-365.
[http://dx.doi.org/10.1093/jat/33.7.356] [PMID: 19796505]

[130]   Samanidou, V.; Stathatos, C.; Njau, S.; Kovatsi, L. Disposable pipette extraction for the simultaneous determination of biperiden and three antipsychotic drugs in human urine by GC-nitrogen phosphorus detection. *Bioanalysis,* **2013**, *5*(1), 21-29.
[http://dx.doi.org/10.4155/bio.12.292] [PMID: 23256469]

[131]   Mozaner Bordin, D.C.; Alves, M.N.; Cabrices, O.G.; de Campos, E.G.; De Martinis, B.S. A rapid assay for the simultaneous determination of nicotine, cocaine and metabolites in meconium using disposable pipette extraction and gas chromatography-mass spectrometry (GC-MS). *J. Anal. Toxicol.,* **2014**, *38*(1), 31-38.
[http://dx.doi.org/10.1093/jat/bkt092] [PMID: 24272386]

[132]   Bordin, D.C.; Alves, M.N.; de Campos, E.G.; De Martinis, B.S. Disposable pipette tips extraction: Fundamentals, applications and state of the art. *J. Sep. Sci.,* **2016**, *39*(6), 1168-1172.
[http://dx.doi.org/10.1002/jssc.201500932] [PMID: 27027593]

[133]   Scheidweiler, K.B.; Newmeyer, M.N.; Barnes, A.J.; Huestis, M.A. Quantification of cannabinoids and their free and glucuronide metabolites in whole blood by disposable pipette extraction and liquid chromatography-tandem mass spectrometry. *J. Chromatogr. A,* **2016**, *1453*, 34-42.
[http://dx.doi.org/10.1016/j.chroma.2016.05.024] [PMID: 27236483]

[134]   Šafaříkováa, M.; Šafařík, I. Magnetic solid-phase extraction. *J. Magn. Magn. Mater.,* **1999**, *194*(1-3), 108-112.
[http://dx.doi.org/10.1016/S0304-8853(98)00566-6]

[135]   Giakisikli, G.; Anthemidis, A.N. Magnetic materials as sorbents for metal/metalloid preconcentration and/or separation. A review. *Anal. Chim. Acta,* **2013**, *789*, 1-16.
[http://dx.doi.org/10.1016/j.aca.2013.04.021] [PMID: 23856225]

[136]   Wierucka, M.; Biziuk, M. Application of magnetic nanoparticles for magnetic solid-phase extraction in preparing biological, environmental and food samples. *Trends Analyt. Chem.,* **2014**, *59*, 50-58.
[http://dx.doi.org/10.1016/j.trac.2014.04.007]

[137]   Kaur, R.; Hasan, A.; Iqbal, N.; Alam, S.; Saini, M.K.; Raza, S.K. Synthesis and surface engineering of magnetic nanoparticles for environmental cleanup and pesticide residue analysis: a review. *J. Sep. Sci.,* **2014**, *37*(14), 1805-1825.
[http://dx.doi.org/10.1002/jssc.201400256] [PMID: 24777942]

[138]   Herrero-Latorre, C.; Barciela-García, J.; García-Martín, S.; Peña-Crecente, R.M.; Otárola-Jiménez, J. Magnetic solid-phase extraction using carbon nanotubes as sorbents: a review. *Anal. Chim. Acta,* **2015**, *892*, 10-26.
[http://dx.doi.org/10.1016/j.aca.2015.07.046] [PMID: 26388472]

[139]   Ibarra, I.S.; Rodriguez, J.A.; Miranda, J.M.; Vega, M.; Barrado, E. Magnetic solid phase extraction based on phenyl silica adsorbent for the determination of tetracyclines in milk samples by capillary electrophoresis. *J. Chromatogr. A,* **2011,** *1218*(16), 2196-2202.
[http://dx.doi.org/10.1016/j.chroma.2011.02.046] [PMID: 21397241]

[140]   He, Z.; Liu, D.; Li, R.; Zhou, Z.; Wang, P. Magnetic solid-phase extraction of sulfonylurea herbicides in environmental water samples by Fe3O4@dioctadecyl dimethyl ammonium chloride@silica magnetic particles. *Anal. Chim. Acta,* **2012,** *747*, 29-35.
[http://dx.doi.org/10.1016/j.aca.2012.08.015] [PMID: 22986132]

[141]   Zhang, N.; Peng, H.; Wang, S.; Hu, B. Fast and selective magnetic solid phase extraction of trace Cd, Mn and Pb in environmental and biological samples and their determination by ICP-MS. *Mikrochim. Acta,* **2011,** *175*, 121-128.
[http://dx.doi.org/10.1007/s00604-011-0659-3]

[142]   Gao, Q.; Luo, D.; Bai, M.; Chen, Z.W.; Feng, Y.Q. Rapid determination of estrogens in milk samples based on magnetite nanoparticles/polypyrrole magnetic solid-phase extraction coupled with liquid chromatography-tandem mass spectrometry. *J. Agric. Food Chem.,* **2011,** *59*(16), 8543-8549.
[http://dx.doi.org/10.1021/jf201372r] [PMID: 21749040]

[143]   Belardi, R.P.; Pawliszyn, J. The application of chemically modified fused silica fibers in the extraction of organics from water matrix samples and their rapid transfer to capillary columns. *Water Pollut. Res. J. Can.,* **1989,** *24*(1), 179-191.

[144]   Arthur, C.L.; Pawliszyn, J. Solid phase microextraction with thermal desorption using fused silica optical fibers. *Anal. Chem.,* **1990,** *62*(19), 2145-2148.
[http://dx.doi.org/10.1021/ac00218a019]

[145]   Pawliszyn, J., Ed. *Solid phase microextraction: Theory and practice*; Wiley-VCH: New York, **1997.**

[146]   Jeleń, H.H.; Wlazły, K.; Wasowicz, E.; Kamiński, E. Solid-phase microextraction for the analysis of some alcohols and esters in beer: Comparison with static headspace method. *J. Agric. Food Chem.,* **1998,** *46*(4), 1469-1473.
[http://dx.doi.org/10.1021/jf9707290]

[147]   Xu, J.; Zheng, J.; Tian, J.; Zhu, F.; Zeng, F.; Su, C.; Ouyang, G. New materials in solid-phase microextraction. *Trends Analyt. Chem.,* **2013,** *47*, 68-83.
[http://dx.doi.org/10.1016/j.trac.2013.02.012]

[148]   Queiroz, M.E.; Melo, L.P. Selective capillary coating materials for in-tube solid-phase microextraction coupled to liquid chromatography to determine drugs and biomarkers in biological samples: a review. *Anal. Chim. Acta,* **2014,** *826*, 1-11.
[http://dx.doi.org/10.1016/j.aca.2014.03.024] [PMID: 24793847]

[149]   Aziz-Zanjani, M.O.; Mehdinia, A. A review on procedures for the preparation of coatings for solid phase microextraction. *Mikrochim. Acta,* **2014,** *181*(11), 1169-1190.
[http://dx.doi.org/10.1007/s00604-014-1265-y]

[150]   Pawliszyn, J., Ed. *Applications of Solid Phase Microextraction*; Royal Society of Chemistry: Cambridge, UK, **1999.**

[151]   Pragst, F. Application of solid-phase microextraction in analytical toxicology. *Anal. Bioanal. Chem.,* **2007,** *388*(7), 1393-1414.
[http://dx.doi.org/10.1007/s00216-007-1289-9] [PMID: 17476482]

[152]   Musteata, M.L.; Musteata, F.M. Analytical methods used in conjunction with solid-phase microextraction: a review of recent bioanalytical applications. *Bioanalysis,* **2009,** *1*(6), 1081-1102.
[http://dx.doi.org/10.4155/bio.09.88] [PMID: 21083077]

[153]   Kumar, A.; Malik, A.K.; Matysik, F.M. Analysis of biological samples using solid-phase microextraction. *Bioanal. Rev.,* **2009,** *1*(1), 35-55.
[http://dx.doi.org/10.1007/s12566-009-0004-z]

[154] Kataoka, H. Recent developments and applications of microextraction techniques in drug analysis. *Anal. Bioanal. Chem.*, **2010**, *396*(1), 339-364.
[http://dx.doi.org/10.1007/s00216-009-3076-2] [PMID: 19727680]

[155] Farhadi, K.; Hatami, M.; Matin, A.A. Microextraction techniques in therapeutic drug monitoring. *Biomed. Chromatogr.*, **2012**, *26*(8), 972-989.
[PMID: 22767149]

[156] Abdulrauf, L.B.; Chai, M.K.; Tan, G.H. Applications of solid-phase microextraction for the analysis of pesticide residues in fruits and vegetables: a review. *J. AOAC Int.*, **2012**, *95*(5), 1272-1290.
[http://dx.doi.org/10.5740/jaoacint.SGE_Abdulrauf] [PMID: 23175958]

[157] Kohler, I.; Schappler, J.; Rudaz, S. Microextraction techniques combined with capillary electrophoresis in bioanalysis. *Anal. Bioanal. Chem.*, **2013**, *405*(1), 125-141.
[http://dx.doi.org/10.1007/s00216-012-6367-y] [PMID: 22965532]

[158] Merkle, S.; Kleeberg, K.K.; Fritsche, J. Recent Developments and Applications of Solid Phase Microextraction (SPME) in Food and Environmental Analysis—A Review. *Chromatography*, **2015**, *2*(3), 293-381.
[http://dx.doi.org/10.3390/chromatography2030293]

[159] Eisert, R.; Pawliszyn, J. Automated in-tube solid-phase microextraction coupled to high-performance liquid chromatography. *Anal. Chem.*, **1997**, *69*(16), 3140-3147.
[http://dx.doi.org/10.1021/ac970319a]

[160] Kataoka, H. Automated sample preparation using in-tube solid-phase microextraction and its application a review. *Anal. Bioanal. Chem.*, **2002**, *373*(1-2), 31-45.
[http://dx.doi.org/10.1007/s00216-002-1269-z] [PMID: 12012170]

[161] Kataoka, H.; Ishizaki, A.; Nonaka, Y.; Saito, K. Developments and applications of capillary microextraction techniques: a review. *Anal. Chim. Acta*, **2009**, *655*(1-2), 8-29.
[http://dx.doi.org/10.1016/j.aca.2009.09.032] [PMID: 19925911]

[162] Rasanen, I.; Viinamäki, J.; Vuori, E.; Ojanperä, I. Headspace in-tube extraction gas chromatography-mass spectrometry for the analysis of hydroxylic methyl-derivatized and volatile organic compounds in blood and urine. *J. Anal. Toxicol.*, **2010**, *34*(3), 113-121.
[http://dx.doi.org/10.1093/jat/34.3.113] [PMID: 20406534]

[163] Campíns-Falcó, P.; Verdú-Andrés, J.; Sevillano-Cabeza, A.; Herráez-Hernández, R.; Molins-Legua, C.; Moliner-Martinez, Y. In-tube solid-phase microextraction coupled by in valve mode to capillary LC-DAD: Improving detectability to multiresidue organic pollutants analysis in several whole waters. *J. Chromatogr. A*, **2010**, *1217*(16), 2695-2702.
[http://dx.doi.org/10.1016/j.chroma.2010.01.018] [PMID: 20102766]

[164] Queiroz, M.E.; Melo, L.P. Selective capillary coating materials for in-tube solid-phase microextraction coupled to liquid chromatography to determine drugs and biomarkers in biological samples: a review. *Anal. Chim. Acta*, **2014**, *826*, 1-11.
[http://dx.doi.org/10.1016/j.aca.2014.03.024] [PMID: 24793847]

[165] Musteata, F.M.; Musteata, M.L.; Pawliszyn, J. Fast *in vivo* microextraction: a new tool for clinical analysis. *Clin. Chem.*, **2006**, *52*(4), 708-715.
[http://dx.doi.org/10.1373/clinchem.2005.064758] [PMID: 16497936]

[166] Cudjoe, E.; Bojko, B.; Togunde, P.; Pawliszyn, J. *In vivo* solid-phase microextraction for tissue bioanalysis. *Bioanalysis*, **2012**, *4*(21), 2605-2619.
[http://dx.doi.org/10.4155/bio.12.250] [PMID: 23173795]

[167] Bojko, B.; Pawliszyn, J. *In vivo* and *ex vivo* SPME: a low invasive sampling and sample preparation tool in clinical bioanalysis. *Bioanalysis*, **2014**, *6*(9), 1227-1239.

[http://dx.doi.org/10.4155/bio.14.91] [PMID: 24946923]

[168] Xie, W.; Mullett, W.M.; Miller-Stein, C.M.; Pawliszyn, J. Automation of in-tip solid-phase microextraction in 96-well format for the determination of a model drug compound in human plasma by liquid chromatography with tandem mass spectrometric detection. *J. Chromatogr. B Analyt. Technol. Biomed. Life Sci.,* **2009**, *877*(4), 415-420.
[http://dx.doi.org/10.1016/j.jchromb.2008.12.036] [PMID: 19144575]

[169] Xie, W.; Chavez-Eng, C.M.; Fang, W.; Constanzer, M.L.; Matuszewski, B.K.; Mullett, W.M.; Pawliszyn, J. Quantitative liquid chromatographic and tandem mass spectrometric determination of vitamin D3 in human serum with derivatization: a comparison of in-tube LLE, 96-well plate LLE and in-tip SPME. *J. Chromatogr. B Analyt. Technol. Biomed. Life Sci.,* **2011**, *879*(17-18), 1457-1466.
[http://dx.doi.org/10.1016/j.jchromb.2011.03.018] [PMID: 21459053]

[170] Xie, W.; Mullett, W.; Pawliszyn, J. High-throughput polymer monolith in-tip SPME fiber preparation and application in drug analysis. *Bioanalysis,* **2011**, *3*(23), 2613-2625.
[http://dx.doi.org/10.4155/bio.11.267] [PMID: 22136050]

[171] Es-haghi, A.; Zhang, X.; Musteata, F.M.; Bagheri, H.; Pawliszyn, J. Evaluation of bio-compatible poly(ethylene glycol)-based solid-phase microextraction fiber for *in vivo* pharmacokinetic studies of diazepam in dogs. *Analyst (Lond.),* **2007**, *132*(7), 672-678.
[http://dx.doi.org/10.1039/b701423f] [PMID: 17592586]

[172] Vuckovic, D.; Cudjoe, E.; Hein, D.; Pawliszyn, J. Automation of solid-phase microextraction in high-throughput format and applications to drug analysis. *Anal. Chem.,* **2008**, *80*(18), 6870-6880.
[http://dx.doi.org/10.1021/ac800936r] [PMID: 18712934]

[173] Mirnaghi, F.S.; Chen, Y.; Sidisky, L.M.; Pawliszyn, J. Optimization of the coating procedure for a high-throughput 96-blade solid phase microextraction system coupled with LC-MS/MS for analysis of complex samples. *Anal. Chem.,* **2011**, *83*(15), 6018-6025.
[http://dx.doi.org/10.1021/ac2010185] [PMID: 21711040]

[174] Mirnaghi, F.S.; Monton, M.R.; Pawliszyn, J. Thin-film octadecyl-silica glass coating for automated 96-blade solid-phase microextraction coupled with liquid chromatography-tandem mass spectrometry for analysis of benzodiazepines. *J. Chromatogr. A,* **2012**, *1246*, 2-8.
[http://dx.doi.org/10.1016/j.chroma.2011.11.030] [PMID: 22197254]

[175] Mirnaghi, F.S.; Pawliszyn, J. Development of coatings for automated 96-blade solid phase microextraction-liquid chromatography-tandem mass spectrometry system, capable of extracting a wide polarity range of analytes from biological fluids. *J. Chromatogr. A,* **2012**, *1261*, 91-98.
[http://dx.doi.org/10.1016/j.chroma.2012.07.012] [PMID: 22824218]

[176] Abdel-Rehim, M. New trend in sample preparation: on-line microextraction in packed syringe for liquid and gas chromatography applications. I. Determination of local anaesthetics in human plasma samples using gas chromatography-mass spectrometry. *J. Chromatogr. B Analyt. Technol. Biomed. Life Sci.,* **2004**, *801*(2), 317-321.
[http://dx.doi.org/10.1016/j.jchromb.2003.11.042] [PMID: 14751801]

[177] Abdel-Rehim, M. Recent advances in microextraction by packed sorbent for bioanalysis. *J. Chromatogr. A,* **2010**, *1217*(16), 2569-2580.
[http://dx.doi.org/10.1016/j.chroma.2009.09.053] [PMID: 19811788]

[178] Blomberg, L.G. Two new techniques for sample preparation in bioanalysis: microextraction in packed sorbent (MEPS) and use of a bonded monolith as sorbent for sample preparation in polypropylene tips for 96-well plates. *Anal. Bioanal. Chem.,* **2009**, *393*(3), 797-807.
[http://dx.doi.org/10.1007/s00216-008-2305-4] [PMID: 18696053]

[179] Rani, S.; Malik, A.K. A novel microextraction by packed sorbent-gas chromatography procedure for the simultaneous analysis of antiepileptic drugs in human plasma and urine. *J. Sep. Sci.,* **2012**, *35*(21), 2970-2977.
[http://dx.doi.org/10.1002/jssc.201200439] [PMID: 22997153]

[180] Alves, G.; Rodrigues, M.; Fortuna, A.; Falcão, A.; Queiroz, J. A critical review of microextraction by packed sorbent as a sample preparation approach in drug bioanalysis. *Bioanalysis,* **2013,** *5*(11), 1409-1442.
[http://dx.doi.org/10.4155/bio.13.92] [PMID: 23742310]

[181] Moein, M.M.; Abdel-Rehim, A.; Abdel-Rehim, M. Microextraction by packed sorbent (MEPS). *Trends Analyt. Chem.,* **2015,** *67,* 34-44.
[http://dx.doi.org/10.1016/j.trac.2014.12.003]

[182] Moein, M.M.; Said, R.; Abdel-Rehim, M. Microextraction by packed sorbent. *Bioanalysis,* **2015,** *7*(17), 2155-2161.
[http://dx.doi.org/10.4155/bio.15.154] [PMID: 26395543]

[183] Baltussen, E.; Sandra, P.; David, F.; Cramers, C. Stir bar sorptive extraction (SBSE), a novel extraction technique for aqueous samples: Theory and principles. *J. Micro.,* **1999,** *11*(10), 737-747.
[http://dx.doi.org/10.1002/(SICI)1520-667X(1999)11:10<737::AID-MCS7>3.0.CO;2-4]

[184] Huang, X.; Qiu, N.; Yuan, D. Direct enrichment of phenols in lake and sea water by stir bar sorptive extraction based on poly (vinylpyridine-ethylene dimethacrylate) monolithic material and liquid chromatographic analysis. *J. Chromatogr. A,* **2008,** *1194*(1), 134-138.
[http://dx.doi.org/10.1016/j.chroma.2008.04.030] [PMID: 18471822]

[185] Hayasaka, Y.; MacNamara, K.; Baldock, G.A.; Taylor, R.L.; Pollnitz, A.P. Application of stir bar sorptive extraction for wine analysis. *Anal. Bioanal. Chem.,* **2003,** *375*(7), 948-955.
[http://dx.doi.org/10.1007/s00216-003-1837-x] [PMID: 12707766]

[186] Kawaguchi, M.; Ito, R.; Saito, K.; Nakazawa, H. Novel stir bar sorptive extraction methods for environmental and biomedical analysis. *J. Pharm. Biomed. Anal.,* **2006,** *40*(3), 500-508.
[http://dx.doi.org/10.1016/j.jpba.2005.08.029] [PMID: 16242285]

[187] Lancas, F.M.; Queiroz, M.E.; Grossi, P.; Olivares, I.R. Recent developments and applications of stir bar sorptive extraction. *J. Sep. Sci.,* **2009,** *32*(5-6), 813-824.
[http://dx.doi.org/10.1002/jssc.200800669] [PMID: 19278005]

[188] Sanchez-Rojas, F.; Bosch-Ojeda, C.; Cano-Pavon, J.M. A review of stir bar sorptive extraction. *Chromatographia,* **2009,** *69*(1), 79-94.
[http://dx.doi.org/10.1365/s10337-008-0687-2]

[189] Prieto, A.; Basauri, O.; Rodil, R.; Usobiaga, A.; Fernández, L.A.; Etxebarria, N.; Zuloaga, O. Stir-bar sorptive extraction: A view on method optimisation, novel applications, limitations and potential solutions. *J. Chromatogr. A,* **2010,** *1217*(16), 2642-2666.
[http://dx.doi.org/10.1016/j.chroma.2009.12.051] [PMID: 20083248]

[190] Soini, H.A.; Klouckova, I.; Wiesler, D.; Oberzaucher, E.; Grammer, K.; Dixon, S.J.; Xu, Y.; Brereton, R.G.; Penn, D.J.; Novotny, M.V. Analysis of volatile organic compounds in human saliva by a static sorptive extraction method and gas chromatography-mass spectrometry. *J. Chem. Ecol.,* **2010,** *36*(9), 1035-1042.
[http://dx.doi.org/10.1007/s10886-010-9846-7] [PMID: 20809147]

[191] Kawaguchi, M.; Takatsu, A.; Ito, R.; Nakazawa, H. Applications of stir-bar sorptive extraction to food analysis. *Trends Analyt. Chem.,* **2013,** *45,* 280-293.
[http://dx.doi.org/10.1016/j.trac.2013.01.007]

[192] Assoumani, A.; Lissalde, S.; Margoum, C.; Mazzella, N.; Coquery, M. *In situ* application of stir bar sorptive extraction as a passive sampling technique for the monitoring of agricultural pesticides in surface waters. *Sci. Total Environ.,* **2013,** *463-464,* 829-835.
[http://dx.doi.org/10.1016/j.scitotenv.2013.06.025] [PMID: 23856404]

[193] Camino-Sánchez, F.J.; Rodríguez-Gómez, R.; Zafra-Gómez, A.; Santos-Fandila, A.; Vílchez, J.L. Stir bar sorptive extraction: recent applications, limitations and future trends. *Talanta,* **2014,** *130,* 388-399.
[http://dx.doi.org/10.1016/j.talanta.2014.07.022] [PMID: 25159426]

[194] He, M.; Chen, B.; Hu, B. Recent developments in stir bar sorptive extraction. *Anal. Bioanal. Chem.,* **2014**, *406*(8), 2001-2026.
[http://dx.doi.org/10.1007/s00216-013-7395-y] [PMID: 24136250]

[195] Hashemi, S.H.; Kaykhaii, M.; Khajeh, M. Molecularly imprinted polymers for stir bar sorptive extraction: synthesis, characterization, and application. *Anal. Lett.,* **2015**, *48*(12), 1815-1829.
[http://dx.doi.org/10.1080/00032719.2014.1003431]

[196] Nazyropoulou, C.; Samanidou, V. Stir bar sorptive extraction applied to the analysis of biological fluids. *Bioanalysis,* **2015**, *7*(17), 2241-2250.
[http://dx.doi.org/10.4155/bio.15.129] [PMID: 26354598]

CHAPTER 2

New Materials Employment in the Electrochemical Sensors Development

Arnaldo C. Pereira[1,*], Daniela N. Silva[1], Débora A. R. Moreira[1] and Juliana F. Giarola[2]

[1] *Departamento de Ciências Naturais, Universidade Federal de São João del Rei ,36301-160, São João del Rei, MG, Brasil*

[2] *Instituto de Química, Universidade Estadual de Campinas, 13083-970, Campinas – SP, Brasil*

Abstract: Electrochemical sensors have been widely used in recent decades because they have many advantages such as high sensitivity, selectivity, stability, with the use of simple instrumentation low cost and can be successfully applied in the samples of clinical, environmental interest, and industry in general. Due to the versatility of these devices, they can be prepared from materials that provide higher reactivity and selectivity. The carbon based carbon materials are widely employed because of the ability to form a wide variety of composite materials. The composites based on graphene and carbon nanotubes have attracted great interest because they have important features, such as high speed in transferring electrons, high surface area, good chemical and mechanical stability, and therefore, they are used in the development of electrochemical sensors. Other promising materials used in the electrodes modification to electroanalytical applications are the Molecularly Imprinted Polymers and / or Ionically (MIP) and (IIP) that have become an important analytical tool due to the biomimetic recognition systems such as the specific antigen-antibody. These materials have many advantages when compared to organic systems such as ease of synthesis, stability in storage for long periods of time and cost. Some inorganic compounds have also been applied as modifiers of electrodes to enhance their electrochemical properties. Among these inorganic compounds, there are the nanoparticles of noble metals, which provide an increased surface area due to the nanoscale and nanostructured metal oxides that present numerous electronic properties and different applications. Thus this study aims to explore the use of new materials used in the development of electrochemical sensors.

Keywords: Carbon nanotubes, Carbon quantum dots, Electrochemical sensors, Fullerene, Graphene, Molecularly Imprinted Polymers and or Ionically (MIP) and (IIP), Nanofibers, Noble metal nanoparticles

* **Corresponding author Arnaldo C. Pereira:** Departamento de Ciências Naturais, Universidade Federal de São João del Rei ,36301-160, São João del Rei, MG, Brasil; Tel: 55 32 988879785; Fax: 55 32 33792483; E-mail: arnaldocsp@yahoo.com.br

INTRODUCTION

The field of material science has contributed much to the development of new materials for electroanalytical applications. Currently, nanomaterials as noble metal nanoparticles and carbon based have attracted interest from the scientific community for their use in chemical sensing, medicine, electronics, agriculture, biotechnology, biomedical and bioanalytical applications. This is because these materials possess unique properties such as excellent electrocatalytic activities, excellent mechanical, optical and conductive properties. With recent advances in nanotechnology, the development of electrochemical sensors have been widely studied because they are fast, sensitive, have a low cost method and low detection and quantification limits, may be used in several analytic determination and it is considered one of the greatest developments in the field of electrochemical devices [1].

Molecularly Imprinted Polymers and / or Ionically (MIP) and (IIP) are also materials used in the electrodes modification to electroanalytical applications. Molecular imprinting is a promising strategy to design a matrix that can be tailor-made materials with high selectivity for a target molecule. In this way, MIPs show an excellent affinity for the template molecule than for other structurally related compounds [2]. The IIP are the materials designed to mimic the binding sites of biological groups and assure an improved recognition of the template species, thereby contributing to increasing selectivity in sensors [3].

The present review discusses the latest advances in sensor technology achieved by the assembly of materials as noble metal nanoparticles, carbon nanotubes, graphene, nanofibers, fullerene, carbon quantum dots, Molecularly Imprinted Polymers and or Ionically (MIP) and (IIP) used in analytical devices.

Noble Metal Nanoparticles

The nanoparticles (NPs) have attracted the attention of researchers for decades, and are promising materials due to their potential applications in nanotechnology [4 - 6]. Nanotechnology is a science and engineering field that refers to materials having dimensions of 100 nm or less order. The concept of nanotechnology was introduced by R. Feynman in 1959, and later described by Taniguchi, 1974 [7].

In recent years, nanomaterials, such as noble metal nanoparticles, have attracted the interest of the scientific community, due to the various possibilities of technological applications. Because they have extremely small dimensions and large specific surface area, the nanoparticles exhibit properties that differ significantly from macrostructural materials [8, 9]. An important feature in the use of nanostructured materials is related to the adjustment and alteration of electronic, optical and mechanical without intervention in the chemical structure. Thus these materials are mainly applied in the field of analytical chemistry for developing electrochemical sensors [10 - 12], which present excellent chemical properties such as; high electrochemical stability, wide working range and potential formation of various oxides, which may be used as catalysts [13, 14].

Among the nanomaterials, the gold nanoparticles have a special role. Brown *et al.* in 1996 described one of the first applications of gold nanoparticles monitoring a reversible electrochemistry reaction of horse heart Cytochrome c (Cc) obtained at SnO_2 electrodes modified with 12-nm-diameter colloidal Au particles [14].

In the last years, the number of published works with electrochemical sensors based on nanoparticles of noble metal have been increasing according to the database to the ScienceDirect® as shown in Fig. (**1**).

In the present review, some of the latest advances in the area of electrochemical sensor development using the following nanoparticles of noble metals: gold, silver, iridium, palladium, platinum and ruthenium are shown.

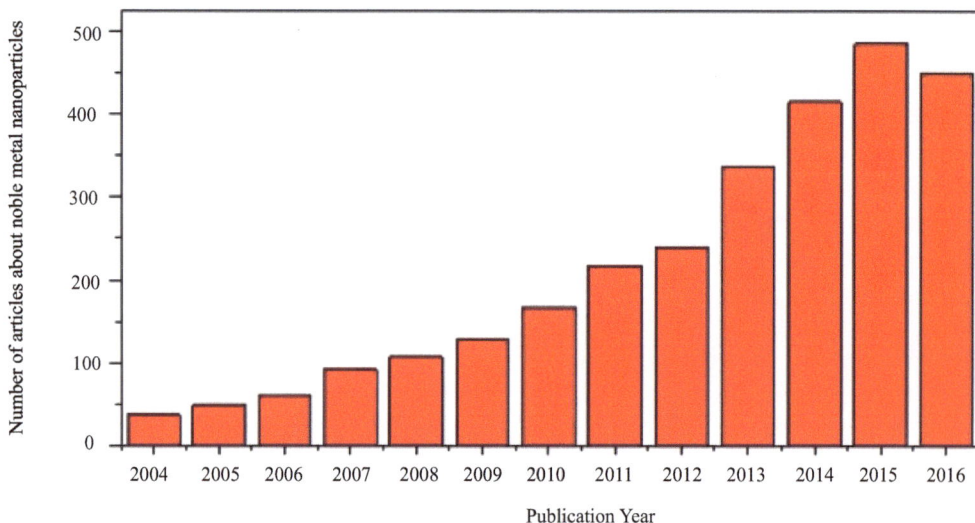

Fig. (1). Number of articles published using electrochemical sensors based on nanoparticles of noble metal in the last 12 years. The values were obtained from ScienceDirect· database in June 2016 using electrochemical sensors based on nanoparticles of noble metal as keywords.

Gold Nanoparticles (AuNPs)

Gold nanoparticles (AuNPs) have played important role in nanoscience and nanotechnology due to characteristics such as high stability, biocompatibility, excellent conductivity. They present effective and electrochemically active surface areas for the adsorption of biomolecules and successfully speed up the electron transfer between the electrode and the analyte to be detected, which can lead to a rapid and sensitive electro-chemical current response. Specific size and morphology of the AuNPs have been the focus of intensive research because of the potential applications in the field of electronic, optical and magnetic devices [15, 16].

In analytical sciences, the nanostructures present several advantages when used as transducers or as a component of the recognition layer in sensing device. When used as transducers, several improvements are achieved such as mass transport, more availability of reactional sites, aggregation/dispersion optical effect and the increase in the optical signal due to surface Plasmon resonance properties of each nanoparticle added [17].

Synthesis of Gold Nanoparticles

The applications and physical properties of nanoparticles can be modified by controlling their size and morphology, thus several chemical and physical methods, such as sol-gel, chemical reduction, laser ablation, thermolysis, among others, are used in the synthesis of different types of nanoparticles [18].

Ionescua *et al.* introduced a new technique that ensures the fabrication of ultra-pure monolayer-capped gold nanoparticles for chemical gas sensing devices. This novel approach involves two steps:(i) physical vapour deposition to make dispersed, ultra-pure, size-controlled, gold nanoparticles (AuNPs), and (ii) coating and functionalizing of the AuNPs with organic ligands. A physical vapour deposition technique was first employed to make dispersed ultra-pure size-controlled gold nanoparticles, and this step was followed by a coating process for functionalization of the gold nanoparticles with an organic ligand, specifically dodecanethiol. Gas sensing experiments demonstrated that these devices are suitable for detecting volatile organic compounds [19].

However, the green synthesis is one of the most popular methods currently used since it is an economic, effective and eco-friendly method. Through this method, it has been possible to obtain bimetallic, semiconductor and oxide nanoparticles, with different applications, such as: nanotechnological, catalytic, biomedical, antimicrobial, among others [20]. Ikram *et al.* in 2016 described a biosynthesises of gold nanoparticles (Fig. **2**) [21]. The study on the synthesis methods and the application fields of gold nanoparticles have been growing due to the contributions of their physicochemical properties and in the fields such as biomedicine, optics, and electronics [20].

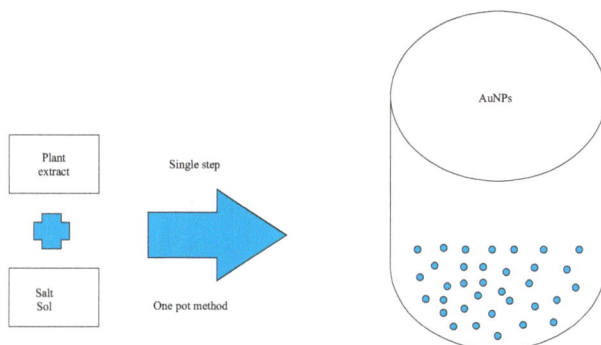

Fig. (2). One step simple method of synthesis of gold nanoparticles using plant extract. (Reference 21).

Application of Gold Nanoparticles

Many analytical methods have been developed in recent years based on gold nanoparticles as shown in Table **1**.

Table 1. Examples of electrochemical sensors based on gold nanoparticles described in the literature.

Analyte	Modifier	Electrode / Transduction	Feature or application	Reference
Bisphenol A	Gold nanoparticles supported carbon nanotubes	Screen printed (gold)/ Voltammetric	LOD = 1.31×10^{-7} mol L^{-1} and application in real plastic samples	22
Clindamycin	Graphene oxide and gold nanoparticles within a film of crosslinked chitosan	Glassy carbon/ Voltammetric	LOD = 2.9×10^{-7} mol L^{-1} and application in pharmaceutical formulations, as well as synthetic urine and river water samples	23
Estrone	Gold nanoparticles (AuNPs) electro-synthesized on a 1-naphtylamine polymer (pNap) film	Glassy carbon/ Voltammetric	LOD = 0.226 pmol L^{-1} and application in water samples	24
Hydrazine	Nitrogen-doped graphene (NG)-polyvinylpyrrolidone (PVP)/gold nanoparticles	Screen printed (gold) / Voltammetric	LOD = 0.07 mol L^{-1} and application in high sugar fruit and vegetable samples	25
Carbofuran	Graphene oxide and gold nanoparticles	Glassy carbon/ Voltammetric	LOD = 0.02×10^{-6} mol L^{-1} and application in real vegetable samples	26

LOD = Limit of detection

Silver Nanoparticles (AgNPs)

The silver nanoparticles (AgNPs) have attracted the attention of the scientific community because of the simple preparation process and advantages, such as biocompatibility, catalysis and low toxicity, making them suitable for improving materials' properties for the preparation of modified electrodes [27, 28].

The electrochemical sensors based on silver nanoparticles are highly preferred for the determination of various analytes as, phenolic compounds, because of their low cost and the conductivity of Ag being higher than other noble metals such as Au, Pd and Pt. The AgNPs acting as nanoelectrocatalyst can facilitate the electron-transfer process during electrocatalysis [29].

Synthesis of Silver Nanoparticles

Silver nanoparticles are synthesized using different methods including: chemical reduction of Ag^+ with or without protecting agent, thermal decomposition, photochemical reduction, and sonochemical reduction. Fungus-mediated synthesis *in situ* generation in polymer film, and microwave- assisted reduction have also been reported [30]. These conventional techniques have various limitations; for instance, complexity, high cost, low stability, toxicity, environmental unfri-endliness. These issues have persuaded researchers to develop simple, safe and reliable methods to synthesize nanoparticles [31].

Ikram *et al.* in 2016, described their work, the green syntheses of silver nanoparticles using plant extracts as illustrated in Fig. (**3**). The use of plants as the production assembly of silver nanoparticles has drawn attention, because of their rapid, ecofriendly, non-pathogenic, and economical protocol providing a single step technique for the biosynthetic processes [32].

Application of Silver Nanoparticles

The most important applications of sensors based on silver nanoparticles recently reported in the literature, are shown in Table **2**.

Iridium Nanoparticles (IrNPs)

The iridium nanoparticles (IrNPs) have been applied in modified electrodes mainly due to excellent electrochemical reversibility at a wide pH range. It is a great candidate for many applications, such as in the preparation of clinical diagnostics devices, power sources, electrochromic devices, pH sensing, and sensors and biosensors production. Among the noble metal nanoparticles, IrNPs have emerged as candidates for catalysis due to their high activity, stability, and selectivity under several reaction conditions. Due to high electron transfer rate constant of Ir(IV)/Ir(III) redox couple, it can be used as an electron transfer mediator for electrocatalytic processes and electroanalysis [38 - 40].

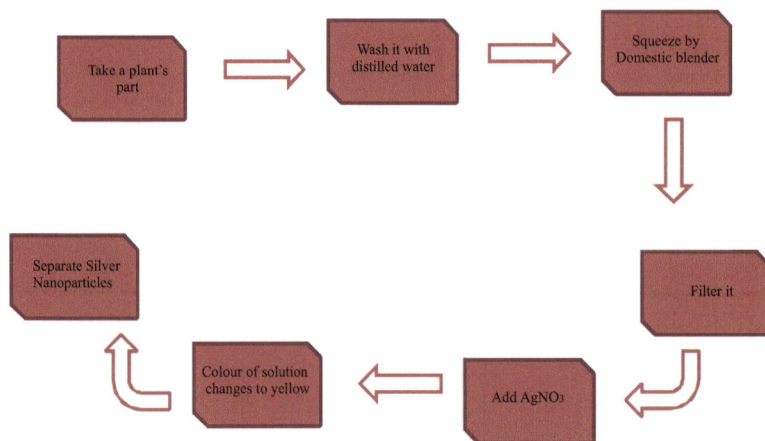

Fig. (3). Protocol for synthesis of silver nanoparticles using plant extract. (Reference 32).

Table 2. Examples of electrochemical sensors based on silver nanoparticles described in the literature.

Analyte	Modifier	Electrode / Transduction	Feature or Application	Reference
Entacapone	Silver (Ag) nanoparticles were electrochemically deposited on the film of a metformin functionalized multiwalled carbon nanotube	Glassy carbon / amperometric	LOD = 15.3 x 10^{-9} mol $L^{-1.}$ Application in pharmaceutical formulations and human urine samples	[33]
Catechol	Trilayers combining a cationic electrolyte, an anionic metallic complex, an anionic phospholipid 1,2-dipalmitoyl-sn-3-glycero-(phosphor-rac-(1-glycerol), DPPG) and silver nanoparticles (AgNPs)	Platinum plate / Voltammetric	LOD = 0.87 x 10^{-6} mol L^{-1} and application in green tea samples	[34]
Glucose	Ag nanoparticles	Carbon paste electrodes/ Voltammetric	LOD = 5.5 x 10^{-6} mol L^{-1} application in human blood serums samples	[35]

(Table 2) contd.....

Analyte	Modifier	Electrode / Transduction	Feature or Application	Reference
Isoniazid	A copolymer of methyl methacrylate and 2-acrylamid--2-methylpropane sulfonic acid (P(MMA-coAMPS)) and silver nanoparticles (Ag NPs)	Glassy carbon / Voltammetric and amperometric	LOD = 10.0 x 10^{-9} mol L^{-1} and application in a pharmaceutical formulation	[36]
4-nitrophenol	Graphene oxide-silver nanocomposite	Glassy carbon / Voltammetric	LOD = 1.2 x 10^{-9} mol L^{-1} and application in real water samples	[37]

LOD = Limit of detection

In general, the behavior of nanomaterials for several applications are highly dependent on the particle size, shape, and size distribution as well as the surrounding environment. This has brought a huge interest in developing methods that allow control over the synthesis parameters and, therefore a better control over the final characteristics of the nanomaterials [41, 42].

Synthesis of Iridium Nanoparticles

In recent years, material scientists have been attempting to synthesize size and morphology controlled Ir nanocatalysts by various synthesis methods, such as thermal decomposition, ionic liquids, solvothermal treatment, microwave, or UV irradiation; however, to do the controlled synthesis of iridium nanoparticles, the synthetic chemists still have a long way to go [43].

Supports are frequently used for the synthesis of nanoparticles, this strategy not only provides nanoparticles with a narrower particle size distribution but it also reduces the tendency of the nanoparticles to aggregate. Furthermore, supports may also provide mechanical strength to the nanostructure and offer additional desired properties for certain applications which would not be obtained, neither by the nanoparticles nor by the support [43]. Seong *et al.* in 2014, showing the fabrication iridium nanoparticles supported in carbon nanotubes (IrNPs-SWCNT) (Fig. **4**) [39]. The SWCNT film electrode was first loaded with iridium nanoparticles (IrNPs) by an easily controllable chronocoulometry technique.The SWCNT film electrode coated with IrNPs was characterized by field emission scanning electron microscopy and electrochemical techniques, like cyclic voltammetry and amperometry.

Application of Iridium Nanoparticles

The recent applications of sensors based on iridium nanoparticles reported in the literature, are shown in Table **3**.

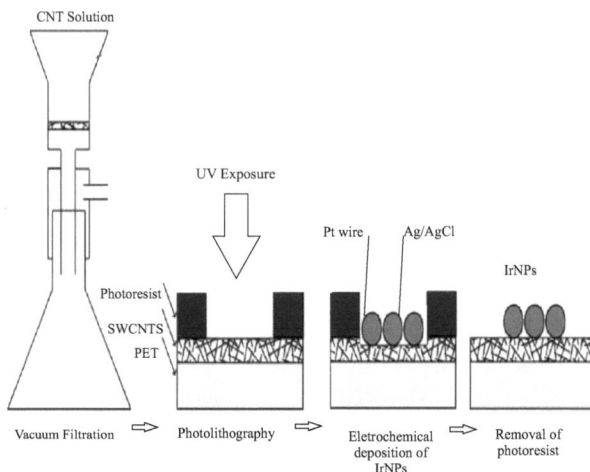

Fig. (4). Schematic diagrams showing the fabrication of IrNPs-SWCNT film electrodes. (Reference 39).

Table 3. Examples of electrochemical sensors based on iridium nanoparticles described in the literature.

Analyte	Modifier	Electrode / Transduction	Feature or Application	Reference
Glucose	Single-walled carbon nanotube film electrodes coated with iridium nanoparticles	Pt wire / Voltammetric and Amperometric	LOD = 17 x 10^{-6} mol L^{-1} and application in human serum	[39]
Butylated hydroxyanisole	Iridium oxide nanoparticles	Glassy carbon / Voltammetric	LOD = 100 x 10^{-9} mol L^{-1} and application in spiked commercial samples	[44]
Ochratoxin A	Polythionine (PTH) and iridium oxide nanoparticles	Screen printed carbon electrode / Voltammetric	LOD = 14 x 10^{-12} mol L^{-1} application in wine samples	[45]
Rutin.	Iridium nanoparticles dispersed in ionic liquid (IL) 1-butyl-3-methylimidazolium hexafluorophosphate	Carbon paste electrodes / Voltammetric	LOD = 0,079 x 10^{-6} mol L^{-1} and application in simulated samples	[46]

(Table 3) contd.....

Analyte	Modifier	Electrode / Transduction	Feature or Application	Reference
Iodate and periodate	Iridium oxide nanoparticles	Glassy carbon / Amperometric	LOD = 36 and 5 x 10^{-9} mol L^{-1} (iodate and periodate) and application in iodate solution is 1.5%	[47]

LOD = Limit of detection

Ruthenium Nanoparticles (RuNPs)

The metal nanoparticles have been widely studied in recent years which possess advantageous characteristics to be used with their unique optical properties [48]. They have excellent physical properties, chemical reactivity applications in electronics and act as catalysts [49].

The ruthenium nanoparticles have been applied to activities as a catalyst in various synthesis processes [50]. As they are the metals with different oxidation states, their versatility is great and can be used in electrochemical sensors acting as mediators in the electron transfer catalysts or redox reactions [51].

In large areas ruthenium catalysis is one of the most important noble metals to be used in catalytic reactions such as ammonia síntesre Fisher-Tropsch synthesis, reduction compounds nitro, oxidation of alcohols and amines among others [52].

Synthesis of Ruthenium Nanoparticles

The synthesis of metals is one of the most important applications in nanotechnology [53]. There are various methods for the synthesis of ruthenium, we can mention physical methods [54] and also the chemicals [55].

A very important factor in the synthesis is the amount of solvent to be used, showing that to determine high performance and adequate size of the particles formed, there is no need for a high amount of solvents, and in these cases it becomes a costly process [52].

The synthesis of nanoparticles in green processes is a major development technology, it is also an important area of research [56], therefore, alternative sources of vegetable can be used for this purpose. We can cite some sources as Moringa oleifera [57], Ocimum sanctum [58] and Syzygium cumini [59].

According to Kannan and Sundrarajan [56], a study was conducted on ruthenium hydroxide precipitation processes using hydrochloric acid solution and the extract of leaves Acalypha Indica. The formed nanoparticles remained stable, obtained by an environmentally friendly process and the process of synthesis is shown in Fig. (**5**).

Application of Ruthenium Nanoparticles

The most important applications of sensors based on ruthenium nanoparticles recently reported in the literature, are shown in Table **4**.

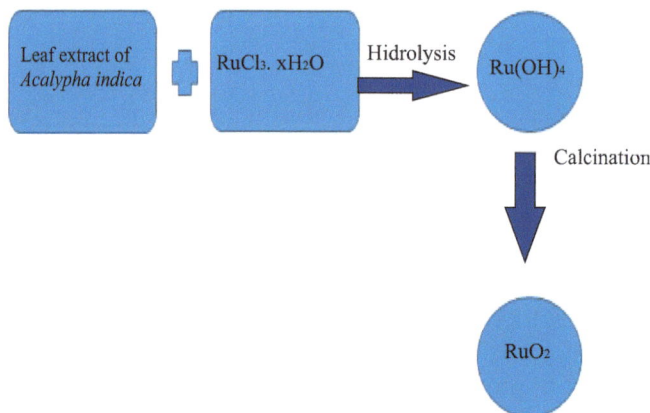

Fig. (5). Synthesis mechanism of ruthenium nanoparticles by green process. (Reference 56).

Table 4. Examples of electrochemical sensors based on ruthenium nanoparticles described in the literature.

Analyte	Modifier	Electrode / Transduction	Feature or Application	Reference
Adrenaline, Uric Acid, and Cysteine	Ruthenium oxide nanoparticles	Glassy Carbon/ Voltammetric	LOD = UA 0.47 x 10^{-6} mol L^{-1} and AD 0.45 µmol L^{-1}, application in human urine sample	[60]
Methyldopa	Pt–Ru nanoparticles on multi-walled carbon nanotubes	Glassy Carbon/ Voltammetric	LOD = 10 x 10^{-9} mol L^{-1} Application in pharmaceutical and clinical preparations	[61]
Glucose	Electrochemiluminescence of poly[tris(N-bipyridylethyl)pyrrole] ruthenium(II) film by Au nanoparticles	Glassy Carbon/ Voltammetric	LOD = 5 x 10^{-6} mol L^{-1} Promising strategy for biomolecules detection	[62]

(Table 4) contd.....

Analyte	Modifier	Electrode / Transduction	Feature or Application	Reference
Iodate and Periodate	Nano-scale islands of ruthenium oxide	Glassy Carbon/ Voltammetric and Amperometric	LOD = IO 0.9 x 10^{-6} mol L^{-1} and PE 0.2 x 10^{-6} mol L^{-1}. Application in water sample.	[63]
Ascorbic acid, Dopamine and N-acetyl-l-cystein	Electrodeposited acetaminophen on ruthenium oxide nanoparticles	Glassy Carbon/ Voltammetric	LOD = 2.84 x 10^{-6} mol L^{-1}. Application in water sample.	[64]

LOD = Limit of detection

Platinum Nanoparticles (PtNPs)

The platinum nanoparticles have attracted the attention of several researchers of nanotechnologies and other areas, due to their numerous applications as catalysts, fuel cell technology and sensors [65, 66].

The chemical or electrochemical deposition of noble metals has an important role in the technological development, because some factors interfere at nanoparticles' formation. Parameters, such as temperature, pH and current density, are important in the production of nanoparticles in general [67].

Synthesis of Platinum Nanoparticles

These include the synthesis by Sonawane *et al.* [68] as already mentioned in other work, the synthesis of platinum nanoparticles was performed using a microreactor flow. Overall synthesis of nanoparticles using a flow microreactor generates nanoparticles of smaller sizes having greater stability and control of the process, when compared with the traditional processes [69].

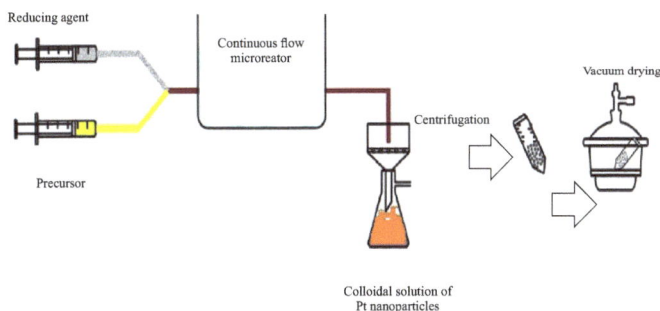

Fig. (6). Experimental set-up for synthesis of Pt nanoparticles in a continuous flow microreactor [68].

In the work of Sonawane *et al.* [68], the synthesis of platinum nanoparticles was performed using a microreactor flow, represented by Fig. (**6**). It was observed that the copper tube was looped in a spiral fashion where each loop had a diameter of 20 mm. Pt precursor and sodium borohydride solutions were injected into the microreactor using two separate peristaltic pumps through Y-connector. At the end of the process, ultra-small particles were obtained, which were controlled by the reaction rate during synthesis.

Application of Platinum Nanoparticles

The recent applications of electrochemical sensors based on platinum nanoparticles are described in the Table **5**.

Table 5. Examples of electrochemical sensors based on platinum nanoparticles described in the literature.

Analyte	Modifier	Electrode / Transduction	Feature or Application	Reference
Isoniazida	Gold@platinum core@shell nanoparticles	Glassy Carbon/ Amperometric	LOD = 26 x 10 $^{-9}$ mol L^{-1}, application in human blood serum and urine samples	[70]
Tetracycline	Platinum nanoparticles supported on carbon	Glassy Carbon/ Voltammetric	LOD = 14.3 x 10 $^{-6}$ mol L^{-1}, application in urine samples	[71]
Xhantine	Reduced graphene oxide-carboxymethylcellulose layered with platinum nanoparticles/PAMAM dendrimer/magnetic nanoparticles Hybrids	Glassy Carbon/ Voltammetric and Amperometric	LOD = 13 x 10 $^{-9}$ mol L^{-1}, application in fish samples	[72]
Paracetamol, Cetirizine and Phenylephrine	Multiwalled carbon nanotube-platinum nanoparticles nanocomposite	Carbon paste/ Voltammetric	LOD = PCT 27.9 x 10 $^{-9}$ mol L^{-1}; Ctz 58.6 x 10 $^{-9}$ mol L^{-1} and Phe 28.3 x 10 $^{-9}$ mol L^{-1}. Application in pharmaceutical formulations, blood serum and urine samples	[73]

LOD = Limit of detection

Palladium Nanoparticles (Pd NPs)

The palladium nanoparticles have received much attention because of their important application in the chemical area where we can highlight, Excellent catalytic performance in the fuel cell, oxidation of various organic compounds and also hydrogen production, purification and storage [74, 75].

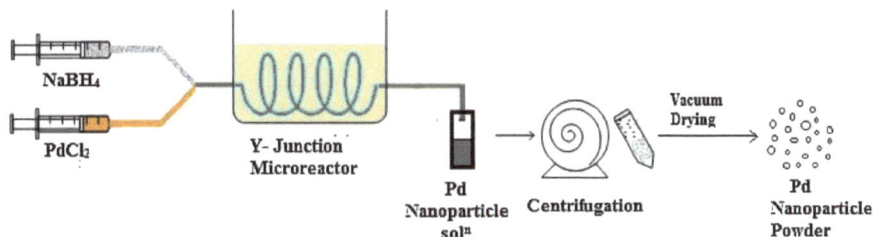

Fig. (7). Experimental setup for stable palladium (Pd) nanoparticle (NP) synthesis using continuous flow microreactor [78].

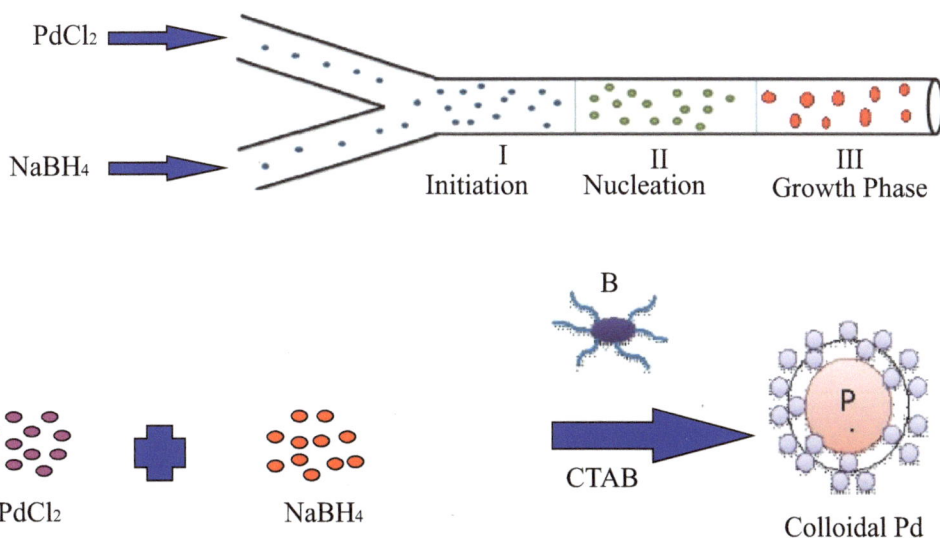

Fig. (8). (a) Mechanism of palladium nanoparticle formation in microreactor. **(b)** Representation of reaction scheme for the formation of metallic Pd nanoparticles in the presence of stabilizing agent CTAB (Reference 78).

Synthesis of Palladium Nanoparticles

In the last decades, the methods for the synthesis of palladium nanoparticles were introduced [75, 76], which observed some important factors, such as size of the particles formed [77]. According to Sharada and colleagues [78], the synthesis of palladium nanoparticles was conducted in a flow microreactor, and the detailed layout as shown in Figs. (**7** & **8**).

Application of Palladium Nanoparticles

Table **6** describes some works shown in the literature, where the palladium nanoparticles are applied in electrochemical sensors, acting as catalysts in the reactions.

Table 6. Examples of electrochemical sensors based on palladium nanoparticles described in the literature.

Analyte	Modifier	Electrode / Transduction	Feature or Application	Reference
Ascorbic Acid and Uric Acid	Electrodeposition of palladium nanoparticles on porous graphitized carbon	Carbon Paste / Amperometric	LOD = AA 0.53 x 10^{-6} mol L^{-1} and UA 0.66 µmol L^{-1} application in human serum sample	[79]
Hydroquinone (HQ) and Bisphenol A (BPA)	Ultrafine Pd nanoparticle@TiO2 functionalized SiC (Pd@TiO2–SiC)	Glassy Carbon/ Voltammetric	LOD = HQ 5.5 x10^{-9} mol L^{-1} and BPA 4.5 x10^{-9} Mol L^{-1}. Application in tap water and wastewater samples	[80]
Isoniazid	Electrodeposited palladium nanoparticles	Carbon Ionic Liquid / Voltammetric	LOD = 4.7x10^{-7} mol L^{-1} application in human blood serum and pharmaceutical samples	[81]
NADH	Palladium nanoparticles in multiwalled carbon nanotubes (PdNPs-MWCNTs)	Glassy Carbon / Amperometric	LOD = 32 x 10^{-9} mol L^{-1}, application in human blood serum.	[82]
Nonenzymatic Glucose	palladium nanoparticles on functional TiO2 nanotubes	Glassy Carbon/ Voltammetric and Amperometric	LOD = 8,0 x 10^{-8} mol L^{-1}, application in human blood serum	[83]

LOD = Limit of detection

As the nanoparticles of noble metals, carbon nanostructured materials have attracted the attention of nanotechnologists recently. The next section describes the features and properties of these materials which are employed in the modification and electrochemical sensors biosensors, with application in different types of samples.

Nanocarbon Materials

Carbon based nanomaterials have been attracted interest from the scientific community over the years because of their features, such as good electrical conductivity, high electrocatalytic effect, strong adsorptive ability and excellent biocompatibility, which make them attractive for a great number of applications in several research fields [84].

Nanocarbon materials were defined to not only determine their primary particle size on a nanometer scale, but their structures and/or textures were also controlled on a nanometer scale. Either the nano-size or nanostructure of the carbon materials had to be deliberately controlled to govern their properties and functions [85].

Different nanostructures of carbon have been widely used in constructing electrochemical sensors or biosensors such as carbon nanotubes (Fernández *et al.*, 2016) [86] graphene (Chen *et al.*, 2010) [87], nanofibers (Yang, 2015) [88], fullerene (Brahman *et al.*, 2016) [84] (Fig. **9**) [89] and carbon quantum dots (Yu *et al.*, 2016) [90] have been attracting increasing attention of nanotechnologists during the past few decades.

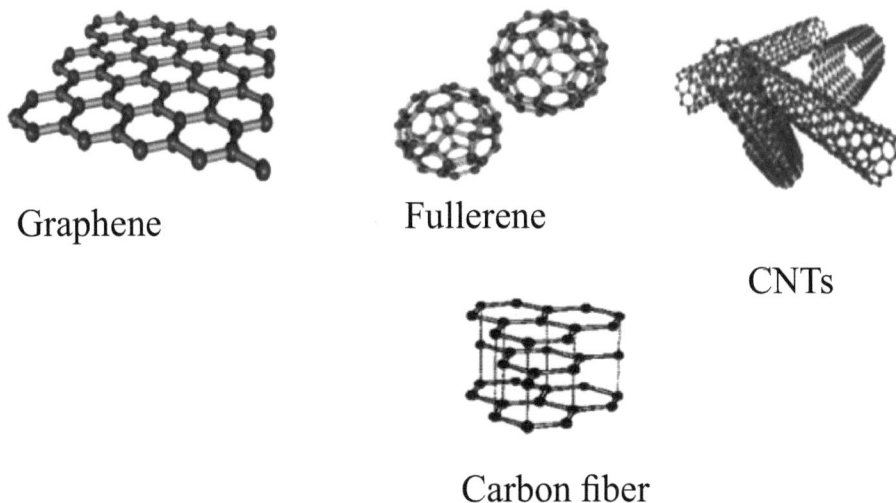

Graphene   Fullerene   CNTs

Carbon fiber

Fig. (9). Allotropes of carbon. (Adaptation of the reference [89]).

The present review highlights some relevant applications in the preparation of sensors and biosensors based on carbon nanomaterials like carbon nanotubes, graphene, nanofibers, fullerene and carbon quantum dots.

Carbon Nanotubes

Since their discovery in 1991, carbon nanotubes have particularly attracted the attention of many scientists in the fields of science and technology as an important component in the realization of nanotechnology. These materials have displayed the combination of superlative mechanical, thermal and electronic properties attributed to them [91, 92].

Carbon nanotubes are a sheet of graphite that is rolled into a tube. Different than diamond, where a 3-D diamond cubic crystal structure is formed with each carbon atom having four nearest neighbors arranged in a tetrahedron, graphite is formed as a 2-D sheet of carbon atoms arranged in a hexagonal array [93].

There are two main types of nanotubes available; single walled nanotubes (SWNT) and multi-walled nanotubes (MWNT). The SWNTs consist of a single sheet of graphene rolled seamlessly to form a cylinder with the diameter of order of 1 nm and length of up to centimeters. It is extremely difficult to separate SWNT from the bundles, making this issue a serious hurdle in the way of real applications. The MWNTs consist of an array of such cylinders formed concentrically and separated by 0.35 nm, similar to the basal plane separation in graphite. MWNTs can have diameters from 2 to 100 nm and lengths of tens of microns [94].

MWCNTs have been widely used in the study of electrochemical sensors due to their unique combination of good mechanical, electrical and electrochemical properties [95]. The CNTs are exhibit π–π conjugating structure with high hydrophobic surface having advantages of enhanced electronic properties, ability to promote electron transfer reaction, a large surface area with enhanced sensitivity and rapid electrode kinetics [96].

Synthesis of Carbon Nanotubes

The most common method for the production for MWNT is undoubtedly chemical vapor deposition (CVD). Nanotubes made from this method generally have very large quantities of defects. This means their structure is very far from the ideal rolled up hexagonal lattice. Their physical properties undergo a change due to the presence of defects with thermal, electronic and mechanical properties deviating significantly from those expected for pristine nanotubes. Therefore MWNT produced CVD are important because they can be produced in very large quantities and are relatively cheap [97].

Elaissari *et al.* described a widely used method for the large scale production of carbon nanotubes, the Chemical vapor deposition (CVD). The method utilizes hydrocarbons like CH_4, acetylene or carbon monoxide as carbon source while high temperature provides sufficient energy for the decomposition of the hydrocarbons to form carbon nanotubes (Fig. **10**) [98].

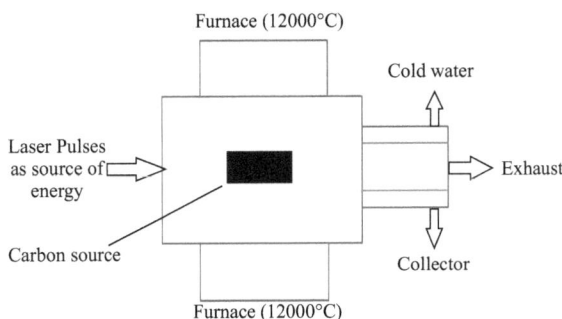

Fig. (10). Schematic illustration of chemical vapor deposition method [98].

Application of Carbon Nanotubes

Table **7** describes the recent publications on modified electrochemical sensors with carbon nanotubes.

Table 7. Examples of electrochemical sensors based on carbon nanotubes described in the literature.

Analyte	Modifier	Electrode / Transduction	Feature or Application	Reference
Adiponectin	Double-walled carbon nanotubes functionalized with 4-aminobenzoic Acid in the presence of isoamylnitrite resulting in the formation of 4-carboxyphenyl- DWCNTs	Screen printed carbon electrodes/ Voltammetric	LOD = 14.5 ng mL^{-1} and application in human serum	[99]
Ceftazidime	Molecularly imprinted polymer immobilized on multiwall carbon nanotube	Glassy carbon / Voltammetric	LOD = 0.55×10^{-9} mol L^{-1} and application in Serum samples	[100]
Propham	Multi-walled carbon nanotubes	Glassy carbon / Voltammetric	LOD = 22.8×10^{-6} mol L^{-1} and application in water samples	[101]
1-aminopyrene	Multi-walled carbon nanotube–sodium dodecyl sulfate/ Nafion (MWCNT–SDS/Nafion)	Glassy carbon / Voltammetric	LOD = 0.02×10^{-9} mol L^{-1} and application in water samples	[102]

(Table 7) contd.....

Analyte	Modifier	Electrode / Transduction	Feature or Application	Reference
Rizatriptan benzoate	Nanocomposite of multiwalled carbon nanotubes (MWCNTs) and Fe_3O_4 nanoparticles (Fe_3O_4/MWCNTs/GCE)	Glassy carbon / Voltammetric	LOD = 0.09 x 10^{-6} mol L^{-1} and application in samples of blood serum and determination in pharmaceutical	[103]
L-tryptophan	Multiwall carbon nanotube	Screen printed gold electrode/ Voltammetric	LOD = 0,36 x 10^{-6} mol L^{-1} and application in cow's milk and human plasma, saliva and urine samples	[104]
Dopamine	Poly-Alizarin red and multiwalled carbon nanotube film	Glassy carbon / Voltammetric	LOD = 1.89 × 10^{-7}mol L^{-1} and application in pharmaceutical formulations	[105]
Citalopram	Multi-walled carbon nanotube (MWCNTS) coated with poly p-aminobenzene sulfonic acid/β-cyclodextrin (p-ABSA/β-CD) film	Glassy carbon / Voltammetric	LOD = 44 x 10^{-9} mol L^{-1} and application in pharmaceutical combinations and human body fluids.	[106]

LOD = Limit of detection

Graphene

Carbon nanomaterials as the nanocomposites graphene have been explored in recent years due to advantageous features, such as high surface area, high conductivity and catalytic ability [107]. Furthermore, it has low production costs and stability [108].

The use of graphene started from 2004 when it was discovered [109]. This material has been applied in different areas such as, electronics [110], battery [111], biological sensors [112] etc. Graphene has interesting properties such as high electron mobility [113, 114], excellent thermal conductivity [115] and mechanical [116].

In recent years, many graphene synthesis processes have been used; for example we can cite mechanical exfoliation [114, 117] epitaxial growth, chemical vapor deposition [118] and also the chemical reduction of graphite oxide [119, 120]. This fact has attracted attention of many researchers due to present simplicity in the synthesis process with satisfactory results [121, 122].

Synthesis of Graphene

Mansha *et al.* [123] synthesized the graphene by a modified Hummer's method. The Fig. (**11**) shows the graphene morphology obtained.

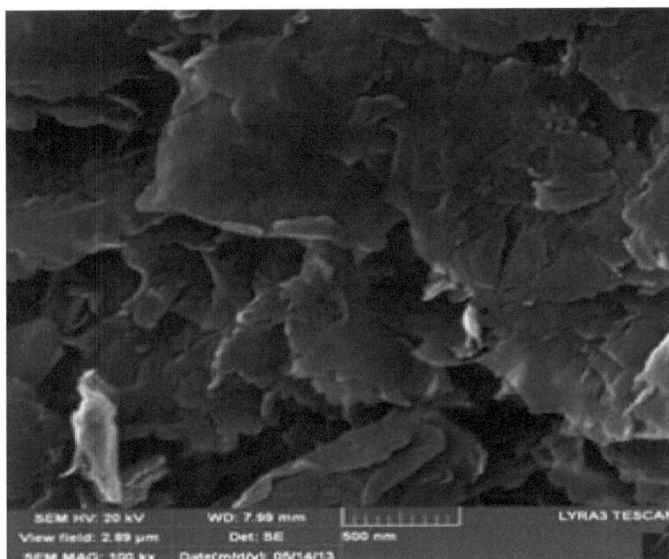

Fig. (11). Low and high magnification FESEM micrographs of Graphene (Reference 123).

Application of Graphene

Because of the wide application, Table **8** describes few works in the literature applying graphene sensors and biosensors electrochemically.

Table 8. Examples of electrochemical sensors based on graphene described in the literature.

Analyte	Modifier	Electrode / Transduction	Feature or Application	Reference
Ascorbic Acid, Dopamine and Uric Acid	Three-dimensional graphene-like carbon frameworks	Glassy Carbon/ Voltammetric	LOD = AA 2 x 10^{-6} mol L^{-1}, DA 10 nmol L^{-1} and UA 10 nmol L^{1}and application in human serum sample	[124]

(Table 8) contd.....

Analyte	Modifier	Electrode / Transduction	Feature or Application	Reference
Nitrite	Graphene Nanoribbons	Glassy Carbon/ Voltammetric and Amperometric	LOD = 0.22 x 10^{-6} mol L^{-1} and application in tap water samples	[107]
Sunset Yellow and Tartrazine. Synthetic Food Dyes	Graphene oxide and multi-walled carbon nanotubes nanocomposite	Glassy Carbon/ Voltammetric.	LOD = SY 0.025 x 10^{-6} mol L^{-1} and TT 56 nmol L^{-1} and application in orange juice	[108]
Sulfate in fine particles	Formation of heteropoly blue at poly-L-lysine-functionalized Grapheme	Glassy Carbon/ Voltammetric	LOD = 0.26 x 10^{-6} mol L^{-1}. Promising applications for the detection of sulfate in environmental fields	[125]
Tert-butylhydroquinone (TBHQ) and Butylated hydroxyanisole (BHA)	Hierarchical triple-shelled porous hollow zinc oxide spheres wrapped in graphene oxide.	Glassy Carbon/ Voltammetric	LOD = TBHQ 0.137 x 10^{-6} mol L^{-1} and BHA 0.0488 µmol L^{-1} and application in edible vegetable oil samples	[126]
Palladium	Polymerized ion-imprinted membranes at graphene	Glassy Carbon/ Amperometric	LOD = 6.4 x 10^{-9} mol L^{-1} and application in palladiumin catalyst samples	[127]
4-chlorophenol	Molecularly imprinted polymer and PDDA-functionalized graphene	Glassy Carbon/ Voltammetric	LOD = 0.3 x 10^{-6} mol L^{-1} and application in real-life water samples.	[128]

LOD = Limit of detection

Carbon Nanofibers (CNFs)

Carbon nanofibers (CNFs) have achieved a prominent relevance in the last decade because they are one of the most versatile materials with high electrical conductivity, chemical stability, flexibility and free standing characteristics, making them promising materials for an extensive range of applications such as energy storage, composites, catalysts in the field of electrochemistry among others [129, 130].

CNFs are graphitic fibers made of graphene layers aligned perpendicular, tilted or parallel to the fiber axis forming different microstructures [131]. CNFs as one-dimensional nanostructures have received considerable interest because they can be simultaneously used as transducers which relay the electrochemical signal and as matrixes for the immobilization of biomolecules [132]. They are mainly produced by the decomposition of hydrocarbons over a transition metal or alloy as catalyst at temperatures in the range 500–1200 °C [130].

Synthesis of Carbon Nanofibers

Several methods can be used to produce CNFs, such as arc-discharge, laser ablation or chemical vapor deposition (CVD) [133]. Hydrothermal methods, template synthesis method and electrospinning method have some incomparable advantages, including convenient operation, low-cost and simplified process to electrodes. As a result, CNFs are considered as the potential electrode materials with supercapacitors [134].

Xiong *et al,.* describe a novel electrochemical catalyst of phosphorus and prepared nitrogen dual-doped cobalt-based carbon nanofibers by a facile and cost-effective electrospinning technique. The CNFs nanomaterial was sintered at 800°C under N_2 atmosphere to prepare the final catalyst [135]. The Fig. (**12**) shows the illustration of the entire procedure for preparing Co-based CNFs catalysts.

Application of Carbon Nanofibers

Table **9** shows recent applications of electrochemical sensors modified with carbon nanofibers.

Table 9. Examples of electrochemical sensors modified with carbon nanofibers described in the literature.

Analyte	Modifier	Electrode / Transduction	Feature or Application	Reference
Microcystin-LR	Carbon nanofibers (CNFs) and gold nanoparticles (AuNPs)	Glassy carbon / Voltammetric	LOD = 1.69 pmol L^{-1} and application in polluted water	[136]
Vanillin	Carbon nanofibers (CNF) and surfactants	Glassy carbon / Voltammetric	LOD = 0.14 x 10^{-6} mol L^{-1} and application in foodstuff (vanilla sugar, vanilla pods, and cream milk powder)	[137]
Uric acid	Carbon nanofibers (CNFs) and carbon nanotubes (CNTs)	Glassy carbon / Voltammetric	LOD = 1.0 x 10^{-6} mol L^{-1} application in the real biological fluids	[138]

(Table 9) contd.....

Analyte	Modifier	Electrode / Transduction	Feature or Application	Reference
Guanine (G) and Adenine (A)	Nickel loaded porous carbon nanofibers (NiCNF)	Glassy carbon / Voltammetric	LOD = 0.03 e 0.03 x 10^{-6} mol l L^{-1} (G and A respectively) and application in fish sperm DNA sample	[139]
Dopamine (DA), uric acid (UA) and ascorbic acid (AA)	Palladium nanoparticle-loaded carbon nanofibers (Pd/CNFs)	Carbon paste electrode / Voltammetric	LOD = 0.2 x 10^{-6} mol L^{-1}, 0.7 x 10^{-6} mol L^{-11} and 15 x 10^{-6} mol L^{-1} for DA, UA and AA, respectively mol L^{-1} and application of DA in injectable medicine and UA in urine sample	[140]

LOD = Limit of detection

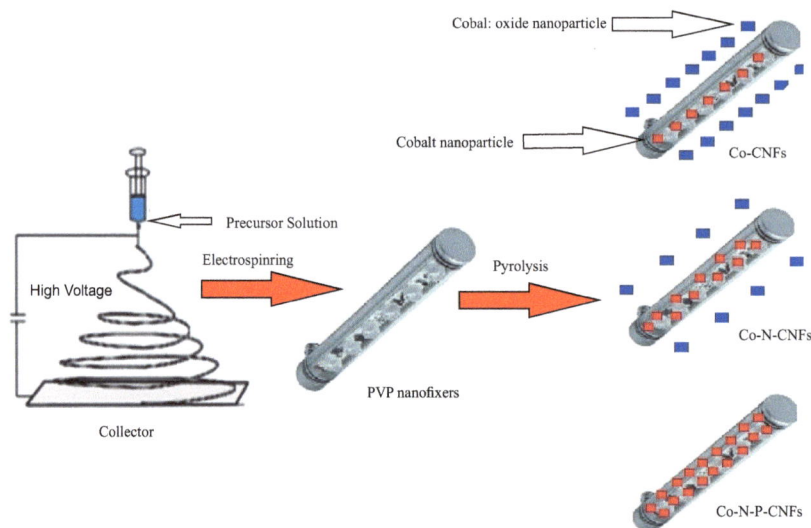

Fig. (12). Illustration of the preparation procedures for the Co-CNFs, CoeN-CNFs and CoeNeP-CNFs samples (Reference 135).

Fullerene

The fullerene (C_{60}) is an allotrope of carbon material that has been applied in different areas due their unique properties [141, 142], being the electrocatalysis, the area of greatest application of fullerene due to advantageous characteristic as the chemical stability [143, 144]. In recent decades many researchers have studied the electrochemical behavior of partially reduced fullerene films, which make them excellent electrical conductors depending on the cation in the reduction process [145 - 147]. Thus the fullerene films can be used as modifying agents which can improve the catalytic process, influencing the sensitivity and selectivity of the sensor during the analysis [148].

Syntesis of Fullerene

Since the discovery of fullerene synthesis [149], new ways have been studied to demonstrate the fullerene curved surfaces [150, 151]. According to Astefanei *et al.* [152], nowadays, fullerenes and fullerene surface functionalized are used in optical applications, electronics, cosmetics and in the field of biomedical. In recent years functionalized fullerene was produced in larger amounts than crystalline fullerenes to create biologically compatible forms. An important feature of fullerenes is that they are hydrophobic, having low solubility in water of about 8ng L^{-1}. Thus several methods have been developed to disperse these materials in the aqueous medium, leading to the formation of a stable colloid. (Fig. **13**) shows the structure of C_{60}, C_{70} and of two surface modified fullerenes.

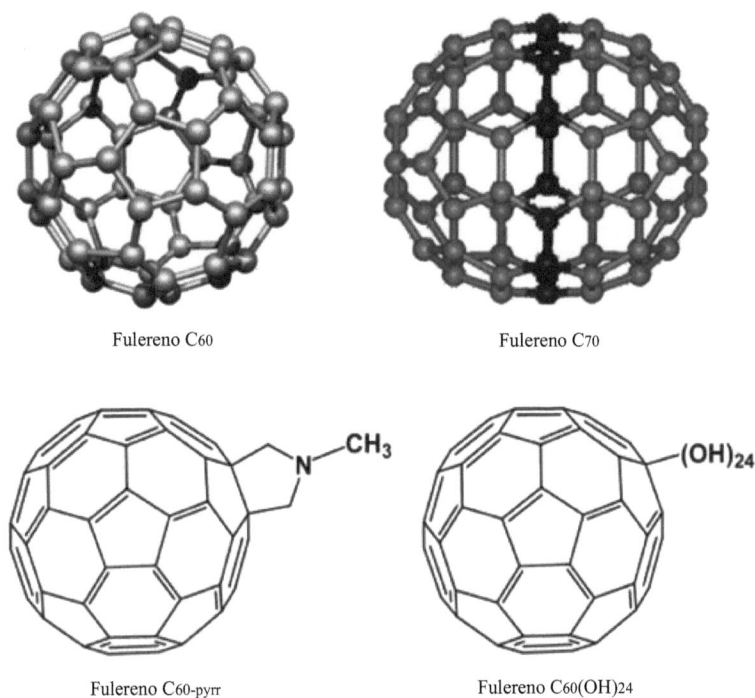

Fulereno C60   Fulereno C70

Fulereno C60-pyrr   Fulereno C60(OH)24

Fig. (13). C_{60}, C_{70}, C_{60}-pyrr and $C_{60}(OH)_{24}$ structures (Reference 152).

Application of Fullerene

Fullerenes have excellent physical and chemical properties when incorporated into polymers, becoming easily processable. Thin films containing fullerenes are of current high interest, owing to the possibility of transferring the interesting fullerene properties to bulk materials by simple surface coating [153].

Table **10** describes the recent publications on the modified electrochemical sensors using fullerene.

Table 10. Examples of electrochemical sensors based on fullerene described in the literature.

Analyte	Modifier	Electrode / Transduction	Feature or Application	Reference
Paracetamol	Nanocomposite film based on CuNPs/fullerene-C60/MWCNTs	Carbon paste/ Voltammetric	LOD = 7.3×10^{-11} molL^{-1} and application in blood serum, plasma and urine samples.	[154]
Norepinephrine (NE), Isoprenaline (IP) and Dopamine (DA)	Fullerene-functionalized carbon nanotubes/ionic liquid	Carbon/ Voltammetric	LOD = 18×10^{-9} mol L^{-1}, NE 22 nmol L^{-1} and 15 nmol L^{-1}DA	[155]
Dopamine	Palladium nanoparticles decorated on activated fullerene	Screen printed carbon/ Voltammetric	LOD = 0.056×10^{-6} mol L^{-1}. Promising application in pharmaceutical samples	[156]
Hydrazine and Hydroxylamine	Fullerene-functionalized carbon nanotubes/ionic liquid nanocomposite	Glassy carbon/ Voltammetric	LOD = 17×10^{-9} mol L^{-1} Hydrazine and 28 nmol L^{-1} Hydroxylamine and application in two water samples	[157]
Pyruvic Acid	Fullerene–C60–MWCNT composite	Glassy carbon/ Voltammetric	LOD = 0.1×10^{-6} mol L^{-1} and and application in biological samples	[158]

LOD = Limit of detection

Carbon Quantum Dots

Carbon quantum dots (CQDs) are of below 10 nm size, have low toxicity, low cost and simple synthetic routes [159]. They were discovered accidentally in a single-walled carbon nanotube purification process [160]. The main composition elements are: carbon, hydrogen and oxygen, but it can also be performed with some doping elements, such as boron [161], nitrogen [162] and sulfur [163], which make powerful methods to improve CQDs properties. In recent years, CQDs have become a promising material because of their diverse applications, such as the catalysis, electro catalysis photo sensors, biosensors *etc.* [164 - 166].

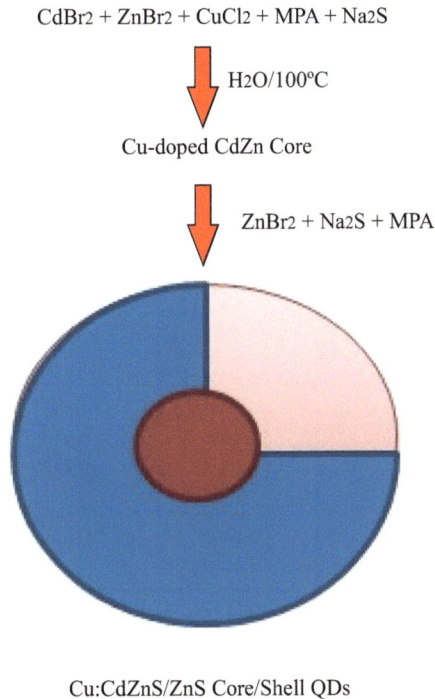

Fig. (14). A schematic illustration of the synthesis of Cu:CdZnS/ZnS QDs (Reference 167).

Synthesis of Carbon Quantum Dots

According to Yakoubi *et al.* [167], there was a low cost aqueous-based synthesis for the production of high quality CdZnS and Cu-doped CdZnS quantum dots (QDs). The scheme synthetic is outlined in Fig. (**14**).

Another widely used method in the synthesis of quantum dots graphene was reported by Martinez et al [168]. Briefly, 2 g of citric acid was placed in a round bottom flask of 5 ml and heated at 200 ° C for 30 min to obtain an orange liquid which was neutralized with NaOH solution. At the end of the process, an aqueous solution of GQDs was obtained. Fig. (**15**) shows the process of synthesis of the quantum dots of graphene.

Application of Carbon Quantum Dots

Table **11** describes some studies in the literature about the different electrochemical applications quantum dots carbon-based sensors.

Table 11. Examples of electrochemical sensors based on carbon quantum dots described in the literature.

Analyte	Modifier	Electrode / Transduction	Feature or Application	Reference
Olanzapine	Immobilization of bisphosphoramidate-derivative and quantum dots onto multi-walled carbon nanotubes	Gold/ Voltammetric and Amperometric	LOD = 6 x 10 $^{-9}$ mol L^{-1} and application in pharmaceuticals and human serum samples	[169]
Puerarin	CdTe quantum Pots-decorated poly(diallyldimethylammonium chloride)-functionalized graphene nanocomposite	Glassy Carbon/ Amperometric	LOD = 0.6 x 10 $^{-9}$ mol L^{-1}. Application in human plasma and pharmaceutical	[170]
Warfarin	Covalent immobilization of quantum dots onto carboxylated multiwalled carbon nanotubes and Chitosan composite film.	Glassy Carbon/ Voltammetric	LOD = 8.5 x 10 $^{-9}$ mol L^{-1} and application in urine serum and milk	[171]
H$_2$O$_2$	Carbon quantum dots via salep hydrothermal treatment as the silver nanoparticles	Glassy Carbon/ Voltammetric and Amperometric	LOD = 80 x 10 $^{-9}$ mol L^{-1}and application in fetal bovine serum	[172]
Dopamine and Epinephrine	Carbon quantum dots	Glassy Carbon/ Voltammetric	LOD = Dop 4,6 x 10 $^{-9}$ mol L^{-1} and Epn 6,1 nmol L^{-1} and application in urine	[166]
L-cysteine	Polypyrrole and graphene quantum dots @ Prussian Blue hybrid film	Graphite felt/ Voltammetric and Amperometric	LOD = 0.15 x 10 $^{-6}$ molL^{-1}. Promising application in medicine, food control and analysis of biological tissues	[173]

LOD = Limit of detection

The next session will address chemical printing technology, where MIP and IP were used for the development of electrochemical sensors. In this case, a great advantage of these materials is that they provide base increased selectivity to the electrode. The characteristics of these materials justify increasing the electroanalysis area.

Molecularly and Ion Imprinted Polymers

The chemical imprinting technology has been widely used as a tool in selective recognition of a given target being used as the substitute of biomolecules that provides recognition [174, 175].

In general, the recognition elements are accountable for specifically recognizing and binding to analyte in a sample way. The transducer then translates the chemical signal generated by this binding or conversion, into an easy quantification output signal. However, the poor stability of biomolecules prevents their use in harsh environments. An alternative for this problem involves the use of biomimetic receptor systems, capable of binding the target molecule or have ion affinities to form a par with natural receptors [174, 176].

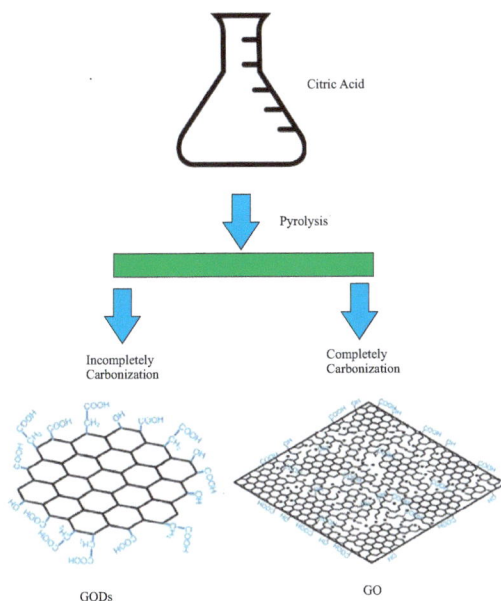

Fig. (15). Synthesis of GQDs and GO. The black dots in GO represent oxygen atoms (Reference 168).

Thus, there is a necessity of the use of molecularly imprinted polymers (MIPs) and ion imprinted polymers (IIPs) as recognition elements (molecules or ions) in the sensors. These are discussed in the next section.

Molecularly Imprinted Polymer (MIP)

The nature of biomolecular interactions has always instigated the interest of researchers in relation to the phenomena which govern the manifestation of biological response and the factors responsible for the inherent selectivity. The concept of molecular imprinting appeared from Pauling's theory for the formation of antibodies, in which an antigen was employed as a template for molding the polypeptide chain, yielding a complementary configuration of the respective antigen-antibodies chain. From there, other works began to appear. In 1972, Wulff and Sarhan [177] synthesized an enantioselective organic polymer for glyceric acid, where the interactions between this molecule and the functional monomers (FM) were of covalent nature. The main characteristic of these polymers was their high selectivity, as a result of good interaction between the FM and the template molecule (TM). However, the removal of MM binding site proved to be very difficult. Thus, in 1981, Arshady and Mosbach [178] reported a MIP that interacted with the MM via non-covalent bonds, making the removal process susceptible to the factors; such as changing the pH, ionic strength, solvent and others. Finally, Whitcombeet al [179], in 1995 proposed a new MIP, where MM and FM interacted by covalent bonds at the time of synthesis and by non-covalent bonds at the time of re-link (other links of MM with the site of MIP). After the synthesis, hydrolysis was necessary to remove the MM of the MIP site, remaining the linker groups capable of interacting non-covalently with the MM in future interactions [180, 181].

The synthesis of the MIP is based on the chemical polymerization of an FM and a cross-linking agent in the presence of a molecule used as a template, often interacting with them by relatively weak interactions, as hydrogen bonds [182]. Therefore, the molecule employed as a template (usually, is used the analyte of interest) interacts by covalent or not with the monomers molecules and, in the sequence, an agent is added in the medium reaction which promotes cross-linking in the polymer to form a rigid matrix. The polymerization reaction is initiated after adding a radical initiator. Finally, the TM is removed from the polymeric matrix using a solvent or by chemical cleavage of the molecule. The resulting polymer will have the size, as well as the capacity of selectively retaining the TM present in a complex sample [181].

The preparation principle of the MIP is quite simple, but without the appropriate combination of reaction components as well as the type of procedure employed in the synthesis, the resulting polymer may gain undesirable characteristics in terms of morphology and uniformity of the particles and the selective sites. In Fig. (**16**), the process of the formation of the MIP is shown [181, 183].

At the end of polymerization and after removal of the TM, a rigid polymer is obtained, containing sites with high affinity for the template, due to acquired shape from the functional groups of the monomer around the TM [182].

MIPs are usually prepared by the polymerization method known as "bulk", where the reaction is made in the homogeneous system. This reaction is carried out in sealed vials containing monomer, analyte, solvent, cross-linking agent and radical initiator. The reaction occurs in the absence of oxygen, under the flow of N_2 or Ar and induced with heat and/or UV radiation. The oxygen must be eliminated from the reaction medium because it retards the reaction of radical polymerization. Finally, the resulting polymeric solid is crushed, screened and subjected to a solvent wash for analyte extraction. This process is illustrated in Fig. (**17**).

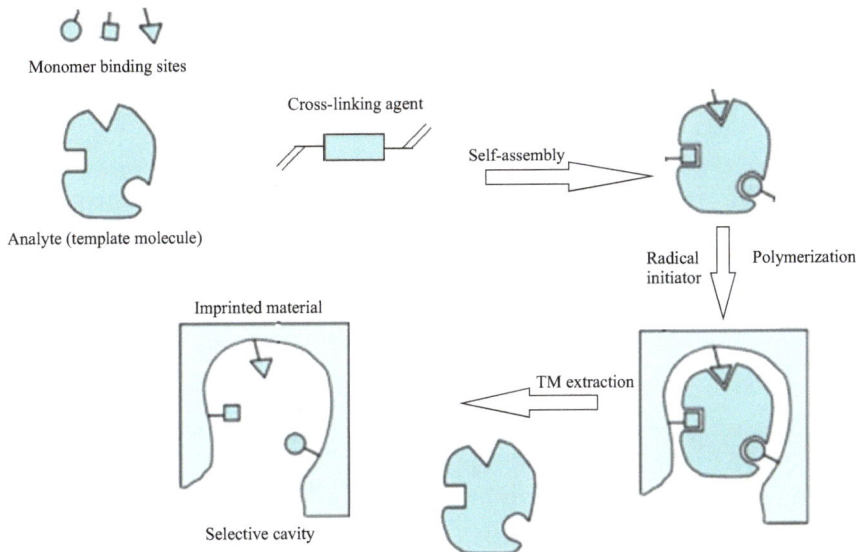

Fig. (16). Schematic representation of the synthesis of MIP (Adaptation of the [181] and [183]).

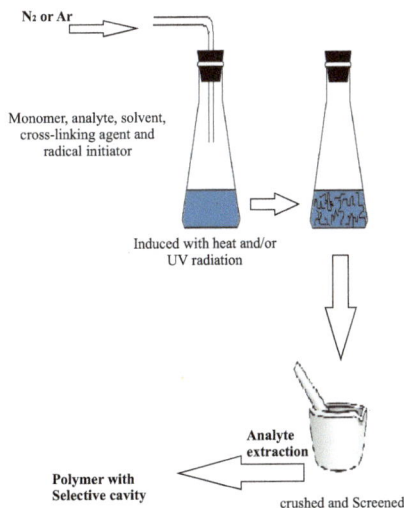

Fig. (17). Scheme of MIP preparation procedure employing bulk polymerization (Adaptation of the [182]).

The synthesis of an MIP is subject to different experimental variables, such as the nature and concentration of TM, the FM, the cross-linking agent, the solvent and the radical initiator. Some rules to these variables can be reported in order to assist new MIP synthesis processes. However, these rules are not always in agreement to the efficient generation of specific sites, requiring individual studies on these variables with regard to the search for selectivity [180]. We will discuss each of them in the following sections.

Template Molecule

Initially, the TM should be evaluated for the presence of groups capable of binding to the FM. Furthermore, this molecule should not have groups which accelerate or retard polymerization reaction (for example, the thiol group) and groups which readily polymerize. That certainly will promote its insertion within the polymeric network and consequently will not form the recognition sites [184].

Functional Monomer

In most of the cases, an FM is employed which is a donor of protons (acid character) when TM is the proton acceptor (basic character) or vice versa. Thus, the methacrylic acid and the 4-vinylpyridine are the most widely used in the MIP synthesis, for basic and acidic molecules, respectively. In terms of concentration, the most appropriate is to use the MF in addition excess to MM (usually 4:1 molar fraction), ensuring the shift of the balance in the formation of the largest possible number of specific sites [180, 181].

Cross-Linking Agent

The cross-linking agent is responsible for controlling morphology of the MIP as well, as for stabilizing the binding sites and the mechanical structure of the polymer. This has arouses the attention towards the high proportion that is required (relative to the FM) to ensure the porosity of the polymer. Several cross-linking agents have been used, such as *p*-divinylbenzene and 1,4-diacrilol. However, the most commonly used agents in molecular imprinting is the ethylene glycol dimethacrylate, due to its ability to form thermally and mechanically stable polymers and allow rapid mass transfer during synthesis [180, 184].

Solvent

The main function of the solvent is to dissolve the synthesis reagents. However, it cannot interfere at the formation of the complex FM-TM, which can result in the formation of selective binding sites. Thus, the suitable solvent must be apolar, aprotic and with low dielectric constant, as chloroform. On the other hand, it is important to note that the thermal capacity and the volume of solvent (compared to other reagents) are directly related to the mechanical stability and the amount of pores of the MIP, which have influence on the molecular recognition capability [180, 181, 184].

Radical Initiator

The function of the radical initiator is to create free radicals to enable the polymerization reaction. However, for the beginning of the reaction, a physical stimulus, such as temperature increasing or the incidence of UV radiation, is needed. This stimulus is decisive in choosing the radical initiator, because the other reagents synthesis may be thermosensitive or photosensitive. Moreover, in the case of hydrogen bonds between the TM and FM, the conducted synthesis at low temperatures is preferable. Some radical initiators have been used in MIP synthesis, highlighting the 2,2'-azobisisobutironitrila, which is used in most cases [180, 181, 184].

Molecularly Imprinted Polymers in Sensors

The combination of specific molecular recognition abilities with stability and robustness are especially attractive to the sensor field [183]. Over the past 10 years, the number of published works is increasing, with about a thousand of articles on the topics of MIP and sensors. Only in the last year, there were about 160 articles according to the Web of Science™ Core Collection.

In the biosensors, a signal is generated with the binding of the analyte to the recognition element. The transducer then translates this signal into a measurable output signal. The same principle applies if a MIP is used as the recognition element as shown in Fig. (**18**).

Similarly to the construction of biosensors, an important issue in the construction of MIP-sensor is to find a suitable way of attaching the synthetic material containing the selective receptor in the transducer. In this sense, the strategy for coupling of the MIP with the transducer depends on the physical nature of the transducer. If it is a solid electrode (gold and glassy carbon), the immobilization can basically be divided into two groups: MIP encapsulation into the support (immobilization by occlusion) and immobilization of the selective material on the transducer's surface (*in situ* polymerization).When the transducer has granular characteristics (powdered graphite and carbon nanotubes), the MIP is easily mixed with the material to form a composite [174].

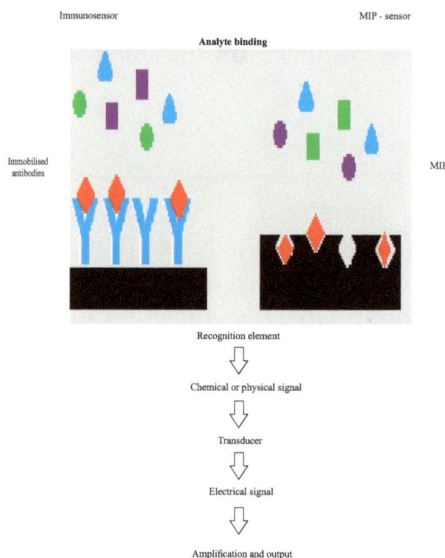

Fig. (18). Representation of a MIP-sensor compared to an immunosensor (Adaptation of the reference 176).

The most important features of biomimetic sensors based on MIP recently reported in the literature, using electrochemical transduction, are shown in Table **12**.

The high selectivity and stability of MIP make them promising alternatives instead of the enzymes, antibodies and the natural receptors used like sensors in the environmental and medical area. However, the factors that limit the development of these sensors are: the lack of a general procedure for preparing MIPs, poor performance in an aqueous medium and difficulty in the conversion of the binding event (MIP-analyte) in a measurable signal. But, with further progress in the polymer science, these problems can be solved [174].

Table 12. Examples of electrochemical sensors based on MIP described in the literature over the last 5 years.

Analyte	Template	Electrode / Transduction	Feature or Application	Reference
Tyramine	Silicic acid tetracthylester and triethoxyphenylsilane	Glassy carbon / Voltammetric	LOD = 57 x 10 $^{-6}$ mol L^{-1} and excellent selectivity	[185]
L-ascorbic acid	Polypyrrole film	Glassy carbon / Potenciometric	LOD = 3 x 10 $^{-6}$ mol L^{-1} and application in orange juice	[186]
Urea	Chitosan and CdS quantum dots	Gold / Impedimetric	LOD = 1.0 x 10 $^{-12}$ mol L^{-1} and application in human blood	[187]

(Table 12) contd.....

Analyte	Template	Electrode / Transduction	Feature or Application	Reference
Dipyridamole	Methacrylic acid, ethylene glycol dimethacrylate and azobisisobutyronitrile	Carbon paste / Voltammetric	LOD = 25.2 x 10^{-6} mol L^{-1} and application in human blood serum	[188]
Myoglobin	PVC-COOH and 1-ethyl-3-(3-dimethyl aminopropyl)carbodiimide	Screen printed (gold) / Voltammetric	LOD = 38.1mol L^{-1} and application in synthetic serum	[189]
Tamoxifen	o-phenylenediamine–resorcinol	Glassy carbon / Voltammetric	Electropolymerization, concentration between 1 to 100 x 10^{-9} mol L^{-1}	[190]
Epinephrine	2,4,6-trisacrylamido-1,3,5-triazine and hemin	Gold disc and chitosan/Nafion / Voltammetric	LOD = 0.12 x 10^{-9} mol L^{-1} and application in human blood serum	[191]
Ciprofloxacin	Methacrylic acid, ethylene glycol dimethacrylate and magnetic multi-walled carbon nanotubes	Carbon paste / Voltammetric	LOD = 1.7 x 10^{-9} mol L^{-1}, application in biological fluid (serum and urine) and in commercialpharmaceuticalsamples	[192]
Metronidazole	poly(anilinomethyltriethoxysilane) and 3-aminopropyltriethoxysilane	Carbon paste / Voltammetric	LOD = 0.91 x 10^{-9} mol L^{-1}, application in commercial tablets, human plasma, serum and urine samples	[193]

LOD = Limit of detection

In the last years, new functionalities to MIPs emerged making them more specific and robust. One approach is the integration of biomolecules, with affinity for a particular protein target, into an MIP, giving rise to a so-called hybrid-MIP system. The hybrid-MIP strategy was first proposed for the detection of lipopolysaccharide (LPS) using the cyclic peptide polymyxin and has also been reported for concanavalin A detection using mannose [194, 195].

Another bioreceptor that has been employed is the DNA aptamer. The DNA aptamers are short and stable oligonucleotide sequences. They have been extensively used as alternatives to antibodies in biosensing applications; which are also known as the 'Aptasensors'. The incorporation of DNA-aptamer within hybrid polymer systems has promoted an increase in the stability [194].

Moreover, polymers can also be electropolymerized on the surface of the transduction platform. This case is what happens in the Quartz Cristal Microbalance (QCM). There are some adhesion problems that, sometimes lead to partly peeling off of the MIP layers produced [195].

Some of these recent advances with hybrid-MIP system employed in the electrochemical detection are shown in Table **13**.

Table 13. Examples of electrochemical sensors based on MIP described in the literature over the last 5 years.

Analyte	Electrode / Transduction	Feature or Application	Reference
Prostate specific antigen	Aptamer–MIP hybrid / gold electrode	LOD = 1 pg/ml	[194]
Concanavalin A	Thiolated oligoethyleneglycol (OEG)/mannose conjugate / gold coated SPR sensor disks	They have approximately 20-fold higher affinity than that obtained from the mannose SAM*	[196]
Glycoprotein 41	Epitope imprinted hydrophilic polymer coated quartz crystal microbalance	LOD = 2 ng/mL and application in human urine	[197]

LOD = Limit of detection * self-assembled monolayer

Ion Imprinted Polymer (IIP)

The IIP represents a peculiarity of the printed polymer with molecules and have the same characteristics of MIPs, including the simplicity and the synthesis. The first work on IIP was published in the 70s, using poly(vinyl pyridine) for metal ion extraction [175]. This conception appears to be similar to the MIPs' by changing the template molecule per metal ion. However, whereas MIPs generally interact with template molecules via classical bonding with functional monomers, a small variation is the resource of the coordination chemistry to the IIPs. The preparation of IIPs generally requires a ligand to form a complex with the metal ion producing selective binding sites after metal leaching.

The synthesis of IIPs can basically be divided into four stages: formation of a complex ion between the metal ions of interest (the analyte) and the monomer; incorporation of the metal ion in the polymeric network *via* the polymerization reaction; removing the metal ion of the printed polymer and, polymer test for selectivity and reconnection of the metal ion of interest. Selectivity is based on the ligand's specificity in relation to the ion in the coordination complex geometry, the coordination number of the ion, and also, on the size and charge of the ion [175, 198]. An example synthesis of an IP can be seen in Fig. (**19**) [199].

The more usual conditions for the synthesis of IIPs include the absence of oxygen, the heat (about 60 °C) and the time of 24 hours of reaction with bulk polymerization. It is important to note that depending on the type of monomer used in the synthesis, the reaction time and temperature can be changed [175].

The polymerization reaction is known as a very complex process, which could be affected by many factors, such as the nature of monomer, cross-linking agent, initiator, temperature and time of polymerization, like the MIP synthesis, and the presence or absence of magnetic field in the case of the IIP. In order to obtain the ideal imprinted polymer, a variety of factors should be optimized. Thus, the synthesis of polymers printed is a time-consuming process and the polymers may be organic, inorganic and hybrid [200].

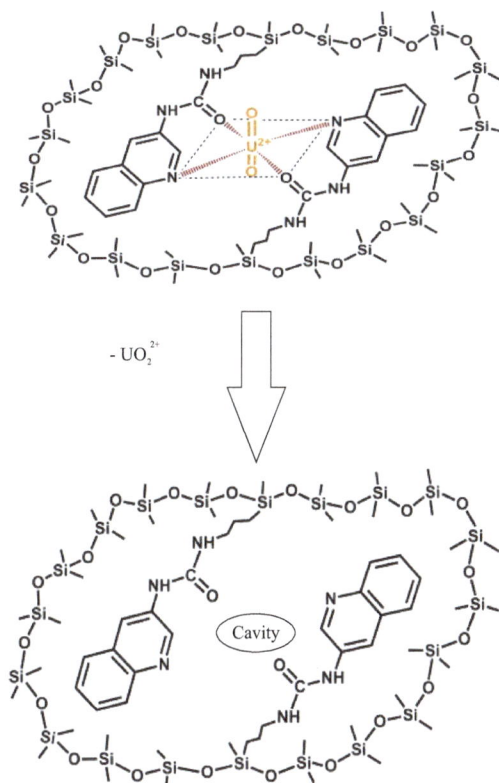

Fig. (19). Example of a synthesis of UO^{2+}/IIP by sol-gel process (Reproduced from [196]).

IIP with Organic Polymer

Most of the works involving the synthesis of ions with organic polymers describe the use of 2,2-azobisisobutironitrila (AIBN) as the radical initiator, and divinylbenzene (DVB) or ethyleneglycoldimethacrylate (EDGMA) as the cross-linking agent. Organic IIPs are used in solid phase extraction (SPE-IIP) and the synthesis can be classified into four approaches, based on how the ligands are immobilized in the polymer matrix: cross-linking of linear polymers, chemical immobilization, superficial impression held in organic interfaces–aqueous and synthesis by the binding of chelating ligands not vinyls [175]. Each of the approaches depends on the desired application of the IIP.

IIP with Inorganic Polymer

These IIPs also include the technological scope of chemical printing. These polymers possess the same purposes as organic polymers, exhibiting a selective adsorption by one ion. Basically, the inorganic polymers are commonly synthesized by the sol-gel process, using tetraethoxysilane (TEOS) as cross-linking agent, which is responsible for formation of the polymer network [175]. The IIPs can also be prepared by the sol-gel process in the presence of surfactants, so called polymers with duplex printing, because besides the connection of the functional precursor with the metal ion of interest, size selective cavity can be controlled by the presence of surfactant, facilitating the analyte access to the selective site of the IIP [175].

IIP with Hybrid (Organic-Inorganic) Polymer

Organic-inorganic hybrid materials are prepared by combining inorganic and organic components, which are alternatives to the production of new multifunctional materials with a wide range of applications. Such materials can be prepared in three ways: only by physical incorporation of the constituents; by chemical bonds between the organic and inorganic components, and a combination of the two types of physical and chemical interactions described above [175].

The process which has been widely used in the synthesis of hybrid polymer is a sol-gel process for ion incorporation into the polymer network using TEOS as a cross-linking agent and organic molecules capable of interacting with the ion of interest [175].

Ion Imprinted Polymers in Sensors

The determination of metal ions using IIPs has been performed by different types of sensors with transduction optical, potentiometric and voltametric. Among them, the cyclic voltammetry method is presented as the most favorable for the determination of metal ions due to its low cost, high sensitivity, ease of operation and the possibility of being portable [175].

Seeking to improve the sensitivity and selectivity of the electrochemical determination, chemically modified electrodes have gained attention. The use of these IIPs in the modified sensors has been shown to be a promising method [175]. Over the past 10 years, the number of published works has increased, with about 250 articles covering the topics of IIP and sensors. Only in the last year, there were about 45 articles according to the Web of ScienceTM Core Collection, so it is an emerging area for advances future.

Among the electrochemical sensors, carbon paste-based electrodes have some advantages, such as simplicity in preparation and in modification and easy regeneration of the sensor surface. For construction of the devices, the polymer should be dispersed and/or dissolved in a suitable solvent and homogenized with graphite or carbon nanotubes. Subsequently, the solvent is evaporated followed by the addition of mineral oil as the binder. The mixture is homogenized to form slurry, which is applied as a thin layer in the cavity of an appropriate electrode, with the excess removed with the aid of a plain paper [175].

Another alternative is the formation of a film on the surface of a solid electrode. In this case, the polymer is synthesized directly on the surface, for example, a glassy carbon electrode. When the film is polymerized on the surface, it produces sites printed selective ion to mold, matching the conductivity of the substrate for easy electronic transfer through the electrode [175].

The chemical printing process with ions is an emerging technology for producing highly selective polymer materials being considered simple, fast, low cost, and presenting various synthetic routes and applications. Furthermore, the IIPs are chemically resistant and may be reusable. The most important features of sensors based on IIPs recently reported in the literature, using electrochemical transduction, are shown in Table **14**.

Table 14. Examples of electrochemical sensors based on IIP described in the literature.

Metallic ion	Template	Electrode / Transduction	Feature or Application	Reference
Cd^{2+}	Vinyl pyridine and quinaldic acid	Carbon paste / Voltammetric	LOD = 0.52 x 10^{-9} molL^{-1} and application in water samples	[201]
Hg^{2+}	5,10,15,20-tetrakis(3-hydroxyphenyl) porphyrin, methacrylic acid, ethyleneglycol-dimethacrylate and 2,2-azobis(isobutyronitrile)	Glassy carbon / Voltammetric	The IIP was prepared with nanobeads and multi-wall carbon nanotubes. Application in water samples LOD = 5.0 x 10^{-9} molL^{-1}	[202]
Eu^{3+}	Vinyl pyridine,methacrylic acid and divinyl benzene	Carbon paste / Voltammetric	LOD = 0.15 x 10^{-6} molL^{-1} and application in synthetic and real water samples	[203]

(Table 14) contd.....

Metallic ion	Template	Electrode / Transduction	Feature or Application	Reference
Pd^{2+}	4-vinylpyridine, ethylene glycol dimethacrylate, eriochrome cyanine R and 2,2'-azobis(isobutyronitrile)	Carbon paste / Voltammetric	The IIP was prepared with nanoparticles. Application in environmental water and soil samples LOD = 3.0 x 10^{-12} mol L^{-1}	[204]
Tl^{3+}	Methacrylic acid, ethyleneglycol-dimethacrylate and 2,2-azobis(isobutyronitrile)	Carbon paste / Voltammetric	The modification was done with IIP and multi-wall carbon nanotubes. Application in water samples and human hair LOD = 0.16 x 10^{-3} mol L^{-1}	[205]
Ce^{3+}	Vinylpyridine,methacrylic acid, divinylbenzene and 2,2-azobis(isobutyronitrile)	Carbon paste / Voltammetric	The modification was done with IIP and multi-wall carbon nanotubes. Application in water samples LOD = 10 x 10^{-12} mol L^{-1}	[206]
UO^{2+}	Tetramethoxysilane, 3-Aminoquinoline and 3-isocyanatopropyl trimethoxysilane	Carbon paste / Voltammetry	LOD = 0.307 x 10^{-9} mol L^{-1} Application in water samples	[199]

LOD = Limit of detection

The number of studies involving IIPs in the preparation of electrochemical sensors is growing and is very attractive. The characteristics of the electrochemical sensors that are inexpensive, possess high selectivity and low detection limit, can be improved with the IIPs, hence making them alternatives to sophisticated instrumental techniques.

CONCLUSION

This review addresses some recent advances in the metal noble nanoparticles, carbon nanotubes, graphene, nanofibers, fullerene, carbon quantum dots, Molecularly Imprinted Polymers and or Ionically (MIP) and (IIP) used in the development of electrochemical sensors.

The metal noble nanoparticles have high surface areas and unique physical-chemical properties that can be easily tuned, making them ideal candidates for developing sensors. In electroanalysis, these materials have been applied for electrode modification, due to advantages, such as their electrocatalytic effect and improved active area and mass transport. These advantages can provide a better analytical performance.

The nanostructures of carbon have been widely used in constructing electro-chemical sensors or biosensors, because of their unique structures and properties; such as good electrical conductivity, high electrocatalytic effect, strong adsorptive ability and excellent biocompatibility, which make them attractive for a great number of applications in various research fields.

Lastly the Molecularly Imprinted Polymers and / or Ionically (MIP) and (IIP) have become an important analytical tool used for the modification of electrodes in the electroanalytical applications due to the biomimetic recognition systems.

Therefore these compounds have contributed to the technological advancement, due presenting amazing features; such as excellent electronic transfer, sensitivity, and selectivity, low cost, low limits of detection and quantification, which are important requisites for a satisfactory analysis.

CONFLICT OF INTEREST

The authors declare no conflict of interest, financial or otherwise.

ACKNOWLEDGEMENTS

UFSJ/DCNAT, FQMat, CNPq, FAPEMIG, RQMG and Capes.

REFERENCES

[1]   Shrivastava, S.; Jadon, N.; Jain, R. Next-generation polymer nanocomposite-based electrochemical sensors and biosensors: A review. *Trends Analyt. Chem.,* **2016**, *82*, 55-67.
         [http://dx.doi.org/10.1016/j.trac.2016.04.005]

[2]   Wei, F.; Xu, G.; Wu, Y.; Wang, X.; Yang, J.; Liu, L.; Zhou, P.; Hu, Q. Molecularly imprinted polymers on dual-color quantum dots for simultaneous detection of norepinephrine and epinephrine. *Sens. Actuators B Chem.,* **2016**, *229*, 38-46.
         [http://dx.doi.org/10.1016/j.snb.2016.01.113]

[3]   Branger, C.; Meouche, W.; Margaillan, A. *Recent advances on ion-imprinted polymers., React. Funct. Polym.,* **2013**, *73*, 859-875.
         [http://dx.doi.org/10.1016/j.reactfunctpolym.2013.03.021]

[4]   Mody, V.V.; Siwale, R.; Singh, A.; Mody, H.R. Introduction to metallic nanoparticles. *J. Pharm. Bioallied Sci.,* **2010**, *2*(4), 282-289.
         [http://dx.doi.org/10.4103/0975-7406.72127] [PMID: 21180459]

[5]   Rai, M.; Ingle, A.P.; Gupta, I.; Brandelli, A. Bioactivity of noble metal nanoparticles decorated with

biopolymers and their application in drug delivery. *Int. J. Pharm.,* **2015**, *496*(2), 159-172.
[http://dx.doi.org/10.1016/j.ijpharm.2015.10.059] [PMID: 26520406]

[6]   Karmaoui, M.; Leonardi, S.G.; Latino, M.; Tobaldi, D.M.; Donato, N.; Pullar, R.C.; Seabra, M.P.; Labrincha, J.A.; Neri, G. Pt-decorated In_2O_3 nanoparticles and their ability as a highly sensitive (<10 ppb) acetone sensor for biomedical applications. *Sens. Actuators B Chem.,* **2016**, *230*, 697-705.
[http://dx.doi.org/10.1016/j.snb.2016.02.100]

[7]   Prakash, J.; Pivin, J.C.; Swart, H.C. Noble metal nanoparticles embedding into polymeric materials: From fundamentals to applications. *Adv. Colloid Interface Sci.,* **2015**, *226*(Pt B), 187-202.

[8]   Islam, N.; Miyazaki, K. Nanotechnology innovation system: Understanding hidden dynamics of nanoscience fusion trajectories. *Technol. Forecast. Soc. Change,* **2009**, *76*, 128-140.
[http://dx.doi.org/10.1016/j.techfore.2008.03.021]

[9]   Pilkington, A.; Lee, L.L.; Cahn, C.K.; Ramahrisna, S. Defining key inventors: A comparison of fuel cell and nanotechnology industries. *Technol. Forecast. Soc. Change,* **2009**, *76*, 118-127.
[http://dx.doi.org/10.1016/j.techfore.2008.03.015]

[10]   Gowthaman, N.S.; Kesavan, S.; John, A. Monitoring isoniazid level in human fluids in the presence of theophylline using gold@platinum core@shell nanoparticles modified glassy carbon electrode. *Sens. Actuators B Chem.,* **2016**, *230*, 157-166.
[http://dx.doi.org/10.1016/j.snb.2016.02.042]

[11]   Mukdasai, S.; Crowley, U.; Mila, P.; He, X.; Nesterenko, E.P.; Nesterenko, P.N.; Paull, B.; Srijaranai, S.; Glennon, J.D.; Moore, E. Electrodeposition of palladium nanoparticles on porous graphitized carbon monolith modified carbon paste electrode for simultaneous enhanced determination of ascorbic acid and uric acid. *Sens. Actuators B Chem.,* **2015**, *218*, 280-288.
[http://dx.doi.org/10.1016/j.snb.2015.04.071]

[12]   Dursun, Z.; Gelmez, B. Simultaneous Determination of Ascorbic Acid, Dopamine and Uric Acid at Pt Nanoparticles Decorated Multiwall Carbon Nanotubes Modified GCE. *Electroanalysis,* **2010**, *22*, 1106-1114.
[http://dx.doi.org/10.1002/elan.200900525]

[13]   Rathod, D.; Dickinson, C.; Egan, D.; Dempsev, E. Platinum nanoparticle decoration of carbon materials with applications in non-enzymatic glucose sensing. *Sens. Actuators B Chem.,* **2010**, *143*, 547-554.
[http://dx.doi.org/10.1016/j.snb.2009.09.064]

[14]   Brown, K.R.; Fox, A.P.; Natan, M.J. Morphology-Dependent Electrochemistry of Cytochrome *c* Au Colloid-Modified SnO_2 Electrodes. *J. Am. Chem. Soc.,* **1996**, *118*, 1154-1157.
[http://dx.doi.org/10.1021/ja952951w]

[15]   Vidotti, M.; Carvalhal, R.F.; Mendes, R.K.; Ferreira, D.C.; Kubota, L.T. Biosensors Based on Gold Nanostructures. *J. Braz. Chem. Soc.,* **2011**, *22*, 3-20.
[http://dx.doi.org/10.1590/S0103-50532011000100002]

[16]   Krithiga, N.; Viswanath, K.B.; Vasantha, V.S.; Jayachitra, A. Specific and selective electrochemical immunoassay for Pseudomonas aeruginosa based on pectin-gold nano composite. *Biosens. Bioelectron.,* **2016**, *79*, 121-129.
[http://dx.doi.org/10.1016/j.bios.2015.12.006] [PMID: 26703990]

[17]   Thanh, T.D.; Balamurugan, J.; Lee, S.H.; Kim, N.H.; Lee, J.H. Effective seed-assisted synthesis of gold nanoparticles anchored nitrogen-doped graphene for electrochemical detection of glucose and dopamine. *Biosens. Bioelectron.,* **2016**, *81*, 259-267.
[http://dx.doi.org/10.1016/j.bios.2016.02.070] [PMID: 26967913]

[18]   Nakamoto, M.; Kashiwagi, Y.; Yamamoto, M. Synthesis and size regulation of gold nanoparticles by controlled thermolysis of ammonium gold(I) thiolate in the absence or presence of amines. *Inorg. Chim. Acta,* **2005**, *358*, 4229-4236.
[http://dx.doi.org/10.1016/j.ica.2005.03.037]

[19]   Ionescua, R.; Cindemir, U.; Welearegaya, T.G.; Calaviaa, R.; Haddi, Z.; Topalianb, Z.; Granqvist, C.G.; Llobet, E. Fabrication of ultra-pure gold nanoparticles capped withdodecanethiol for Schottky-diode chemical gas sensing devicesRadu. *Sens. Actuators B Chem.,* **2017**, *239*, 455-461. [http://dx.doi.org/10.1016/j.snb.2016.07.182]

[20]   Alvarez, R.A.; Cortez-Valadez, M.; Bueno, L.O.; Hurtado, R.B.; Rocha-Rocha, O. Delgado-Beleno, **Y.; Martinez-Nuñez, C.E.; Serrano-Corrales, L.I.; Arizpe-Chávez, H.; Flores-Acosta, M. Vibrational properties of gold nanoparticles obtained by green synthesis.** *Physica E,* **2016**, *84*, 191-195. [http://dx.doi.org/10.1016/j.physe.2016.04.024]

[21]   Ahmed, S.; Annu, I.S.; Ikram, S.; Yudha S, S. Biosynthesis of gold nanoparticles: A green approach. *J. Photochem. Photobiol. B,* **2016**, *161*, 141-153. [http://dx.doi.org/10.1016/j.jphotobiol.2016.04.034] [PMID: 27236049]

[22]   Li, H.; Wang, W.; Lv, Q.; Xi, G.; Bai, H.; Zhang, Q. Disposable paper-based electrochemical sensor based on stacked gold nanoparticles supported carbon nanotubes for the determination of bisphenol A. *Electrochem. Commun.,* **2016**, *68*, 104-107. [http://dx.doi.org/10.1016/j.elecom.2016.05.010]

[23]   Wong, A.; Razzino, C.A.; Silva, T.A.; Fatibello-Filho, O. Square-wave voltammetric determination of clindamycin using a glassy carbon electrode modified with graphene oxide and gold nanoparticles within a crosslinked chitosan film. *Sens. Actuators B Chem.,* **2016**, *231*, 183-193. [http://dx.doi.org/10.1016/j.snb.2016.03.014]

[24]   Monerris, M.J.; D'Eramo, F.; Arévalo, F.J.; Fernández, H.; Zon, M.A.; Molina, P.G. Electrochemical immunosensor based on gold nanoparticles deposited on a conductive polymer to determine estrone in water samples. *Microchem. J.,* **2016**, *129*, 71-77. [http://dx.doi.org/10.1016/j.microc.2016.06.001]

[25]   Saengsookwaow, C.; Rangkupan, R.; Chailapakul, O.; Rodthongkum, N. Nitrogen-doped graphene–polyvinylpyrrolidone/gold nanoparticles modified electrode as a novel hydrazine sensor. *Sens. Actuators B Chem.,* **2016**, *227*, 524-532. [http://dx.doi.org/10.1016/j.snb.2015.12.091]

[26]   Tan, X.; Hu, Q.; Wu, J.; Li, X.; Li, P.; Yu, H.; Li, X.; Lei, F. Electrochemical sensor based on molecularly imprinted polymer reduced graphene oxide and gold nanoparticles modified electrode for detection of carbofuran. *Sens. Actuators B Chem.,* **2015**, *220*, 216-221. [http://dx.doi.org/10.1016/j.snb.2015.05.048]

[27]   Xiao, X.; Bard, A.J. Observing single nanoparticle collisions at an ultramicroelectrode by electrocatalytic amplification. *J. Am. Chem. Soc.,* **2007**, *129*(31), 9610-9612. [http://dx.doi.org/10.1021/ja072344w] [PMID: 17630740]

[28]   Jia, M.; Wang, T.; Liang, F. Hu. A Novel Process for the Fabrication of a Silver-Nanoparticl--Modified Electrode and Its Application in Nonenzymatic Glucose Sensing. *Electroanalysis,* **2012**, *24*, 1864-1868. [http://dx.doi.org/10.1002/elan.201200273]

[29]   Maduraiveeran, G.; Ramaraj, R. Potential sensing platform of silver nanoparticles embedded in functionalized silicate shell for nitroaromatic compounds. *Anal. Chem.,* **2009**, *81*(18), 7552-7560. [http://dx.doi.org/10.1021/ac900781d] [PMID: 19691270]

[30]   Hebeish, A.; Shaheen, T.I.; El-Naggar, M.E. Solid state synthesis of starch-capped silver nanoparticles. *Int. J. Biol. Macromol.,* **2016**, *87*, 70-76. [http://dx.doi.org/10.1016/j.ijbiomac.2016.02.046] [PMID: 26902893]

[31]   Mohammadi, S.; Pourseyedi, S.; Amini, A. Green synthesis of silver nanoparticles with a long lasting stability using colloidal solution of cowpea seeds (*Vigna sp. L*). *J. Environ. Chem. Eng.,* **2016**, *4*, 2023-2032. [http://dx.doi.org/10.1016/j.jece.2016.03.026]

[32]   Ahmed, S.; Ahmad, M.; Swami, B.L.; Ikram, S. A review on plants extract mediated synthesis of silver nanoparticles for antimicrobial applications: A green expertise. *J. Adv. Res.,* **2016**, *7*(1), 17-28.
[http://dx.doi.org/10.1016/j.jare.2015.02.007] [PMID: 26843966]

[33]   Baghayeri, M.; Tehrani, M.B.; Amiri, A.; Maleki, B.; Farhadi, S. A novel way for detection of antiparkinsonism drug entacapone via electrodeposition of silver nanoparticles/functionalized multi-walled carbon nanotubes as an amperometric sensor. *Mater. Sci. Eng. C,* **2016**, *66*, 77-83.
[http://dx.doi.org/10.1016/j.msec.2016.03.077] [PMID: 27207040]

[34]   Alessio, P.; Martin, C.S.; Saja, J.A.; Rodriguez-Mendez, M.L. Mimetic biosensors composed by layer-by-layer films of phospholipid, phthalocyanine and silver nanoparticles to polyphenol detection. *Sens. Actuators B Chem.,* **2016**, *233*, 654-666.
[http://dx.doi.org/10.1016/j.snb.2016.04.139]

[35]   Ghiaci, M.; Tghizadeh, M.; Ensafi, A.A.; Zandi-Atashbar, N.; Rezaei, B. Silver nanoparticles decorated anchored type ligands as new electrochemical sensors for glucose detection. *Journal of the Taiwan Institute of Chemical Engineers,* **2016**, *63*, 39-45.
[http://dx.doi.org/10.1016/j.jtice.2016.03.013]

[36]   Rastogi, P.K.; Ganesan, V.; Azad, U.P. Electrochemical determination of nanomolar levels of isoniazid in pharmaceutical formulation using silver nanoparticles decorated copolymer. *Electrochim. Acta,* **2016**, *188*, 818-824.
[http://dx.doi.org/10.1016/j.electacta.2015.12.058]

[37]   Ikhsan, N.I.; Rameshkumar, P.; Huang, N.M. Controlled synthesis of reduced graphene oxide supported silver nanoparticles for selective and sensitive electrochemical detection of 4-nitrophenol. *Electrochim. Acta,* **2016**, *192*, 392-399.
[http://dx.doi.org/10.1016/j.electacta.2016.02.005]

[38]   Kundu, S.; Liang, H. Shape-selective formation and characterization of catalytically active iridium nanoparticles. *J. Colloid Interface Sci.,* **2011**, *354*(2), 597-606.
[http://dx.doi.org/10.1016/j.jcis.2010.11.032] [PMID: 21144533]

[39]   Irfan, M.; Pham, X-H.; Han, K.N.; Li, C.A.; Hong, M.H.; Seong, G.H. Decoration of carbon nanotube films with iridium nanoparticles and their electrochemical characterization. *Journal BioChip,* **2014**, *8*, 129-136.
[http://dx.doi.org/10.1007/s13206-014-8208-x]

[40]   Salimi, A.; Hallaj, R.; Kavosi, B.; Hagighi, B. Highly sensitive and selective amperometric sensors for nanomolar detection of iodate and periodate based on glassy carbon electrode modified with iridium oxide nanoparticles. *Anal. Chim. Acta,* **2010**, *661*(1), 28-34.
[http://dx.doi.org/10.1016/j.aca.2009.12.005] [PMID: 20113712]

[41]   Pachon, L.D. Rothenberg. G. Transition-metal nanoparticles: synthesis, stability and the leaching issue. *Appl. Organomet. Chem.,* **2008**, *22*, 288-299.
[http://dx.doi.org/10.1002/aoc.1382]

[42]   Rojas, J.V.; Higgins, M.M.; Gonzalez, M.T.; Castano, C.E. Single step radiolytic synthesis of iridium nanoparticles onto graphene oxide. *Appl. Surf. Sci.,* **2015**, *357*, 2087-2093.
[http://dx.doi.org/10.1016/j.apsusc.2015.09.190]

[43]   Zhang, T.; Li, S-C.; Zhu, W.; Ke, J.; Yu, J-W.; Zhang, Z-P.; Dai, L-X.; Gu, J.; Zhang, Y-W. Iridium ultrasmall nanoparticles, worm-like chain nanowires, and porous nanodendrites: One-pot solvothermal synthesis and catalytic CO oxidation activity. *Surf. Sci.,* **2016**, *648*, 319-327.
[http://dx.doi.org/10.1016/j.susc.2015.10.007]

[44]   Roushani, M.; Sarabaegi, M. Electrochemical detection of butylated hydroxyanisole based on glassy carbon electrode modified by iridium oxide nanoparticles. *J. Electroanal. Chem.,* **2014**, *717-718*, 147-152.
[http://dx.doi.org/10.1016/j.jelechem.2014.01.013]

[45]   Rivas, L.; Mayorga-Martinez, C.C.; Quesada-González, D.; Zamora-Gálvez, A.; de la Escosura-Muñiz, A.; Merkoçi, A. Label-free impedimetric aptasensor for ochratoxin-A detection using iridium oxide nanoparticles. *Anal. Chem.,* **2015,** *87*(10), 5167-5172.
[http://dx.doi.org/10.1021/acs.analchem.5b00890] [PMID: 25901535]

[46]   da Silva, C.P.; Franzoi, A.C.; Fernandes, S.C.; Dupont, J.; Vieira, I.C. Development of biosensor for phenolic compounds containing PPO in β-cyclodextrin modified support and iridium nanoparticles. *Enzyme Microb. Technol.,* **2013,** *52*(4-5), 296-301.
[http://dx.doi.org/10.1016/j.enzmictec.2012.12.001] [PMID: 23540933]

[47]   Salimi, A.; Hallaj, R.; Kavosi, B.; Hagighi, B. Highly sensitive and selective amperometric sensors for nanomolar detection of iodate and periodate based on glassy carbon electrode modified with iridium oxide nanoparticles. *Anal. Chim. Acta,* **2010,** *661*(1), 28-34.
[http://dx.doi.org/10.1016/j.aca.2009.12.005] [PMID: 20113712]

[48]   Kumar, P.S.; Manievel, A.; Anandan, S.; Zhou, M.; Grieser, F.; Ashokkumar, M. Sonochemical synthesis and characterization of gold–ruthenium bimetallic nanoparticles. *Colloids Surf. A Physicochem. Eng. Asp.,* **2010,** *356,* 140-144.
[http://dx.doi.org/10.1016/j.colsurfa.2010.01.004]

[49]   Tsukatani, T.; Fujihara, H. New method for facile synthesis of amphiphilic thiol-stabilized ruthenium nanoparticles and their redox-active ruthenium nanocomposite. *Langmuir,* **2005,** *21*(26), 12093-12095.
[http://dx.doi.org/10.1021/la052332t] [PMID: 16342978]

[50]   Li, H.; Wang, R.; Hong, Q.; Chen, L.; Zhong, Z.; Koltypin, Y.; Calderon-Moreno, J.; Gedanken, A. Ultrasound-assisted polyol method for the preparation of SBA-15-supported ruthenium nanoparticles and the study of their catalytic activity on the partial oxidation of methane. *Langmuir,* **2004,** *20*(19), 8352-8356.
[http://dx.doi.org/10.1021/la049290d] [PMID: 15350113]

[51]   Zare, H.R.; Hashemi, S.H.; Benvidi, A. Electrodeposited nano-scale islands of ruthenium oxide as a bifunctional electrocatalyst for simultaneous catalytic oxidation of hydrazine and hydroxylamine. *Anal. Chim. Acta,* **2010,** *668*(2), 182-187.
[http://dx.doi.org/10.1016/j.aca.2010.04.028] [PMID: 20493296]

[52]   Garcia-Pena, N.G.; Redon, R.; Herrera-Gomez, A.; Fernandez-Osorio, A.L.; Bravo-Sanchez, M.; Gomez-Sosa, G. Solventless synthesis of ruthenium nanoparticles. *Appl. Surf. Sci.,* **2015,** *340,* 25-34.
[http://dx.doi.org/10.1016/j.apsusc.2015.02.186]

[53]   Cuenya, B.R. Synthesis and catalytic properties of metal nanoparticles: Size, shape, support, composition, and oxidation state effects. *Thin Solid Films,* **2010,** *518,* 3127-3150.
[http://dx.doi.org/10.1016/j.tsf.2010.01.018]

[54]   Yu, H.D.; Regulacio, M.D.; Ye, E.; Han, M.Y. Chemical routes to top-down nanofabrication. *Chem. Soc. Rev.,* **2013,** *42*(14), 6006-6018.
[http://dx.doi.org/10.1039/c3cs60113g] [PMID: 23653019]

[55]   Yang, J.; Lee, J.Y.; Deivaraj, T.C.; Too, H.P. A highly efficient phase transfer method for preparing alkylamine-stabilized Ru, Pt, and Au nanoparticles. *J. Colloid Interface Sci.,* **2004,** *277*(1), 95-99.
[http://dx.doi.org/10.1016/j.jcis.2004.03.074] [PMID: 15276043]

[56]   Kannan, S.K.; Sundrarajan, M. Green synthesis of ruthenium oxide nanoparticles: Characterization and its antibacterial activity. *Adv. Powder Technol.,* **2005,** *26,* 1505-1511.
[http://dx.doi.org/10.1016/j.apt.2015.08.009]

[57]   Prasad, T.N.; Elumalai, E.K. Biofabrication of Ag nanoparticles using Moringa oleifera leaf extract and their antimicrobial activity. *Asian Pac. J. Trop. Biomed.,* **2011,** *1*(6), 439-442.
[http://dx.doi.org/10.1016/S2221-1691(11)60096-8] [PMID: 23569809]

[58]   Mallikarjun, K.; Narsimba, G.; Dillip, G. Green synthesis of silver nanoparticles using Ocimum leaf extract and their characterization. *Dig. J. Nanomater. Biostruct.,* **2011,** *6,* 181-186.

[59]   Banerjee, L.; Nerendhirakannan, R. Biosynthesis of silver nanoparticles from syzygium cumini (l.) Seed extract and evaluation of their in vitro antioxidant activities. *Dig. J. Nanomater. Biostruct.,* **2011**, *6*(3), 961-968.

[60]   Zare, H.R.; Ghanbari, Z.; Nasirizadeh, N.; Benvidi, A. Simultaneous determination of adrenaline, uric acid, and cysteine using bifunctional electrocatalyst of ruthenium oxide nanoparticles. *C. R. Chim.,* **2013**, *16*, 287-295.
[http://dx.doi.org/10.1016/j.crci.2013.01.004]

[61]   Shahrokhian, S.; Rastgar, S. Electrodeposition of Pt–Ru nanoparticles on multi-walled carbon nanotubes: Application in sensitive voltammetric determination of methyldopa. *Electrochim. Acta,* **2011**, *58*, 125-133.
[http://dx.doi.org/10.1016/j.electacta.2011.09.023]

[62]   Xia, J.; Ding, S.N.; Gao, B.H.; Sun, Y.M.; Wang, Y.H.; Cosnier, S.; Guo, X. A biosensing application based on quenching the enhanced electrochemiluminescence of poly[tris(N-bipyridylethyl)pyrrole] ruthenium(II) film by Au nanoparticles. *J. Electroanal. Chem.,* **2013**, *692*, 60-65.
[http://dx.doi.org/10.1016/j.jelechem.2013.01.002]

[63]   Chatraei, F.; Zare, H.R. Nano-scale islands of ruthenium oxide as an electrochemical sensor for iodate and periodate determination. *Mater. Sci. Eng. C,* **2013**, *33*(2), 721-726.
[http://dx.doi.org/10.1016/j.msec.2012.10.024] [PMID: 25427479]

[64]   Jahanbakhshi, M.; Habibi, B. A novel and facile synthesis of carbon quantum dots via salep hydrothermal treatment as the silver nanoparticles support: Application to electroanalytical determination of H2O2 in fetal bovine serum. *Biosens. Bioelectron.,* **2016**, *81*, 143-150.
[http://dx.doi.org/10.1016/j.bios.2016.02.064] [PMID: 26943787]

[65]   Radziuk, K.; Mohywald, H.; Shchukin, D. Ultrasonic Activation of Platinum Catalysts. *J. Phys. Chem. C,* **2008**, *112*, 19257-19262.
[http://dx.doi.org/10.1021/jp806508t]

[66]   Teranishi, T.; Hosoe, M.; Tanaka, T.; Miyake, M. Size Control of Monodispersed Pt Nanoparticles and Their 2D Organization by Electrophoretic Deposition. *J. Phys. Chem. B,* **1999**, *103*, 3818-3827.
[http://dx.doi.org/10.1021/jp983478m]

[67]   Rao, C.R.; Trivedi, D.C. Chemical and electrochemical depositions of platinum group metals and their applications. *Coord. Chem. Rev.,* **2005**, *249*, 613-631.
[http://dx.doi.org/10.1016/j.ccr.2004.08.015]

[68]   Suryawanshi, P.L.; Gumfekar, S.P.; Kumar, P.R.; Kale, B.B.; Sonawane, S.H. Synthesis of ultra-small platinum nanoparticles in a continuous flow microreactor. *Colloid and Interface Science Communications,* **2016**, *13*, 6-9.
[http://dx.doi.org/10.1016/j.colcom.2016.05.001]

[69]   Ravi Kumar, D.V.; Prasad, B.L.; Kulkarni, A.A. Segmented flow synthesis of Ag nanoparticles in spiral microreactor: Role of continuous and dispersed phase. *Chem. Eng. J.,* **2012**, *192*, 357-368.
[http://dx.doi.org/10.1016/j.cej.2012.02.084]

[70]   Gowthaman, N.S.; Srinivasan, K.; John, S.A. Monitoring isoniazid level in human fluids in the presence of theophylline using gold@platinum core@shell nanoparticles modified glassy carbon electrode. *Sens. Actuators B Chem.,* **2016**, *230*, 157-166.
[http://dx.doi.org/10.1016/j.snb.2016.02.042]

[71]   Kushikawa, R.T.; Silva, M.R.; Angelo, A.C.; Teixeira, M.F. Construction of an electrochemical sensing platform based on platinum nanoparticles supported on carbon for tetracycline determination. *Sens. Actuators B Chem.,* **2016**, *228*, 207-213.
[http://dx.doi.org/10.1016/j.snb.2016.01.009]

[72]   Borisova, B.; Sanchez, A.; Jimenez-Falcao, S.; Martin, M.; Salazar, P.; Parrado, C.; Pingarron, J.M.; Villalong, R. Reduced graphene oxide-carboxymethylcellulose layered with platinum

nanoparticles/PAMAM dendrimer/magnetic nanoparticles hybrids. Application to the preparation of enzyme electrochemical biosensors. *Sens. Actuators B Chem.,* **2016**, *232*, 84-90.
[http://dx.doi.org/10.1016/j.snb.2016.02.106]

[73] Kalambate, P.K.; Srivastava, A.K. Simultaneous voltammetric determination of paracetamol, cetirizine and phenylephrine using a multiwalled carbon nanotube-platinum nanoparticles nanocomposite modified carbon paste electrode. *Sens. Actuators B Chem.,* **2016**, *233*, 237-248.
[http://dx.doi.org/10.1016/j.snb.2016.04.063]

[74] Chen, A.; Ostrom, C. Palladium-Based Nanomaterials: Synthesis and Electrochemical Applications. *Chem. Rev.,* **2015**, *115*(21), 11999-12044.
[http://dx.doi.org/10.1021/acs.chemrev.5b00324] [PMID: 26402587]

[75] Alonso, F.; Beletskaya, I.P.; Yus, M. Metal-mediated reductive hydrodehalogenation of organic halides. *Chem. Rev.,* **2002**, *102*(11), 4009-4091.
[http://dx.doi.org/10.1021/cr0102967] [PMID: 12428984]

[76] Zhang, H.; Jin, M.; Xiong, Y.; Lim, B.; Xia, Y. Shape-controlled synthesis of Pd nanocrystals and their catalytic applications. *Acc. Chem. Res.,* **2013**, *46*(8), 1783-1794.
[http://dx.doi.org/10.1021/ar300209w] [PMID: 23163781]

[77] Xiong, Y.; Xia, Y. Shape-Controlled Synthesis of Metal Nanostructures: The Case of Palladium. *Adv. Mater.,* **2007**, *19*, 3385-3391.
[http://dx.doi.org/10.1002/adma.200701301]

[78] Sharada, S.; Suryawanshi, P.L.; Rajesh, K.P.; Gumfekar, S.P.; Narsaiah, T.B.; Sonawane, S.H. Synthesis of palladium nanoparticles using continuous flow microreactor. *Colloids Surf. A Physicochem. Eng. Asp.,* **2016**, *498*, 297-304.
[http://dx.doi.org/10.1016/j.colsurfa.2016.03.068]

[79] Mukdasai, S.; Crowley, U.; Pravda, M.; He, X.; Nesterenko, E.P.; Nesterenko, P.N.; Paull, B.; Srijaranai, S.; Gennon, J.D.; Moore, E. Electrodeposition of palladium nanoparticles on porous graphitized carbon monolith modified carbon paste electrode for simultaneous enhanced determination of ascorbic acid and uric acid. *Sens. Actuators B Chem.,* **2015**, *218*, 280-288.
[http://dx.doi.org/10.1016/j.snb.2015.04.071]

[80] Yang, L.; Zhao, H.; Fan, S.; Li, B.; Li, C.P. A highly sensitive electrochemical sensor for simultaneous determination of hydroquinone and bisphenol A based on the ultrafine Pd nanoparticle@TiO2 functionalized SiC. *Anal. Chim. Acta,* **2014**, *852*, 28-36.
[http://dx.doi.org/10.1016/j.aca.2014.08.037] [PMID: 25441876]

[81] Absalan, G.; Akhond, M.; Soleimani, M.; Ershadifar, H. Efficient electrocatalytic oxidation and determination of isoniazid on carbon ionic liquid electrode modified with electrodeposited palladium nanoparticles. *J. Electroanal. Chem.,* **2016**, *761*, 1-7.
[http://dx.doi.org/10.1016/j.jelechem.2015.11.041]

[82] Hamidi, H.; Haghighi, B. Fabrication of a sensitive amperometric sensor for NADH and H2O2 using palladium nanoparticles-multiwalled carbon nanotube nanohybrid. *Mater. Sci. Eng. C,* **2016**, *62*, 423-428.

[83] Chen, X.; Li, G.; Zhang, G.; Hou, K.; Pan, H.; Du, M. Self-assembly of palladium nanoparticles on functional TiO2 nanotubes for a nonenzymatic glucose sensor. *Mater. Sci. Eng. C,* **2016**, *62*, 323-328.

[84] Brahman, P.K.; Suresh, L.; Lokesh, V.; Nizamuddin, S. Fabrication of highly sensitive and selective nanocomposite film based on CuNPs/fullerene-C60/MWCNTs: An electrochemical nanosensor for trace recognition of paracetamol. *Anal. Chim. Acta,* **2016**, *917*, 107-116.
[http://dx.doi.org/10.1016/j.aca.2016.02.044] [PMID: 27026607]

[85] Inagaki, M.; Kaneko, K.; Nishizawa, T. Nanocarbons—recent research in Japan. *Carbon,* **2004**, *42*, 1401-1417.
[http://dx.doi.org/10.1016/j.carbon.2004.02.032]

[86]   Tarditto, L.V.; Arévalo, F.J.; Zon, M.A.; Ovando, H.G.; Vettorazzi, N.R.; Fernández, H. Electrochemical sensor for the determination of enterotoxigenic *Escherichia coli* in swine feces using glassy carbon electrodes modified with multi-walled carbon nanotubes. *Microchem. J.,* **2016**, *127*, 220-225.
[http://dx.doi.org/10.1016/j.microc.2016.03.011]

[87]   Sakthinathan, S.; Lee, H.F.; Chen, S-M.; Tamizhdurai, P. Electrocatalytic oxidation of dopamine based on non-covalent functionalization of manganese tetraphenylporphyrin/reduced graphene oxide nanocomposite. *J. Colloid Interface Sci.,* **2016**, *468*, 120-127.
[http://dx.doi.org/10.1016/j.jcis.2016.01.014] [PMID: 26835582]

[88]   Li, L.; Zhou, T.; Sun, G.; Li, Z.; Yang, W.; Jia, J.; Yang, G. Ultrasensitive electrospun nickel-doped carbon nanofibers electrode for sensing paracetamol and glucose. *Electrochim. Acta,* **2015**, *152*, 31-37.
[http://dx.doi.org/10.1016/j.electacta.2014.11.048]

[89]   Hanaei, H.; Assadi, M.K.; Saidur, R. Highly efficient antireflective and self-cleaning coatings that incorporate carbon nanotubes (CNTs) into solar cells: A review. *Renew. Sustain. Energy Rev.,* **2016**, *59*, 620-635.
[http://dx.doi.org/10.1016/j.rser.2016.01.017]

[90]   Lin, B.; Yu, Y.; Li, R.; Cao, Y.; Guo, M. Turn-on sensor for quantification and imaging of acetamiprid residues based on quantum dots functionalized with aptamer. *Sens. Actuators B Chem.,* **2016**, *229*, 100-109.
[http://dx.doi.org/10.1016/j.snb.2016.01.114]

[91]   Baughman, R.H.; Zakhidov, A.A.; de Heer, W.A. Carbon nanotubes--the route toward applications. *Science,* **2002**, *297*(5582), 787-792.
[http://dx.doi.org/10.1126/science.1060928] [PMID: 12161643]

[92]   Cao, J.; Wang, Q.; Dai, H. Electromechanical properties of metallic, quasimetallic, and semiconducting carbon nanotubes under stretching. *Phys. Rev. Lett.,* **2003**, *90*(15), 157601.
[http://dx.doi.org/10.1103/PhysRevLett.90.157601] [PMID: 12732069]

[93]   Thostenson, E.T.; Ren, Z.; Chou, T-W. Advances in the science and technology of carbon nanotubes and their composites: a review. *Compos. Sci. Technol.,* **2001**, *61*, 1899-1912.
[http://dx.doi.org/10.1016/S0266-3538(01)00094-X]

[94]   Iijima, S.; Ichihashi, T. Single-shell carbon nanotubes of 1-nm diameter. *Nature,* **1993**, *363*, 603-605.
[http://dx.doi.org/10.1038/363603a0]

[95]   Fang, B.; Feng, Y.H.; Wang, G.F.; Zhang, C.H.; Gu, A.X.; Liu, M. A uric acid sensor based on electrodeposition of nickel hexacyanoferrate nanoparticles on an electrode modified with multi-walled carbon nanotubes. *Mikrochim. Acta,* **2011**, *173*, 27-32.
[http://dx.doi.org/10.1007/s00604-010-0509-8]

[96]   Yogeswaran, U.; Chen, S.M. Separation and concentration effect of *f*-MWCNTs on electrocatalytic responses of ascorbic acid, dopamine and uric acid at *f*-MWCNTs incorporated with poly (neutral red) composite films. *Electrochim. Acta,* **2006**, *52*, 5985-5996.
[http://dx.doi.org/10.1016/j.electacta.2007.03.047]

[97]   Coleman, J.N.; Khan, U.; Blau, W.J.; Gun'ko, Y.K. Small but strong: A review of the mechanical properties of carbon nanotube–polymer composites. *Carbon,* **2006**, *44*, 1624-1652.
[http://dx.doi.org/10.1016/j.carbon.2006.02.038]

[98]   Sajid, M.I.; Jamshaid, U.; Jamshaid, T.; Zafar, N.; Fessi, H.; Elaissari, A. Carbon nanotubes from synthesis to in vivo biomedical applications. *Int. J. Pharm.,* **2016**, *501*(1-2), 278-299.
[http://dx.doi.org/10.1016/j.ijpharm.2016.01.064] [PMID: 26827920]

[99]   Ojeda, I.; Barrejón, M.; Arellano, L.M.; González-Cortés, A.; Yáñez-Sedeño, P.; Langa, F.; Pingarrón, J.M. Grafted-double walled carbon nanotubes as electrochemical platforms for immobilization of antibodies using a metallic-complex chelating polymer: Application to the determination of

adiponectin cytokine in serum. *Biosens. Bioelectron.*, **2015**, *74*, 24-29.
[http://dx.doi.org/10.1016/j.bios.2015.06.001] [PMID: 26093125]

[100] Torkashvand, M.; Gholivand, M.B. Malekzadeh.Gh. Construction of a new electrochemical sensor based on molecular imprinting recognition sites on multiwall carbon nanotube surface for analysis of ceftazidime in real samples. *Sens. Actuators B Chem.*, **2016**, *231*, 759-767.
[http://dx.doi.org/10.1016/j.snb.2016.03.061]

[101] Leniart, A.; Brycht, M.; Burnat, B.; Skrzypek, S. Voltammetric determination of the herbicide propham on glassy carbon electrode modified with multi-walled carbon nanotubes. *Sens. Actuators B Chem.*, **2016**, *231*, 54-63.
[http://dx.doi.org/10.1016/j.snb.2016.02.126]

[102] Zarei, K.; Helli, H. Electrochemical determination of aminopyrene on glassy carbon electrode modified with multi-walled carbon nanotube–sodium dodecyl sulfate/Nafion composite film. *J. Electroanal. Chem.*, **2015**, *749*, 10-15.
[http://dx.doi.org/10.1016/j.jelechem.2015.04.027]

[103] Madrakian, T.; Maleki, S.; Heidari, M.; Afkhami, A. An electrochemical sensor for rizatriptan benzoate determination using Fe_3O_4 nanoparticle/multiwall carbon nanotube-modified glassy carbon electrode in real samples. *Mater. Sci. Eng. C*, **2016**, *63*, 637-643.
[http://dx.doi.org/10.1016/j.msec.2016.03.041] [PMID: 27040259]

[104] Majidi, M.R.; Omidi, Y.; Karami, P.; Johari-Ahar, M. Reusable potentiometric screen-printed sensor and label-free aptasensor with pseudo-reference electrode for determination of tryptophan in the presence of tyrosine. *Talanta*, **2016**, *150*, 425-433.
[http://dx.doi.org/10.1016/j.talanta.2015.12.064] [PMID: 26838426]

[105] Reddaiah, K.; Madhusudana Reddy, T.; Venkata Ramana, D.K.; Subba Rao, Y. Poly-Alizarin red S/multiwalled carbon nanotube modified glassy carbon electrode for the boost up of electrocatalytic activity towards the investigation of dopamine and simultaneous resolution in the presence of 5-HT: A voltammetric study. *Mater. Sci. Eng. C*, **2016**, *62*, 506-517.
[http://dx.doi.org/10.1016/j.msec.2015.12.036] [PMID: 26952453]

[106] Gholivand, M-B.; Akbari, A. A novel voltammetric sensor for citalopram based on multiwall carbon nanotube/(poly(*p*-aminobenzene sulfonic acid)/β-cyclodextrin). *Mater. Sci. Eng. C*, **2016**, *62*, 480-488.
[http://dx.doi.org/10.1016/j.msec.2016.01.066] [PMID: 26952450]

[107] Mehmeti, E.; Stanković, D.M.; Hajrizi, A.; Kalcher, K. The use of graphene nanoribbons as efficient electrochemical sensing material for nitrite determination. *Talanta*, **2016**, *159*, 34-39.
[http://dx.doi.org/10.1016/j.talanta.2016.05.079] [PMID: 27474276]

[108] Qiu, X.; Lu, L.; Leng, J.; Yu, Y.; Wang, W.; Jiang, M.; Bai, L. An enhanced electrochemical platform based on graphene oxide and multi-walled carbon nanotubes nanocomposite for sensitive determination of Sunset Yellow and Tartrazine. *Food Chem.*, **2016**, *190*, 889-895.
[http://dx.doi.org/10.1016/j.foodchem.2015.06.045] [PMID: 26213053]

[109] Novoselov, K.S.; Geim, A.K.; Morozov, S.V.; Jiang, D.; Zhang, Y.; Dubonos, S.V.; Grigorieva, I.V.; Firsov, A.A. Electric field effect in atomically thin carbon films. *Science*, **2004**, *306*(5696), 666-669.
[http://dx.doi.org/10.1126/science.1102896] [PMID: 15499015]

[110] Roy-Mayhew, J.D.; Bozym, D.J.; Punckt, C.; Aksay, I.A. Functionalized graphene as a catalytic counter electrode in dye-sensitized solar cells. *ACS Nano*, **2010**, *4*(10), 6203-6211.
[http://dx.doi.org/10.1021/nn1016428] [PMID: 20939517]

[111] Hiralal, P.; Imaizumi, S.; Unalan, H.E.; Matsumoto, H.; Minagawa, M.; Rouvala, M.; Tanioka, A.; Amaratunga, G.A. Nanomaterial-enhanced all-solid flexible zinc--carbon batteries. *ACS Nano*, **2010**, *4*(5), 2730-2734.
[http://dx.doi.org/10.1021/nn901391q] [PMID: 20415426]

[112] Dong, X.; Shi, Y.; Huang, W.; Chen, P.; Li, L.J. Electrical detection of DNA hybridization with single-base specificity using transistors based on CVD-grown graphene sheets. *Adv. Mater.*, **2010**,

22(14), 1649-1653.
[http://dx.doi.org/10.1002/adma.200903645] [PMID: 20496398]

[113] Novoselov, K.S.; Geim, A.K.; Morozov, S.V.; Jiang, D.; Katsnelson, M.I.; Grigorieva, I.V.; Dubonos, S.V.; Firsov, A.A. Two-dimensional gas of massless Dirac fermions in graphene. *Nature,* **2005,** *438*(7065), 197-200.
[http://dx.doi.org/10.1038/nature04233] [PMID: 16281030]

[114] Novoselov, K.S.; Geim, A.K.; Morozov, S.V.; Jiang, D.; Zhang, Y.; Dubonos, S.V.; Grigorieva, I.V.; Firsov, A.A. Electric field effect in atomically thin carbon films. *Science,* **2004,** *306*(5696), 666-669.
[http://dx.doi.org/10.1126/science.1102896] [PMID: 15499015]

[115] Balandin, A.A.; Ghosh, S.; Bao, W.; Calizo, I.; Teweldebrhan, D.; Miao, F.; Lau, C.N. Superior thermal conductivity of single-layer graphene. *Nano Lett.,* **2008,** *8*(3), 902-907.
[http://dx.doi.org/10.1021/nl0731872] [PMID: 18284217]

[116] Lee, C.; Wei, X.; Kysar, J.W.; Hone, J. Measurement of the elastic properties and intrinsic strength of monolayer graphene. *Science,* **2008,** *321*(5887), 385-388.
[http://dx.doi.org/10.1126/science.1157996] [PMID: 18635798]

[117] Berger, C.; Song, Z.; Li, X.; Wu, X.; Brown, N.; Naud, C.; Mayou, D.; Li, T.; Hass, J.; Marchenkov, A.N.; Conrad, E.H.; First, P.N.; de Heer, W.A. Electronic confinement and coherence in patterned epitaxial graphene. *Science,* **2006,** *312*(5777), 1191-1196.
[http://dx.doi.org/10.1126/science.1125925] [PMID: 16614173]

[118] Reina, A.; Jia, X.; Ho, J.; Nezich, D.; Son, H.; Bulovic, V.; Dresselhaus, M.S.; Kong, J. Large area, few-layer graphene films on arbitrary substrates by chemical vapor deposition. *Nano Lett.,* **2009,** *9*(1), 30-35.
[http://dx.doi.org/10.1021/nl801827v] [PMID: 19046078]

[119] Zhuo, Q.Q.; Gao, J.; Peng, M.F.; Bai, L.L.; Deng, J.J.; Xia, Y.J. Large-scale synthesis of graphene by the reduction of graphene oxide at room temperature using metal nanoparticles as catalyst. *Carbon,* **2013,** *52,* 559-564.
[http://dx.doi.org/10.1016/j.carbon.2012.10.014]

[120] Zhang, H.; Kuila, T.; Kim, N.H.; Yu, D.S.; Lee, J.H. Simultaneous reduction, exfoliation, and nitrogen doping of graphene oxide via a hydrothermal reaction for energy storage electrode materials. *Carbon,* **2014,** *69,* 66-78.
[http://dx.doi.org/10.1016/j.carbon.2013.11.059]

[121] Park, S.; Ruoff, R.S. Chemical methods for the production of graphenes. *Nat. Nanotechnol.,* **2009,** *4*(4), 217-224.
[http://dx.doi.org/10.1038/nnano.2009.58] [PMID: 19350030]

[122] Bai, H.; Li, C.; Shi, G. Functional composite materials based on chemically converted graphene. *Adv. Mater.,* **2011,** *23*(9), 1089-1115.
[http://dx.doi.org/10.1002/adma.201003753] [PMID: 21360763]

[123] Mansha, M.; Qurashi, A.; Ullah, N.; Bakare, F.O.; Khan, I.; Yamani, Z.H. Synthesis of In_2O_3/graphene heterostructure and their hydrogen gas sensing properties. *Ceram. Int.,* **2016,** *42,* 11490-11495.
[http://dx.doi.org/10.1016/j.ceramint.2016.04.035]

[124] Deng, W.; Yuan, X.; Tan, Y.; Ma, M.; Xie, Q. Three-dimensional graphene-like carbon frameworks as a new electrode material for electrochemical determination of small biomolecules. *Biosens. Bioelectron.,* **2016,** *85,* 618-624.
[http://dx.doi.org/10.1016/j.bios.2016.05.065] [PMID: 27240008]

[125] Yu, L.; Xu, Q.; Jin, D.; Zhang, Q.; Mao, A.; Shu, Y.; Yan, B.; Iiu, X. Highly sensitive electrochemical determination of sulfate in $PM_{2.5}$ based on the formation of heteropoly blue at poly-l-lysi-e-functionalized graphene modified glassy carbon electrode in the presence of cetyltrimethylammonium bromide. *Chem. Eng. J.,* **2016,** *294,* 122-131.
[http://dx.doi.org/10.1016/j.cej.2016.02.063]

[126] Gan, T.; Zhao, A.X.; Wang, S.H.; Lv, Z.; Sun, J.Y. Hierarchical triple-shelled porous hollow zinc oxide spheres wrapped in graphene oxide as efficient sensor material for simultaneous electrochemical determination of synthetic antioxidants in vegetable oil. *Sens. Actuators B Chem.*, **2016**, *235*, 707-716.
[http://dx.doi.org/10.1016/j.snb.2016.05.137]

[127] Bai, H.; Wang, S.; Liu, P.; Xiong, C.; Zhang, K.; Cao, Q. Electrochemical sensor based on in situ polymerized ion-imprinted membranes at graphene modified electrode for palladium determination. *J. Electroanal. Chem.*, **2016**, *771*, 29-36.
[http://dx.doi.org/10.1016/j.jelechem.2016.04.013]

[128] Wang, B.; Okoth, O.K.; Yan, K.; Zhang, J. A highly selective electrochemical sensor for 4-chlorophenol determination based on molecularly imprinted polymer and PDDA-functionalized graphene. *Sens. Actuators B Chem.*, **2016**, *236*, 294-303.
[http://dx.doi.org/10.1016/j.snb.2016.06.017]

[129] Arshad, S.N.; Naraghi, M.; Chasiotis, I. Strong carbon nanofibers from electrospun polyacrylonitrile. *Carbon*, **2011**, *49*, 1710-1719.
[http://dx.doi.org/10.1016/j.carbon.2010.12.056]

[130] Cuesta, N.; Cameán, I.; Ramos, A.; Llobet, S.; García, A.B. Graphitic nanomaterials from biogas-derived carbon nanofibers. *Fuel Process. Technol.*, **2016**, *152*, 1-6.
[http://dx.doi.org/10.1016/j.fuproc.2016.05.043]

[131] Ramos, A.; Cameán, I.; García, A.B. Graphitization thermal treatment of carbon nanofibers. *Carbon*, **2013**, *59*, 2-32.
[http://dx.doi.org/10.1016/j.carbon.2013.03.031]

[132] Adabi, M.; Saber, R.; Faridi-Majidi, R.; Faridbod, F. Performance of electrodes synthesized with polyacrylonitrile-based carbon nanofibers for application in electrochemical sensors and biosensors. *Mater. Sci. Eng. C*, **2015**, *48*, 673-678.
[http://dx.doi.org/10.1016/j.msec.2014.12.051] [PMID: 25579970]

[133] Mamun, A.A.; Ahmed, Y.M.; Muyibi, S.A.; Al-Khatib, M.F.; Jameel, A.T.; AlSaadi, M.A. Synthesis of carbon nanofibers on impregnated powdered activated carbon as cheap substrate. *Arab. J. Chem.*, **2016**, *9*, 532-536.
[http://dx.doi.org/10.1016/j.arabjc.2013.09.001]

[134] Sheng, J.; Ma, C.; Ma, Y.; Zhang, H.; Wang, R.; Xie, Z.; Shi, J. Synthesis of microporous carbon nanofibers with high specific surface using tetraethyl orthosilicate template for supercapacitors. *Int. J. Hydrogen Energy*, **2016**, *41*, 9383-9393.
[http://dx.doi.org/10.1016/j.ijhydene.2016.04.076]

[135] Wang, Z.; Zuo, P.; Fan, L.; Han, J.; Xiong, Y.; Yin, G. Facile electrospinning preparation of phosphorus and nitrogen dual-doped cobalt-based carbon nanofibers as bifunctional electrocatalyst. *J. Power Sources*, **2016**, *311*, 68-80.
[http://dx.doi.org/10.1016/j.jpowsour.2016.02.012]

[136] Zhang, J.; Sun, Y.; Dong, H.; Zhang, X.; Wang, W.; Chen, Z. An electrochemical non-enzymatic immunosensor for ultrasensitive detection of microcystin-LR using carbon nanofibers as the matrix. *Sens. Actuators B Chem.*, **2016**, *233*, 624-632.
[http://dx.doi.org/10.1016/j.snb.2016.04.145]

[137] Ziyatdinova, G.; Kozlova, E.; Ziganshina, E.; Budnikov, H. Surfactant/carbon nanofibers-modified electrode for the determination of vanillin. *Monatsh. Chem.*, **2016**, *147*, 191-200.

[138] Zhang, H.; Zhang, J.; Zheng, J. Electrochemical behavior of modified electrodes with carbon nanotubes and nanofibers: Application to the sensitive measurement of uric acid in the presence of ascorbic acid. *Measurement*, **2015**, *59*, 177-183.
[http://dx.doi.org/10.1016/j.measurement.2014.09.044]

[139] Lu, Y.; Luo, L.; Ding, Y.; Wang, Y.; Zhou, M.; Zhou, T.; Zhu, D.; Li, X. Electrospun nickel loaded

porous carbon nanofibers for simultaneous determination of adenine and guanine. *Electrochim. Acta,* **2015**, *174*, 191-198.
[http://dx.doi.org/10.1016/j.electacta.2015.05.165]

[140] Huang, J.; Liu, Y.; Hou, H.; You, T. Simultaneous electrochemical determination of dopamine, uric acid and ascorbic acid using palladium nanoparticle-loaded carbon nanofibers modified electrode. *Biosens. Bioelectron.,* **2008**, *24*(4), 632-637.
[http://dx.doi.org/10.1016/j.bios.2008.06.011] [PMID: 18640024]

[141] Bettencourt-Dias, A.D.; Winkler, K.; Fawcett, W.R.; Balch, A.L. The influence of electroactive solutes on the properties of electrochemically formed fullerene C_{60}-based films. *J. Electroanal. Chem.,* **2003**, *549*, 109-117.
[http://dx.doi.org/10.1016/S0022-0728(03)00265-1]

[142] Grodzka, E.; Pieta, P.; Dłuzewski, P.; Kutner, W.; Winkler, K. Formation and electrochemical properties of composites of the C_{60}–Pd polymer and multi-wall carbon nanotubes. *Electrochim. Acta,* **2009**, *54*, 5621-5628.
[http://dx.doi.org/10.1016/j.electacta.2009.04.066]

[143] Xiao, L.; Wildgoose, G.G.; Compton, R.G. Exploring the origins of the apparent "electrocatalysis" observed at C_{60} film-modified electrodes. *Sens. Actuators B Chem.,* **2009**, *138*, 524-531.
[http://dx.doi.org/10.1016/j.snb.2009.02.006]

[144] Tan, W.T.; Bond, A.M.; Ngooi, S.W.; Lim, E.B.; Goh, J.K. Electrochemical oxidation of l-cysteine mediated by a fullerene-C_{60}-modified carbon electrode. *Anal. Chim. Acta,* **2003**, *491*, 181-191.
[http://dx.doi.org/10.1016/S0003-2670(03)00791-8]

[145] Szucs, A.; Loix, A.; Nagy, J.B.; Lamberts, L. Fullerene film electrodes in aqueous solutions Part 1. Preparation and electrochemical characterization. *J. Electroanal. Chem.,* **1995**, *397*, 191-203.
[http://dx.doi.org/10.1016/0022-0728(95)04180-5]

[146] Goyal, R.N.; Singh, S.P. Voltammetric determination of paracetamol at C_{60}-modified glassy carbon electrode. *Electrochim. Acta,* **2006**, *51*, 3008-3012.
[http://dx.doi.org/10.1016/j.electacta.2005.08.036]

[147] Goyal, R.N.; Singh, S.P. Voltammetric determination of atenolol at C(60)-modified glassy carbon electrodes. *Talanta,* **2006**, *69*(4), 932-937.
[http://dx.doi.org/10.1016/j.talanta.2005.11.041] [PMID: 18970660]

[148] Goyal, R.N.; Gupta, V.K.; Bachheti, N.; Sharma, R.A. Electrochemical Sensor for the Determination of Dopamine in Presence of High Concentration of Ascorbic Acid Using a Fullerene-C_{60} Coated Gold Electrode. *Electroanalysis,* **2008**, *20*, 757-764.
[http://dx.doi.org/10.1002/elan.200704073]

[149] Kratschmer, W.; Lamb, L.D.; Fostiropoulos, K.; Huffman, D.R. Solid C_{60}: a new form of carbono. *Nature,* **1990**, *347*, 354-358.
[http://dx.doi.org/10.1038/347354a0]

[150] Diederich, C.; Thilgen, C. Covalent Fullerene Chemistry. *Science,* **1996**, *271*, 317-323.
[http://dx.doi.org/10.1126/science.271.5247.317]

[151] Prato, M.; Maggini, M. Fulleropyrrolidines: A Family of Full-Fledged Fullerene Derivatives. *Acc. Chem. Res.,* **1998**, *31*, 519-530.
[http://dx.doi.org/10.1021/ar970210p]

[152] Astefanei, A.; Núñez, O.; Galceran, M.T. Characterisation and determination of fullerenes: A critical review. *Anal. Chim. Acta,* **2015**, *882*, 1-21.
[http://dx.doi.org/10.1016/j.aca.2015.03.025] [PMID: 26043086]

[153] Prato, M. [60]Fullerene chemistry for materials science applications. *J. Mater. Chem. A Mater. Energy Sustain.,* **1997**, *7*, 1097-1109.

[154] Brahman, P.K.; Suresh, L.; Lokesh, V.; Nizamuddin, S. Fabrication of highly sensitive and selective

nanocomposite film based on CuNPs/fullerene-C60/MWCNTs: An electrochemical nanosensor for trace recognition of paracetamol. *Anal. Chim. Acta,* **2016**, *917*, 107-116.
[http://dx.doi.org/10.1016/j.aca.2016.02.044] [PMID: 27026607]

[155] Mazloum-Ardakeni, M.; Khoshroo, A. High performance electrochemical sensor based on fullerene-functionalized carbon nanotubes/ionic liquid: Determination of some catecholamines. *Electrochem. Commun.,* **2014**, *42*, 9-12.
[http://dx.doi.org/10.1016/j.elecom.2014.01.026]

[156] Palanisamy, S.; Thirumalraj, B.; Chen, S.M.; Ali, M.A.; Al-Hemaid, F.M. Palladium nanoparticles decorated on activated fullerene modified screen printed carbon electrode for enhanced electrochemical sensing of dopamine. *J. Colloid Interface Sci.,* **2015**, *448*, 251-256.
[http://dx.doi.org/10.1016/j.jcis.2015.02.013] [PMID: 25744858]

[157] Mazlou-Ardakani, M.; Khoshroo, A.; Hosseinzadeh, L. Simultaneous determination of hydrazine and hydroxylamine based on fullerene-functionalized carbon nanotubes/ionic liquid nanocomposite. *Sens. Actuators B Chem.,* **2015**, *214*, 132-137.
[http://dx.doi.org/10.1016/j.snb.2015.03.010]

[158] Brahman, P.K.; Pandey, N.; Topkaya, S.N.; Singhai, R. Fullerene-C60-MWCNT composite film based ultrasensitive electrochemical sensing platform for the trace analysis of pyruvic acid in biological fluids. *Talanta,* **2015**, *134*, 554-559.
[http://dx.doi.org/10.1016/j.talanta.2014.10.054] [PMID: 25618707]

[159] Shen, L.M.; Liu, J. New development in carbon quantum dots technical applications. *Talanta,* **2016**, *156-157*, 245-256.
[http://dx.doi.org/10.1016/j.talanta.2016.05.028] [PMID: 27260460]

[160] Xu, X.; Ray, R.; Gu, Y.; Ploehn, H.J.; Gearheart, L.; Raker, K.; Scrivens, W.A. Electrophoretic analysis and purification of fluorescent single-walled carbon nanotube fragments. *J. Am. Chem. Soc.,* **2004**, *126*(40), 12736-12737.
[http://dx.doi.org/10.1021/ja040082h] [PMID: 15469243]

[161] Bourlinos, A.B.; Trivizas, G.; Karakassides, M.A.; Baikousi, M.; Kouloumpis, A.; Gourni, D.; Bakandritsos, A.; Hola, K.; Kozak, O.; Zboril, R.; Papagiannouli, I.; Aloukos, P.; Couris, S. Green and simple route toward boron doped carbon dots with significantly enhanced non-linear optical properties. *Carbon,* **2005**, *83*, 173-179.

[162] Shen, L.; Zhang, L.; Chen, M.; Chen, X.; Wang, J. The production of pH-sensitive photoluminescent carbon nanoparticles by the carbonization of polyethylenimine and their use for bioimaging. *Carbon,* **2013**, *55*, 343-349.
[http://dx.doi.org/10.1016/j.carbon.2012.12.074]

[163] Li, L.; Yu, B.; You, T. Nitrogen and sulfur co-doped carbon dots for highly selective and sensitive detection of Hg (II) ions. *Biosens. Bioelectron.,* **2015**, *74*, 263-269.
[http://dx.doi.org/10.1016/j.bios.2015.06.050] [PMID: 26143466]

[164] Qian, Z.S.; Chai, L.J.; Huang, Y.Y.; Tang, C.; Shen, J.J.; Chen, J.R.; Feng, H. A real-time fluorescent assay for the detection of alkaline phosphatase activity based on carbon quantum dots. *Biosens. Bioelectron.,* **2015**, *68*, 675-680.
[http://dx.doi.org/10.1016/j.bios.2015.01.068] [PMID: 25660658]

[165] Yang, Y.; Liu, N.; Qiao, S.; Liu, R.; Huang, H.; Liu, Y. Silver modified carbon quantum dots for solvent-free selective oxidation of cyclohexane. *New J. Chem.,* **2015**, *39*, 2815-2821.
[http://dx.doi.org/10.1039/C4NJ02256D]

[166] Canevari, T.C.; Nakamura, M.; Cincotto, F.H.; Melo, F.M.; Toma, H.E. High performance electrochemical sensors for dopamine and epinephrine using nanocrystalline carbon quantum dots obtained under controlled chronoamperometric conditions. *Electrochim. Acta,* **2016**, *209*, 464-470.
[http://dx.doi.org/10.1016/j.electacta.2016.05.108]

[167] Yakoubi, A.; Chaabane, T.B.; Aboulaich, A.; Mahiou, R.; Balan, L.; Medjahdi, G.; Schineider, R.

Aqueous synthesis of Cu-doped CdZnS quantum dots with controlled and efficient photoluminescence. *J. Lumin.*, **2016**, *175*, 193-202.
[http://dx.doi.org/10.1016/j.jlumin.2016.02.035]

[168] Benitez-Martinez, S.; Valcarcel, M. Graphene quantum dots in analytical science. *Trends Analyt. Chem.*, **2015**, *72*, 93-113.
[http://dx.doi.org/10.1016/j.trac.2015.03.020]

[169] Mohammadi-Behzad, L.; Gholivand, M.B.; Shamsipur, M.; Gholivand, K.; Barati, A.; Gholami, A. Highly sensitive voltammetric sensor based on immobilization of bisphosphoramidate-derivative and quantum dots onto multi-walled carbon nanotubes modified gold electrode for the electrocatalytic determination of olanzapine. *Mater. Sci. Eng. C*, **2016**, *60*, 67-77.
[http://dx.doi.org/10.1016/j.msec.2015.10.068] [PMID: 26706508]

[170] Yang, R.; Miao, D.; Liang, Y.; Qu, L.; Li, J.; Harrington, P.B. Ultrasensitive electrochemical sensor based on CdTe quantum dots-decorated poly(diallyldimethylammonium chloride)-functionalized graphene nanocomposite modified glassy carbon electrode for the determination of puerarin in biological samples. *Electrochim. Acta*, **2015**, *173*, 839-846.
[http://dx.doi.org/10.1016/j.electacta.2015.05.139]

[171] Gholivand, M.B.; Mohammadi-Behzad, L. An electrochemical sensor for warfarin determination based on covalent immobilization of quantum dots onto carboxylated multiwalled carbon nanotubes and chitosan composite film modified electrode. *Mater. Sci. Eng. C*, **2015**, *57*, 77-87.
[http://dx.doi.org/10.1016/j.msec.2015.07.020] [PMID: 26354242]

[172] Jahanbakhshi, M.; Habibi, B. A novel and facile synthesis of carbon quantum dots via salep hydrothermal treatment as the silver nanoparticles support: Application to electroanalytical determination of H2O2 in fetal bovine serum. *Biosens. Bioelectron.*, **2016**, *81*, 143-150.
[http://dx.doi.org/10.1016/j.bios.2016.02.064] [PMID: 26943787]

[173] Wang, L.; Tricard, S.; Yue, P.; Zhao, J.; Fang, J.; Shen, W. Polypyrrole and graphene quantum dots @ Prussian Blue hybrid film on graphite felt electrodes: Application for amperometric determination of l-cysteine. *Biosens. Bioelectron.*, **2016**, *77*, 1112-1118.
[http://dx.doi.org/10.1016/j.bios.2015.10.088] [PMID: 26569441]

[174] Tarley, C.R.; Sotomayor, M.D.; Kubota, L.T. Polímeros Biomiméticos em Química Analítica. Parte 2: Preparo e Aplicações de MIP ("Molecularly Imprinted Polymers") em Técnicas de Extração e Separação. *Quim. Nova*, **2005**, *28*(6), 1087-1101.
[http://dx.doi.org/10.1590/S0100-40422005000600025]

[175] Marestoni, L.D.; Sotomayor, M.D.; Segatelli, M.G.; Sartori, L.R.; Tarley, C.R. Polímeros impressos com íons: fundamentos, estratégias de preparo e aplicações em química analítica. *Quim. Nova*, **2013**, *36*(8), 1194-1207.
[http://dx.doi.org/10.1590/S0100-40422013000800018]

[176] Haupt, K.; Mosbach, K. Molecularly imprinted polymers and their use in biomimetic sensors. *Chem. Rev.*, **2000**, *100*(7), 2495-2504.
[http://dx.doi.org/10.1021/cr990099w] [PMID: 11749293]

[177] Wulff, G.; Sarhan, A. Über die Anwendung von enzymanalog gebauten Polymeren zur Racemattrennung. *Angew. Chem.*, **1972**, *84*, 364.
[http://dx.doi.org/10.1002/ange.19720840838]

[178] Arshady, R.; Mosbach, M. Synthesis of substrate-selective polymers by host-guest polymerization. *Macromol. Chem. Phys.*, **1981**, *182*, 687-692.
[http://dx.doi.org/10.1002/macp.1981.021820240]

[179] Whitcombe, M.J.; Rodriguez, M.E.; Villar, P.; Vulfson, E.N. A New Method for the Introduction of Recognition Site Functionality into Polymers Prepared by Molecular Imprinting: Synthesis and Characterization of Polymeric Receptors for Cholesterol. *J. Am. Chem. Soc.*, **1995**, *117*, 7105-7111.
[http://dx.doi.org/10.1021/ja00132a010]

[180] Figueiredo, E. D.; Dias, A. C. B.; Arruda, M. A. Z. Molecular Imprinting: A Promising Strategy In Matrices Elaboration For Drug Delivery Systems. *Revista Brasileira de Ciências Farmacêuticas,* **2008**, *44*, 361-375. [impressão Molecular: Uma Estratégia Promissora Na Elaboração De Matrizes Para A Liberação Controlada De Fármacos].

[181] Tarley, C.R.; Sotomayor, M.D.; Kubota, L.T. Polímeros Biomiméticos em Química Analítica. Parte 1: Preparo e Aplicações de MIP ("Molecularly Imprinted Polymers") em Técnicas de Extração e Separação. *Quim. Nova,* **2005**, *28*, 1076-1086.
[http://dx.doi.org/10.1590/S0100-40422005000600024]

[182] Malitesta, C.; Losito, I.; Zambonin, P.G. Molecularly imprinted electrosynthesized polymers: new materials for biomimetic sensors. *Anal. Chem.,* **1999**, *71*(7), 1366-1370.
[http://dx.doi.org/10.1021/ac980674g] [PMID: 21662960]

[183] Sellergren, B. *Molecularly imprinted polymers man-made mimics of antibodies and their applications in analytical chemistry*; Elsevier: Amsterdam, **2001**.

[184] Cormack, P.A.; Elorza, A.Z. Molecularly imprinted polymers: synthesis and characterisation. *J. Chromatogr. B Analyt. Technol. Biomed. Life Sci.,* **2004**, *804*(1), 173-182.
[http://dx.doi.org/10.1016/j.jchromb.2004.02.013] [PMID: 15093171]

[185] Huang, J. Xing. X., Zhang, X., He, X., Lin, Q., Lian, W., Zhu, H. A molecularly imprinted electrochemical sensor based on multiwalled carbon nanotube-gold nanoparticle composites and chitosan for the detection of tyramine. *Food Res. Int.,* **2011**, *44*, 276-281.
[http://dx.doi.org/10.1016/j.foodres.2010.10.020]

[186] Tonelli, D.; Ballarin, B.; Guadagnini, L.; Mignani, A.; Scavetta, E. A novel potentiometric sensor for l-ascorbic acid based on molecularly imprinted polypyrrole. *Electrochim. Acta,* **2011**, *56*, 7149-7154.
[http://dx.doi.org/10.1016/j.electacta.2011.05.076]

[187] Lian, H.T.; Liu, B.; Chen, Y.P.; Sun, X.Y. A urea electrochemical sensor based on molecularly imprinted chitosan film doping with CdS quantum dots. *Anal. Biochem.,* **2012**, *426*(1), 40-46.
[http://dx.doi.org/10.1016/j.ab.2012.03.024] [PMID: 22484037]

[188] Javanbakht, M.; Fathollahi, F.; Divsar, F.; Ganjali, M.R.; Norouzi, P. A selective and sensitive voltammetric sensor based on molecularly imprinted polymer for the determination of dipyridamole in pharmaceuticals and biological fluids. *Sens. Actuators B Chem.,* **2013**, *182*, 362-367.
[http://dx.doi.org/10.1016/j.snb.2013.02.097]

[189] Moreira, F.T.; Dutra, R.A.; Noronha, J.P.; Sales, M.G. Electrochemical biosensor based on biomimetic material for myoglobin detection. *Electrochim. Acta,* **2013**, *107*, 481-487.
[http://dx.doi.org/10.1016/j.electacta.2013.06.061]

[190] Yarman, A.; Scheller, F.W. The first electrochemical MIP sensor for tamoxifen. *Sensors (Basel),* **2014**, *14*(5), 7647-7654.
[http://dx.doi.org/10.3390/s140507647] [PMID: 24776936]

[191] Tadi, K.K.; Motghare, R.V.; Ganesh, V. Electrochemical detection of epinephrine using a biomimic made up of hemin modified molecularly imprinted microspheres. *RSC Advances,* **2015**, *5*, 99115-99124.
[http://dx.doi.org/10.1039/C5RA16636E]

[192] Bagheri, H.; Khoshsafar, H.; Amidi, S.; Ardakani, H. Fabrication of an electrochemical sensor based on magnetic multi-walled carbon nanotubes for the determination of ciprofloxacin. *Anal. Methods,* **2016**, *8*, 3383-3390.
[http://dx.doi.org/10.1039/C5AY03410H]

[193] Xiao, N.; Deng, J.; Cheng, J.; Ju, S.; Zhao, H.; Xie, J.; Qian, D.; He, J. Carbon paste electrode modified with duplex molecularly imprinted polymer hybrid film for metronidazole detection. *Biosens. Bioelectron.,* **2016**, *81*, 54-60.
[http://dx.doi.org/10.1016/j.bios.2016.02.041] [PMID: 26921552]

[194]   Jolly, P.; Tamboli, V.; Harniman, R.L.; Estrela, P.; Allender, C.J.; Bowen, J.L. Aptamer-MIP hybrid receptor for highly sensitive electrochemical detection of prostate specific antigen. *Biosens. Bioelectron.,* **2016,** *75,* 188-195.
[http://dx.doi.org/10.1016/j.bios.2015.08.043] [PMID: 26318788]

[195]   Vasapollo, G.; Sole, R.D.; Mergola, L.; Lazzoi, M.R.; Scardino, A.; Scorrano, S.; Mele, G. Molecularly imprinted polymers: present and future prospective. *Int. J. Mol. Sci.,* **2011,** *12*(9), 5908-5945.
[http://dx.doi.org/10.3390/ijms12095908] [PMID: 22016636]

[196]   Dechtrirat, D.; Eichelmann, N.G.; Bier, F.F.; Scheller, F.W. Hybrid material for protein sensing based on electrosynthesized mip on a mannose terminated selfassembled monolayer. *Adv. Funct. Mater.,* **2014,** *24,* 2233-2239.
[http://dx.doi.org/10.1002/adfm.201303148]

[197]   Lu, C.H.; Zhang, Y.; Tang, S.F.; Fang, Z.B.; Yang, H.H.; Chen, X.; Chen, G.N. Sensing HIV related protein using epitope imprinted hydrophilic polymer coated quartz crystal microbalance. *Biosens. Bioelectron.,* **2012,** *31*(1), 439-444.
[http://dx.doi.org/10.1016/j.bios.2011.11.008] [PMID: 22143073]

[198]   Say, R.; Birlik, E.; Ersoz, A.; Yilmaz, F.; Gedikbey, T.; Denizli, A. Preconcentration of copper on ion-selective imprinted polymer microbeads. *Anal. Chim. Acta,* **2003,** *480,* 251-258.
[http://dx.doi.org/10.1016/S0003-2670(02)01656-2]

[199]   Güney, S.; Güney, O. A novel electrochemical sensor for selective determination of uranyl ion based on imprinted polymer sol–gel modified carbon paste electrode. *Sens. Actuators B Chem.,* **2016,** *231,* 45-53.
[http://dx.doi.org/10.1016/j.snb.2016.02.119]

[200]   Ramstrom, O.; Mosbach, K. Synthesis and catalysis by molecularly imprinted materials. *Current Opinion in Chemical Biology,* 3 (6), 759-764, 1999.

[201]   Alizadeh, T.; Ganjali, M.R.; Nourozi, P.; Zare, M.; Hoseini, M. A carbon paste electrode impregnated with Cd2+ imprinted polymer as a new and high selective electrochemical sensor for determination of ultra-trace Cd2+ in water samples. *J. Electroanal. Chem.,* **2011,** *657,* 98-106.
[http://dx.doi.org/10.1016/j.jelechem.2011.03.029]

[202]   Rajabi, H.R.; Roushani, M.; Shamsipur, M. Development of a highly selective voltammetric sensor for nanomolar detection of mercury ions using glassy carbon electrode modified with a novel ion imprinted polymeric nanobeads and multi-wall carbon nanotubes. *J. Electroanal. Chem.,* **2013,** *693,* 16-22.
[http://dx.doi.org/10.1016/j.jelechem.2013.01.003]

[203]   Alizadeh, T.; Amjadi, S. Synthesis of nano-sized Eu³□-imprinted polymer and its application for indirect voltammetric determination of europium. *Talanta,* **2013,** *106,* 431-439.
[http://dx.doi.org/10.1016/j.talanta.2013.01.019] [PMID: 23598148]

[204]   Bojdi, M.K.; Behbahani, M.; Sahragard, A.; Amin, B.G.; Fakhari, A.; Bagheri, A. A palladium imprinted polymer for highly selective and sensitive electrochemical determination of ultra-trace of palladium ions. *Electrochim. Acta,* **2014,** *149,* 108-116.
[http://dx.doi.org/10.1016/j.electacta.2014.10.096]

[205]   Nasiri-Majd, M.; Taher, M.A.; Fazelirad, H. Synthesis and application of nano-sized ionic imprinted polymer for the selective voltammetric determination of thallium. *Talanta,* **2015,** *144,* 204-209.
[http://dx.doi.org/10.1016/j.talanta.2015.05.058] [PMID: 26452811]

[206]   Alizadeh, T.; Ganjali, M.R.; Akhoundian, M.; Norouzi, P. Voltammetric determination of ultratrace levels of cerium(III) using a carbon paste electrode modified with nano-sized cerium-imprinted polymer and multiwalled carbon nanotubes. *Mikrochim. Acta,* **2016,** *183*(3), 1123-1130.
[http://dx.doi.org/10.1007/s00604-015-1702-6]

CHAPTER 3

Monitoring Circulating Tumor Cells by *In Vivo* Flow Cytometry in Cancer Therapy

Xunbin Wei[1,*], **Yuanzhen Suo**[1], **Pengfei Hai**[2] and **Xi Zhu**[1]

[1] *Med-X Research Institute and School of Biomedical Engineering, Shanghai Jiao Tong University, Shanghai, China*

[2] *Department of Biomedical Engineering, Washington University in St. Louis, St. Louis, Missouri, USA*

Abstract: The *in vivo* flow cytometry is an emerging tool to monitor circulating tumor cells (CTCs) *in vivo*. As a non-invasive method, the *in vivo* flow cytometry allows long-term monitoring of CTCs without changing their native biological environment. It has been applied for monitoring CTC dynamics in extensive biomedical researches on cancer therapy. In this chapter, the principles of fluorescence based and photoacoustic based *in vivo* flow cytometry are introduced. A number of studies concerning basic research in cancer therapy with *in vivo* flow cytometry are reviewed. Potential applications and disadvantages of *in vivo* flow cytometry in cancer therapy are also discussed.

Keywords: Cancer metastasis, Circulating tumor cells, Coherent Raman scattering, Fluorescence detection, *in vivo* flow cytometry, Liver cancer, Melanoma, Optical tweezers, Photoacoustic detection, Tumor resection, Tumor prognosis.

INTRODUCTION

In Vivo Flow Cytometry

Flow cytometry (FCM) is a fundamental biotechnology in cell quantification, biochemical analysis and cell sorting. It is routinely used in both basic biomedical research and clinical practice, including monitoring tumor cells in cancer therapy. FCM is usually referred as the conventional *in vitro* FCM. In the conventional FCM, cells of interest are sampled from patients or experimental animals. Then they are labeled with fluorescent tracers and pumped into a sheath flow. In the sheath flow, cells pass in single file through a laser beam and get excited. Emitted

---

* **Corresponding author Xunbin Wei:** Med-X Research Institute and School of Biomedical Engineering, Shanghai Jiao Tong University, Shanghai, China; Tel: +86 21-6293-3693; E-mail: xwei01@sjtu.edu.cn

Atta-ur-Rahman, Sibel A. Ozkan & Rida Ahmed (Eds.)

fluorescence from the cells are detected and analyzed to provide the physical and biochemical information.

Nevertheless, the conventional FCM can only test cells *in vitro* and requires blood drawing from patients or experimental animals. This process of making samples might change the native biological environment of the cells while the blood sample volume is limited. Red blood cell lysis, cell centrifuge and cell suspension during this process can cause cell loss and change the adhesions between cells. To overcome these limitations of *in vitro* FCM, *in vivo* flow cytometry has been developed. The basic principle of *in vivo* flow cytometry is to use the natural blood flow as the sheath flow in the conventional FCM. As cells, *e.g.* circulating tumor cells (CTCs), pass through a focused laser slit across a blood vessel of a living animal, exogenous labeling agents or endogenous chemical contrasts are excited (Fig. **1**). Generated signals are acquired. Thus biological information can be obtained during a long period of time without blood drawing. There are various exogenous or endogenous contrasts that can provide signals for *in vivo* flow cytometry. Fluorescence and photoacoustic signals are two types of signals that are commonly used in *in vivo* flow cytometry. Other types include photothermal signals, high-speed image, scattered light and Raman scattering.

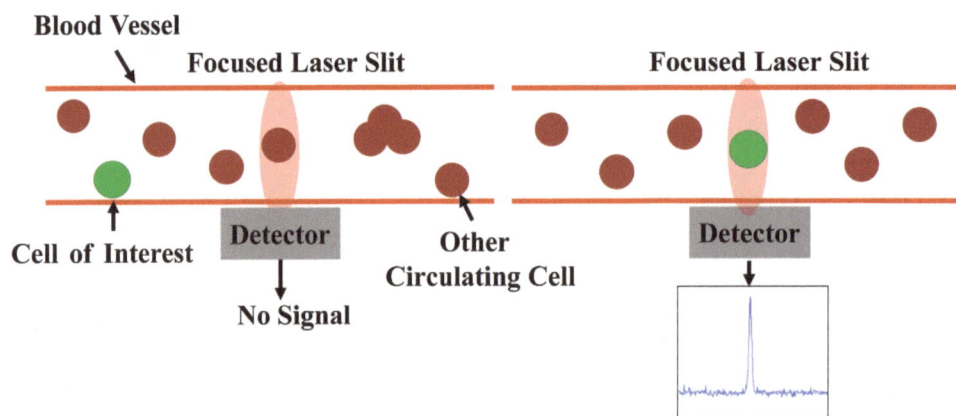

Fig. (1). Principle of monitoring circulating cells with *in vivo* flow cytometry. When a labeled cell passes through a focused laser slit and gets excited, its emitted signals are detected.

Fluorescence Based *In Vivo* Flow Cytometry

The principle of fluorescence based *in vivo* flow cytometry is to detect fluorescently labeled circulating cells by detecting the fluorescence excited by the laser. For example, to detect CTCs in animal models, researchers can transfect tumor cells with fluorescent proteins and introduce them to mice to build tumor models. At various time points of interest, researchers can quantify CTCs by detecting the fluorescence from them.

The fluorescence based *in vivo* flow cytometry was first developed by Lin's group at Massachusetts General Hospital in 2004 [1]. In the first *in vivo* flow cytometer Fig. (**2**), a 632 nm beam from a He-Ne laser was reshaped as a slit and focused onto a blood vessel of a mouse ear. When blood cells labeled with DiD (a lipo-philic dye binding to cell membranes) went through the laser slit across the vessel, they emitted fluorescence signals, which were detected by the photo-multiplier tubes. It was the first demonstration of *in vivo* flow cytometry to quantify the number of fluorescently labeled circulating cells in a reproducible manner. Lin's group also discussed its applications in monitoring circulating tumor cells. From then on, the fluorescence based *in vivo* flow cytometry was further developed by a number of groups including Lin's group, Wei's group at Shanghai Jiao Tong University, Zharov's group at University of Arkansas and Georgakoudi's group at Tufts University. The *in vivo* flow cytometer has now been developed to up to 6 channels, which is able to cover most of the fluorophores in the visible spectrum and the near infrared spectrum [2].

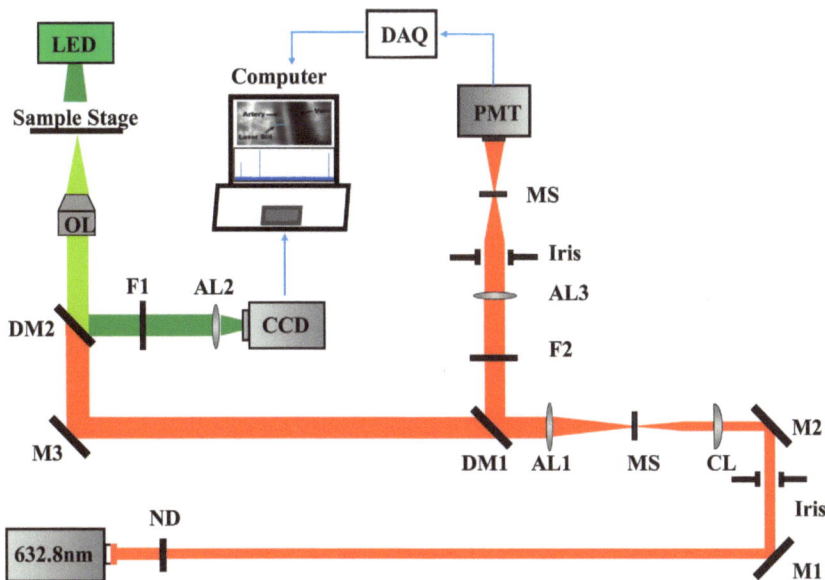

Fig. (2). Schematic of the first fluorescence based *in vivo* flow cytometer developed by Lin's group. AL1-AL3, achromatic lens; BPF1, BPF2, band pass filters; CCD, charge-coupled device; CL, cylindrical lens; DM1, DM2, dichroic mirrors; ND, neutral-density filter; M1-M3, mirrors; DAQ, data acquisition; PMT, photomultiplier tube.

Photoacoustic *In Vivo* Flow Cytometry

The *in vivo* flow cytometry can also be achieved based on photoacoustic effect, namely photoacoustic flow cytometry (PAFC) [3]. For photoacoustic effect, biological tissue is usually illuminated by a short laser pulse. Light absorption by molecules in the biological tissue induces a transient thermoelastic expansion that launches ultrasonic waves, which are detected by ultrasonic transducers [4]. Photoacoustic based imaging and sensing techniques have been demonstrated to be capable of providing anatomical, functional, molecular, metabolic and mechanical information of biological tissue [5 - 9]. In PAFC, a high-repetition short-pulsed laser slit is focused onto the target blood vessel. When a cell of interest passes through the laser slit, it gets excited and generates photoacoustic waves, which are detected by an ultrasonic transducer placed confocally with the objective to maximize the detection sensitivity (Fig. **3**).

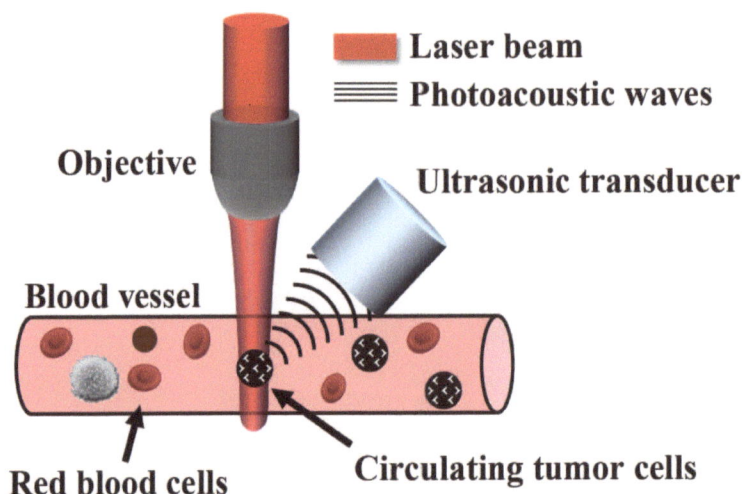

Fig. (3). Principle of monitoring CTCs with PAFC. When a CTC passes through the focused laser slit, it absorbs light and generates photoacoustic waves, which are detected by an ultrasonic transducer placed confocally with the objective.

Using endogenous contrast, PAFC can achieve label-free detection of CTCs with high optical absorption such as melanoma cells due to melanin absorption [10]. PAFC can also utilize exogenous contrast such as magnetic nanoparticles to improve the detection sensitivity and specificity of CTCs [11]. Benefiting from the weak acoustic attenuation in biological tissue, PAFC can penetrate deeper and monitor blood vessels up to 1-3 mm depth compared to 500 μm penetration depth for fluorescence based *in vivo* flow cytometry [12]. However, CTCs with lower or comparable optical absorption as red blood cells (RBCs) are difficult for PAFC to detect.

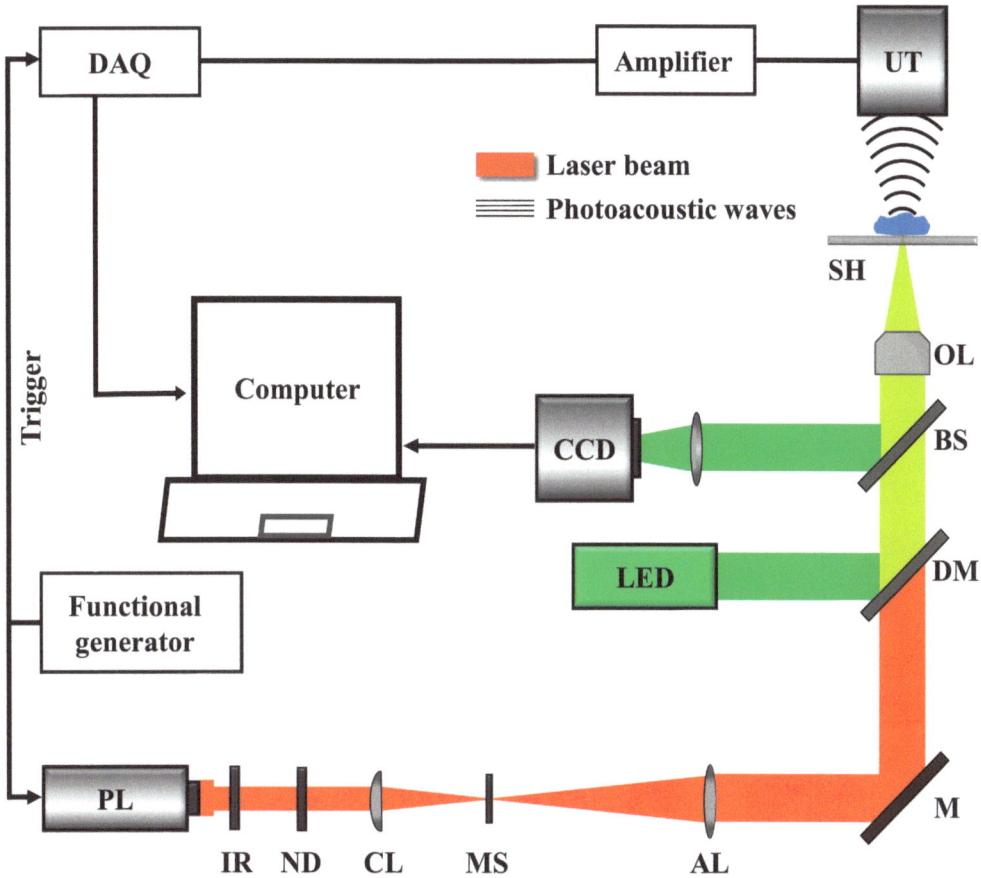

Fig. (4). Schematic of a typical setup for PAFC. AL, achromatic lens; BS, beam splitter; CCD, charge-coupled device ; CL, cylindrical lens; DAQ, data acquisition; DM, dichroic mirror; IR, iris; M, mirror; MS, mechanical slit; ND, neural density filter; OL, objective lens; PL, pulsed laser; UT, ultrasonic transducer.

The schematic of a typical setup for PAFC is shown in Fig. (**4**). A nanosecond or picosecond pulsed laser is usually used for photoacoustic excitation. Near-infrared light is usually chosen for excitation to achieve deep penetration in biological tissue and low background photoacoustic signals from red blood cells [13]. The laser beam is usually reshaped by an iris and attenuated by a neural density filter. Then it is expanded in one direction to cover the entire cross section of a blood vessel. The excitation laser beam is combined with a green LED illumination light by a dichroic mirror to visualize the blood vessels. An illumination light of 532 nm is chosen for the strong blood absorption at such wavelength, which gives strong blood to background ratio. A charge-coupled device (CCD) camera is used to capture the back-reflected light and provide an image to locate blood vessels. The excitation laser beam is focused onto a blood vessel as a slit to cover the entire cross section. An ultrasonic transducer placed confocally with the objective detects the generated photoacoustic waves with maximized sensitivity. The photoacoustic signal is then amplified and recorded by a data acquisition (DAQ) system. The entire PAFC system is synchronized by a functional generator that sends triggers to both the laser and DAQ.

To be able to detect CTCs without missing them, the laser pulse repetition rate R in PAFC is a critical parameter. It also determines the throughput of PAFC. The time for a CTC to flow through the excitation laser beam t_c is calculated by

$$t_c = \frac{D_c + D_l}{V} \tag{1}$$

Here D_c is the cell diameter while D_l is the laser beam width. V represents the blood flow speed at the point of detection. To be able to excite the CTC, the lower limit of laser pulse repetition rate R is determined as

$$R \geq \frac{1}{t_c} \tag{2}$$

For a typical B16 melanoma cell, the average diameter D_c is ~10 μm. The typical laser beam width used in PAFC D_l is also ~10 μm. For a typical blood vessel with ~10 mm/s (mouse ear artery) and ~50 mm/s (human arm artery) flow speed, the required minimum laser pulse repetition rates are 500 Hz and 2.5 kHz. Ideally, the laser pulse repetition rate should be set to several times of the minimum required value, allowing detection of the same CTC multiple times to confirm the detection of CTC and increase of the signal to noise ratio by averaging. The maximum laser pulse repetition rate is limited by the acoustic propagation time in the biological tissue and can be up to 500 kHz. The laser pulse repetition rate determines another important parameter, throughput of PAFC in CTC detection and monitoring. The throughput of PAFC, T, can be calculated by the following equation when we match the laser pulse repetition rate with the blood flow velocity so that no blood is excited multiple times.

$$T = V \times R = \pi \times (D_v / 2)^2 \times D_l \times R \tag{3}$$

Here, V is the blood volume examined per laser pulse while D_v is the diameter of the blood vessel monitored by PAFC. The throughput of PAFC can be up to 10 mL/min [3].

The detection sensitivity is also critical for PAFC. To maximize the detection sensitivity, the excitation wavelength is typically between 640 nm and 910 nm, within which low optical absorption of hemoglobin keeps the background signal low [13]. The detection sensitivity of PAFC can be further improved by using two illumination wavelengths and spectrally unmixing the photoacoustic signals [14].

It is worth noting that apart from CTCs in the blood circulation. PAFC is capable of detecting and monitoring other kinds of cells and contrast agents in various blood fluids, among which many are closely related to cancer metastasis and therapy. Individual circulating normal cells such as erythrocytes and leukocytes at different functional states can be detected by PAFC in both blood and lymph flows *in vivo* in real time [15]. PAFC can also probe single circulating nanoparticles, bacteria, and dyes *in vivo* with high sensitivity [16].

APPLICATIONS OF *IN VIVO* FLOW CYTOMETRY IN MONITORING CIRCULATING TUMOR CELLS

In vivo flow cytometry has broad applications in basic researches and clinical trials, in which monitoring CTCs related to cancer therapy is the most representative application.

Applications of Fluorescence Based *In Vivo* Flow Cytometry

Lin's group reported the first demonstration of enumerating circulating tumor cells with *in vivo* flow cytometry. They examined the circulation kinetics of two prostate cancer cell lines with different metastatic potentials in mice and rats [17]. It was found that the cell line and the host environment affected the circulation kinetics of prostate cancer cells, with the intrinsic cell line properties determining the initial rate of cell depletion from the circulation and the host environment affecting cell circulation at later time points. This work was a starting point that indicated the great potential of *in vivo* flow cytometry for monitoring CTCs.

In 2005, Lin's group collaborated with Scadden's group at Massachusetts General Hospital to investigate the molecular mechanism of CTC homing for leukemia [18]. Using *in vivo* flow cytometry, they found out that Nalm-6 cells (an acute lymphoblastic leukemia cell line) were blocked from bone marrow homing by stromal-cell-derived factor 1 (SDF-1) induced CXCR4 desensitization or a single dose of AMD3100 remaining in the peripheral circulation. Combined with intravital bone marrow imaging, they discovered that CXCR4 and SDF-1 was an important pair of molecules associated with bone marrow metastasis. The results were of great value for developing molecular targeted drugs in the therapy of leukemia.

In 2012, Wei's group at Shanghai Jiao Tong University reported their work under clinically relevant oncological condition of cancer therapy: they monitored CTC dynamics after tumor resection with *in vivo* flow cytometry [19]. A human hepatocellular carcinoma cell line HCCLM3 [20] with high metastatic potential was obtained from the Liver Cancer Institute, Zhongshan Hospital of Fudan University. They established an orthotopic metastatic tumor model of hepatocellular carcinoma by implanting a small cube of HCCLM3-GFP tumor into the liver of a mouse. By comparing the CTC dynamics in both models, it was found out that CTCs in the orthotopic tumor model increased significantly with the increase of the primary tumor size while CTCs in the subcutaneous tumor model did not increase with the increase of the primary tumor size. It meant that the orthotopic tumor model could describe the CTC dynamics in clinical situations more accurately. Following tumor resection in the orthotopic tumor model, CTCs and early metastasis decreased significantly. Tumor resection is usually considered as an important way in cancer therapy. *in vivo* flow cytometry provides direct evidence that resection prominently restricts hematogenous metastasis and distant metastasis in liver cancer therapy. It can serve as a general method to monitor CTCs in other kinds of cancer therapies.

A recent application of fluorescence based *in vivo* flow cytometry is to detect CTC clusters in animal models. CTC clusters were found out to have up to 50-fold metastatic potential when compared with single CTCs by Haber's group at Massachusetts General Hospital, which might change our knowledge of the role that CTCs play in cancer metastasis and cancer therapy. In this work, Lin's group injected *in vitro* cultured CTC clusters and single CTCs labeled with a fluorescent dye, DiD. Using fluorescence based *in vivo* flow cytometry, they found that the clearance rate of CTC clusters was at least 3 times faster than single CTCs [21]. In 2016, Wei's group published their finding that the proportion of CTC clusters kept growing during cancer metastasis in an orthotopic liver cancer and a subcutaneous prostate cancer model of mice. They distinguished CTC clusters and single CTCs by analyzing the signal patterns of the CTC events, in which a CTC cluster had multiple peaks while a single CTC had only one peak [22].

Applications of Photoacoustic Based *In Vivo* Flow Cytometry

First demonstrated by Zharov's group at University of Arkansas, PAFC has been widely applied for monitoring CTCs in cancer diagnosis and therapy [23]. A number of major technical improvements have been achieved for PAFC, including ultrafast CTC detection with high laser pulse repetition rate, two-color and multispectral PAFC with improved CTC detection sensitivity, *in vivo* magnetic enrichment and detection of CTCs, minimally invasive fiber delivery of laser radi-ation to vessels and time-of-flight velocity measurement of a single CTC [24 - 28].

As we discussed above, high laser pulse repetition rate in PAFC would allow CTC detection in larger blood vessels with high flow velocity and gain signal to noise ratio by averaging. An ultrafast PAFC was achieved by employing an Yb-doped fiber laser at 1064 nm, with laser pulse repetition rate of 500 kHz, pulse duration of 10 ns, and pulse energy up to 100 nJ [29]. The ultrafast PAFC enabled CTC detection in blood vessels with flow velocity up to 2.5 m/s and greatly improved the throughput of CTC detection.

By employing two laser pulses at wavelengths of 865 nm and 639 nm, two-color PAFC can detect melanoma CTCs with higher sensitivity by spectral identification [14]. Due to the weak optical absorption of hemoglobin, blood generated much weaker photoacoustic signals at 639 nm than at 865 nm, while melanoma CTCs generated stronger photoacoustic signals at 639 nm than at 865 nm. The distinctive features would allow detection of melanoma CTCs in blood flow with higher sensitivity.

The CTC detection efficiency and specificity by PAFC was improved by an *in vivo* magnetic enrichment method [26]. In this method, magnetic nanoparticles with specific targeting capability to the urokinase plasminogen activator receptors, which were usually overexpressed in CTCs, were injected to the blood circulation and quickly enriched CTCs *in vivo*. To further improve the CTC detection sensitivity and specificity, a second contrast agent, golden carbon nanotubes were also used and targeted at the overexpressed folate acid receptors in CTCs. For the two contrast agents, two wavelengths, 639 nm and 900 nm, were used to further differentiate the CTCs and improved the detection specificity.

To be able to detect CTCs with high sensitivity by PAFC, effective delivery of light to the region of interest is important. In the typical PAFC setup described above, confocal external optical illumination and acoustic detection was used to maximize the detection sensitivity. However, illumination laser beams still suffer from strong optical attenuation in biological tissue, especially for animals with high skin pigmentation level. To overcome this issue, a minimally invasive fiber delivery method for illumination was developed [27]. This method utilized a quartz fiber with 100 μm diameter inside a needle and inserted the fiber to blood vessels with diameters about 300 μm. This method allowed effective delivery of illumination light to the blood vessel to achieve CTC detection with high sensitivity and overcome the strong optical attenuation in the skin at the expense of spatial resolution.

POTENTIAL APPLICATIONS OF *IN VIVO* FLOW CYTOMETRY IN CANCER THERAPY

Label Free *In Vivo* Flow Cytometry

Fluorescence based and photoacoustic based *in vivo* flow cytometry are two practical types of *in vivo* flow cytometry for monitoring circulating tumor cells. It should be noted that photoacoustic based *in vivo* flow cytometry is a kind of label free *in vivo* flow cytometry, but only for cells with enough endogenous photoacoustic contrast agents, *e.g.* melanoma cells, which contain melanin). In most cases, the CTCs have to be labeled with fluorophores or photoacoustic enhancement agents. These fluorophores and photoacoustic enhancement agents might perturb the biological system, especially for small molecules with 'big' labels. Despite of numerous fluorophores, only several of them are approved for clinical use by Food and Drug Administration. Therefore, label-free *in vivo* flow cytometry can be a method to monitor CTCs in clinical studies or trials [30].

There are two possible ways for label-free *in vivo* flow cytometry. One is based on the detection of autofluorescence from CTCs. It is reported that tumor cells have more porphyrin, NADH and other chemical agents than normal blood cells [31 - 33]. However, the autofluorescence from CTCs is in short wavelength range, and its intensity is weak. It is very difficult to distinguish them from a complicated background. The other way is coherent Raman scattering based *in vivo* flow cytometry, which is sensitive to the vibrational signatures of molecules, typically the nuclear vibrations of chemical bonds. It is reported that CTCs have more lipid than normal cells, especially for prostate cancer [34, 35]. Coherent Raman scattering based *in vivo* flow cytometry can potentially monitor CTCs *in vivo* based on the chemical characters. Once *in vivo* flow cytometry achieves label-free detection and monitoring of CTCs with high robustness, it can serve as an important tool for clinical cancer therapy.

In Vivo Theranostic of CTCs by *In Vivo* Flow Cytometry

CTCs are considered the "seed" for cancer metastasis [36]. Killing CTCs *in vivo* would reduce metastasis and contribute to cancer therapy. With the capability to detect CTCs in real time [37], *in vivo* flow cytometry can achieve *in vivo* theranostic of CTCs. It is reported that ultrafast laser is able to induce the apoptosis of HeLa cells [38]. The laser is also able to modulate cell signaling in many ways [39, 40]. Zharov's group has demonstrated detection and eradication of melanoma CTCs *in vivo* with PAFC guided photothermal ablation [14]. By increasing the illumination laser energy, they observed a decrease in the CTC counts, demonstrating *in vivo* eradication of CTCs. Effective as this method is, it might cause damage to the surrounding RBCs. A potential improvement of this technique would be a real time feedback *in vivo* flow cytometry-guided photothermal therapy system. Once a CTC is detected by the *in vivo* flow cytometry, a trigger is sent to a therapeutic laser that focuses a strong laser pulse on the CTC to ablate it. An alternative way to achieve CTC ablation *in vivo* is to capture a CTC once it is detected by *in vivo* flow cytometry. Wei's group and Li's group have successfully trapped red blood cells in living animals using optical tweezers [29]. This technique could be adopted and combined with *in vivo* flow cytometry for *in vivo* detecting, trapping, and ablation of CTCs.

In Vivo Monitoring of CTCs During Medical Interventions

During cancer therapy, many medical interventions usually take place, including tumor biopsy, tumor compression and laser treatment. However, these medical procedures may stimulate tumor metastasis and increase the CTC counts in the blood stream. *in vivo* flow cytometry, with the ability to monitor and enumerate CTCs, is an ideal tool to study the CTC fluctuation during medical intervention and guide cancer therapy. In one study, tumor bearing mice were treated with various medical procedures including biopsy, physical compression of tumor and laser surgery [41]. A multimodal *in vivo* flow cytometry, both on photoacoustic and fluorescence methods, was used to monitor the CTC released from the tumors. Experimental data showed that CTC counts in the blood flow had greatly increased after the above medical intervention, resulting in greater risk of tumor metastasis. The results have demonstrated the capability of IVFC to detect the intervention-amplified CTCs in real time and its potential to guide such medical intervention during cancer therapy. The results also indicate that IVFC, as an independent tool, can be used to evaluate the responses to treatment by monitoring CTC fluctuation after cancer therapy.

LIMITATIONS OF *IN VIVO* FLOW CYTOMETRY IN CANCER THERAPY

The *in vivo* flow cytometry is an optical method, in which lasers play an important role. It has all the limitations as other optical methods when applied in biomedical research and clinic, *i.e.* penetration depth. In *in vivo* flow cytometry, the laser can only detect a maximum depth of 1 mm underneath the skin, due to the scattering, absorption and reflection of light. It is a much greater depth when compared with laser scanning confocal microscopy with a detection depth of about 300 μm. However, this depth limits its potential in clinical applications, because most large human vessels are deeper than 1 mm.

For the safety of *in vivo* flow cytometry, one limitation is the use of fluorescent or photoacoustic labels, in which only quite a few fluorescent dyes are approved for clinical use. Although photoacoustic *in vivo* flow cytometry has the potential of label-free detection, it only applies to detect cells with high photoacoustic contrast agents, *e.g.* melanoma cells with high content of melanin. Another limitation is that the laser power of *in vivo* flow cytometry is relatively high. The biosafety of the lasers in *in vivo* flow cytometry has not yet been studied systemically.

Although with a number of limitations for clinical use, the *in vivo* flow cytometry is still a powerful tool to monitor circulating cells dynamically in biomedical research, in which the requirement of detection depth and biosafety is not that strict. More and more discoveries have been made with *in vivo* flow cytometry. There is no doubt that these findings would enhance the development of cancer therapies.

CONFLICT OF INTEREST

The authors declare no conflict of interest, financial or otherwise.

ACKNOWLEDGEMENTS

Declared none.

REFERENCES

[1]   Novak, J.; Georgakoudi, I.; Wei, X.; Prossin, A.; Lin, C.P. *In vivo* flow cytometer for real-time detection and quantification of circulating cells. *Opt. Lett.,* **2004**, *29*(1), 77-79.
       [http://dx.doi.org/10.1364/OL.29.000077] [PMID: 14719666]

[2]   Suo, Y.; Liu, T.; Xie, C.; Wei, D.; Tan, X.; Wu, L.; Wang, X.; He, H.; Shi, G.; Wei, X.; Shi, C. Near infrared *in vivo* flow cytometry for tracking fluorescent circulating cells. *Cytometry A,* **2015**, *87*(9), 878-884.
       [http://dx.doi.org/10.1002/cyto.a.22711] [PMID: 26138257]

[3]   Galanzha, E.I.; Zharov, V.P. Photoacoustic flow cytometry. *Methods,* **2012**, *57*(3), 280-296.
       [http://dx.doi.org/10.1016/j.ymeth.2012.06.009] [PMID: 22749928]

[4]   Wang, L.V.; Hu, S. Photoacoustic tomography: *in vivo* imaging from organelles to organs. *Science,* **2012**, *335*(6075), 1458-1462.
       [http://dx.doi.org/10.1126/science.1216210] [PMID: 22442475]

[5]   Wang, X.; Pang, Y.; Ku, G.; Xie, X.; Stoica, G.; Wang, L.V. Noninvasive laser-induced photoacoustic tomography for structural and functional *in vivo* imaging of the brain. *Nat. Biotechnol.,* **2003**, *21*(7), 803-806.
       [http://dx.doi.org/10.1038/nbt839] [PMID: 12808463]

[6]   Hai, P.; Yao, J.; Maslov, K.I.; Zhou, Y.; Wang, L.V. Near-infrared optical-resolution photoacoustic microscopy. *Opt. Lett.,* **2014**, *39*(17), 5192-5195.
       [http://dx.doi.org/10.1364/OL.39.005192] [PMID: 25166107]

[7]   Zhang, H.F.; Maslov, K.; Stoica, G.; Wang, L.V. Functional photoacoustic microscopy for high-resolution and noninvasive *in vivo* imaging. *Nat. Biotechnol.,* **2006**, *24*(7), 848-851.
       [http://dx.doi.org/10.1038/nbt1220] [PMID: 16823374]

[8]   Hai, P.; Zhou, Y.; Liang, J.; Li, C.; Wang, L.V. Photoacoustic tomography of vascular compliance in humans. *J. Biomed. Opt.,* **2015**, *20*(12), 126008.
       [http://dx.doi.org/10.1117/1.JBO.20.12.126008] [PMID: 26720875]

[9]   Wang, L.V. Multiscale photoacoustic microscopy and computed tomography. *Nat. Photonics,* **2009**, *3*(9), 503-509.
        [http://dx.doi.org/10.1038/nphoton.2009.157] [PMID: 20161535]

[10]   Galanzha, E.I.; Zharov, V.P. Circulating tumor cell detection and capture by photoacoustic flow cytometry *in vivo* and *ex vivo. Cancers (Basel),* **2013**, *5*(4), 1691-1738.
        [http://dx.doi.org/10.3390/cancers5041691] [PMID: 24335964]

[11]   Kim, J-W.; Galanzha, E.I.; Shashkov, E.V.; Moon, H-M.; Zharov, V.P. Golden carbon nanotubes as multimodal photoacoustic and photothermal high-contrast molecular agents. *Nat. Nanotechnol.,* **2009**, *4*(10), 688-694.
        [http://dx.doi.org/10.1038/nnano.2009.231] [PMID: 19809462]

[12]   Tuchin, V.V.; Tárnok, A.; Zharov, V.P. *In vivo* flow cytometry: a horizon of opportunities. *Cytometry A,* **2011**, *79*(10), 737-745.
        [http://dx.doi.org/10.1002/cyto.a.21143] [PMID: 21915991]

[13]   Nedosekin, D.A.; Sarimollaoglu, M.; Ye, J.H.; Galanzha, E.I.; Zharov, V.P. *In vivo* ultra-fast photoacoustic flow cytometry of circulating human melanoma cells using near-infrared high-pulse rate lasers. *Cytometry A,* **2011**, *79*(10), 825-833.
        [http://dx.doi.org/10.1002/cyto.a.21102] [PMID: 21786417]

[14]   Galanzha, E.I.; Shashkov, E.V.; Spring, P.M.; Suen, J.Y.; Zharov, V.P. *In vivo*, noninvasive, label-free detection and eradication of circulating metastatic melanoma cells using two-color photoacoustic flow cytometry with a diode laser. *Cancer Res.,* **2009**, *69*(20), 7926-7934.
        [http://dx.doi.org/10.1158/0008-5472.CAN-08-4900] [PMID: 19826056]

[15]   Zharov, V.P.; Galanzha, E.I.; Shashkov, E.V.; Kim, J-W.; Khlebtsov, N.G.; Tuchin, V.V. Photoacoustic flow cytometry: principle and application for real-time detection of circulating single nanoparticles, pathogens, and contrast dyes *in vivo. J. Biomed. Opt.,* **2007**, *12*(5), 051503-051514.
        [http://dx.doi.org/10.1117/1.2793746] [PMID: 17994867]

[16]   Galanzha, E.I.; Shashkov, E.V.; Tuchin, V.V.; Zharov, V.P. *In vivo* multispectral, multiparameter, photoacoustic lymph flow cytometry with natural cell focusing, label-free detection and multicolor nanoparticle probes. *Cytometry A,* **2008**, *73*(10), 884-894.
        [http://dx.doi.org/10.1002/cyto.a.20587] [PMID: 18677768]

[17]   Georgakoudi, I.; Solban, N.; Novak, J.; Rice, W.L.; Wei, X.; Hasan, T.; Lin, C.P. *In vivo* flow cytometry: a new method for enumerating circulating cancer cells. *Cancer Res.,* **2004**, *64*(15), 5044-5047.
        [http://dx.doi.org/10.1158/0008-5472.CAN-04-1058] [PMID: 15289300]

[18]   Sipkins, D.A.; Wei, X.; Wu, J.W.; Runnels, J.M.; Côté, D.; Means, T.K.; Luster, A.D.; Scadden, D.T.; Lin, C.P. *In vivo* imaging of specialized bone marrow endothelial microdomains for tumour engraftment. *Nature,* **2005**, *435*(7044), 969-973.
        [http://dx.doi.org/10.1038/nature03703] [PMID: 15959517]

[19]   Fan, Z-C.; Yan, J.; Liu, G-D.; Tan, X-Y.; Weng, X-F.; Wu, W-Z.; Zhou, J.; Wei, X.B. Real-time monitoring of rare circulating hepatocellular carcinoma cells in an orthotopic model by *in vivo* flow cytometry assesses resection on metastasis. *Cancer Res.,* **2012**, *72*(10), 2683-2691.
        [http://dx.doi.org/10.1158/0008-5472.CAN-11-3733] [PMID: 22454286]

[20]   Li, Y.; Tang, Y.; Ye, L.; Liu, B.; Liu, K.; Chen, J.; Xue, Q. Establishment of a hepatocellular carcinoma cell line with unique metastatic characteristics through *in vivo* selection and screening for metastasis-related genes through cDNA microarray. *J. Cancer Res. Clin. Oncol.,* **2003**, *129*(1), 43-51.
        [PMID: 12618900]

[21]   Aceto, N.; Bardia, A.; Miyamoto, D.T.; Donaldson, M.C.; Wittner, B.S.; Spencer, J.A.; Yu, M.; Pely, A.; Engstrom, A.; Zhu, H.; Brannigan, B.W.; Kapur, R.; Stott, S.L.; Shioda, T.; Ramaswamy, S.; Ting, D.T.; Lin, C.P.; Toner, M.; Haber, D.A.; Maheswaran, S. Circulating tumor cell clusters are oligoclonal precursors of breast cancer metastasis. *Cell,* **2014**, *158*(5), 1110-1122.

[http://dx.doi.org/10.1016/j.cell.2014.07.013] [PMID: 25171411]

[22]   Suo, Y.; Xie, C.; Zhu, X.; Fan, Z.; Yang, Z.; He, H. Proportion of circulating tumor cell clusters increases during cancer metastasis. *Cytometry Part A*, **2017**, *91*(3), 250-253.
[http://dx.doi.org/10.1002/cyto.a.23037] [PMID: 28009470]

[23]   Zharov, V.P.; Galanzha, E.I.; Shashkov, E.V.; Khlebtsov, N.G.; Tuchin, V.V. *In vivo* photoacoustic flow cytometry for monitoring of circulating single cancer cells and contrast agents. *Opt. Lett.,* **2006**, *31*(24), 3623-3625.
[http://dx.doi.org/10.1364/OL.31.003623] [PMID: 17130924]

[24]   Proskurnin, M.A.; Zhidkova, T.V.; Volkov, D.S.; Sarimollaoglu, M.; Galanzha, E.I.; Mock, D.; Nedosekin, D.A.; Zharov, V.P. *In vivo* multispectral photoacoustic and photothermal flow cytometry with multicolor dyes: a potential for real-time assessment of circulation, dye-cell interaction, and blood volume. *Cytometry A*, **2011**, *79*(10), 834-847.
[http://dx.doi.org/10.1002/cyto.a.21127] [PMID: 21905207]

[25]   Nedosekin, D.A.; Sarimollaoglu, M.; Shashkov, E.V.; Galanzha, E.I.; Zharov, V.P. Ultra-fast photoacoustic flow cytometry with a 0.5 MHz pulse repetition rate nanosecond laser. *Opt. Express,* **2010**, *18*(8), 8605-8620.
[http://dx.doi.org/10.1364/OE.18.008605] [PMID: 20588705]

[26]   Galanzha, E.I.; Shashkov, E.V.; Kelly, T.; Kim, J-W.; Yang, L.; Zharov, V.P. *In vivo* magnetic enrichment and multiplex photoacoustic detection of circulating tumour cells. *Nat. Nanotechnol.,* **2009**, *4*(12), 855-860.
[http://dx.doi.org/10.1038/nnano.2009.333] [PMID: 19915570]

[27]   Galanzha, E.I.; Kokoska, M.S.; Shashkov, E.V.; Kim, J-W.; Tuchin, V.V.; Zharov, V.P. *In vivo* fiber-based multicolor photoacoustic detection and photothermal purging of metastasis in sentinel lymph nodes targeted by nanoparticles. *J. Biophotonics,* **2009**, *2*(8-9), 528-539.
[http://dx.doi.org/10.1002/jbio.200910046] [PMID: 19743443]

[28]   Sarimollaoglu, M.; Nedosekin, D.A.; Simanovsky, Y.; Galanzha, E.I.; Zharov, V.P. *In vivo* photoacoustic time-of-flight velocity measurement of single cells and nanoparticles. *Opt. Lett.,* **2011**, *36*(20), 4086-4088.
[http://dx.doi.org/10.1364/OL.36.004086] [PMID: 22002394]

[29]   Zhong, M-C.; Wei, X-B.; Zhou, J-H.; Wang, Z-Q.; Li, Y-M. Trapping red blood cells in living animals using optical tweezers. *Nat. Commun.,* **2013**, *4*, 1768.
[http://dx.doi.org/10.1038/ncomms2786] [PMID: 23612309]

[30]   Zeng, Y.; Xu, J.; Li, D.; Li, L.; Wen, Z.; Qu, J.Y. Label-free *in vivo* flow cytometry in zebrafish using two-photon autofluorescence imaging. *Opt. Lett.,* **2012**, *37*(13), 2490-2492.
[http://dx.doi.org/10.1364/OL.37.002490] [PMID: 22743431]

[31]   Li, C.; Pastila, R.K.; Pitsillides, C.; Runnels, J.M.; Puorishaag, M.; Côté, D.; Lin, C.P. Imaging leukocyte trafficking *in vivo* with two-photon-excited endogenous tryptophan fluorescence. *Opt. Express*, **2010**, *18*(2), 988-999.
[http://dx.doi.org/10.1364/OE.18.000988] [PMID: 20173920]

[32]   Masilamani, V.; Alsalhi, M.S.; Vijmasi, T.; Govindarajan, K.; Rathan Rai, R.; Atif, M.; Prasad, S.; Aldwayyan, A.S. Fluorescence spectra of blood and urine for cervical cancer detection. *J. Biomed. Opt.,* **2012**, *17*(9), 98001-1.
[http://dx.doi.org/10.1117/1.JBO.17.9.098001] [PMID: 23085927]

[33]   Harris, D.M.; Werkhaven, J. Endogenous porphyrin fluorescence in tumors. *Lasers Surg. Med.,* **1987**, *7*(6), 467-472.
[http://dx.doi.org/10.1002/lsm.1900070605] [PMID: 3123828]

[34]   Yue, S.; Li, J.; Lee, S-Y.; Lee, H.J.; Shao, T.; Song, B.; Cheng, L.; Masterson, T.A.; Liu, X.; Ratliff, T.L.; Cheng, J.X. Cholesteryl ester accumulation induced by PTEN loss and PI3K/AKT activation underlies human prostate cancer aggressiveness. *Cell Metab.,* **2014**, *19*(3), 393-406.

[http://dx.doi.org/10.1016/j.cmet.2014.01.019] [PMID: 24606897]

[35]   Le, T.T.; Huff, T.B.; Cheng, J-X. Coherent anti-Stokes Raman scattering imaging of lipids in cancer metastasis. *BMC Cancer,* **2009**, *9*(1), 42.
[http://dx.doi.org/10.1186/1471-2407-9-42] [PMID: 19183472]

[36]   Paget, S. The distribution of secondary growths in cancer of the breast. *Lancet,* **1889**, *133*(3421), 571-573.
[http://dx.doi.org/10.1016/S0140-6736(00)49915-0] [PMID: 2673568]

[37]   Boutrus, S.; Greiner, C.; Hwu, D.; Chan, M.; Kuperwasser, C.; Lin, C.P.; Georgakoudi, I. Portable two-color *in vivo* flow cytometer for real-time detection of fluorescently-labeled circulating cells. *J. Biomed. Opt.,* **2007**, *12*(2), 020507-3.
[http://dx.doi.org/10.1117/1.2722733] [PMID: 17477705]

[38]   He, H.; Chan, K.T.; Kong, S.K. Role of nuclear tubule on the apoptosis of HeLa cells induced by femtosecond laser. *Appl. Phys. Lett.,* **2010**, *96*(22), 223701.
[http://dx.doi.org/10.1063/1.3447365] [PMID: 21200442]

[39]   Echevarría, W.; Leite, M.F.; Guerra, M.T.; Zipfel, W.R.; Nathanson, M.H. Regulation of calcium signals in the nucleus by a nucleoplasmic reticulum. *Nat. Cell Biol.,* **2003**, *5*(5), 440-446.
[http://dx.doi.org/10.1038/ncb980] [PMID: 12717445]

[40]   He, H.; Li, S.; Wang, S.; Hu, M.; Cao, Y.; Wang, C. Manipulation of cellular light from green fluorescent protein by a femtosecond laser. *Nat. Photonics,* **2012**, *6*(10), 651-656.
[http://dx.doi.org/10.1038/nphoton.2012.207]

[41]   Juratli, M.A.; Siegel, E.R.; Nedosekin, D.A.; Sarimollaoglu, M.; Jamshidi-Parsian, A.; Cai, C.; Menyaev, Y.A.; Suen, J.Y.; Galanzha, E.I.; Zharov, V.P. *In vivo* Long-Term Monitoring of Circulating Tumor Cells Fluctuation during Medical Interventions. *PLoS One,* **2015**, *10*(9), e0137613.
[http://dx.doi.org/10.1371/journal.pone.0137613] [PMID: 26367280]

CHAPTER 4

Advances in Glow Discharge Spectroscopy for Depth Profile Analytical Applications

Lara Lobo and Rosario Pereiro[*]

Department of Physical and Analytical Chemistry, Faculty of Chemistry, University of Oviedo, 33006 Oviedo, Spain

Abstract: Glow discharges (GDs), either coupled to optical emission or to mass spectrometry, have been widely investigated during the last three decades for a high variety of direct solid analytical applications. The intrinsic characteristics of these techniques, *e.g.* low matrix effects, multi-elemental capabilities, analytical sensitivity and good depth resolution, explain the continuous effort towards new developments aiming at further broadening their applications field and, so, the interesting applications reported so far.

In this chapter, a brief description of the basics of the GDs is given first. Most recent instrumental advances will be then described in detail, both for optical emission and mass spectrometry, together with the analytical improvements that these instrumental progresses have allowed. Particular interest is paid to GD time-of-flight mass spectrometry, as this instrumentation (commercially launched in 2014) has proved to be promising in terms of high depth resolution, fast acquisition rates and time-gated detection, showing also interesting capabilities to obtain both elemental and molecular information.

Finally, recent applications of GD techniques are described (focusing on the last five years). Special attention is given to the characterization of advanced materials such as multilayers, thin film solar cells and polymers.

Keywords: Depth profile analysis, Direct solid analysis, Elemental analysis, Glow discharge, Mass spectrometry, Multi-matrix calibration, Multilayers, Optical emission spectrometry, Thin films, Time of flight.

INTRODUCTION

Glow discharges (**GDs**) operated at low pressure (1-10 torr) are atomization / excitation / ionization sources with interesting applications in analytical chemistry for direct solid analysis of materials. GDs for analytical purposes can be operated

[*] **Corresponding Author Rosario Pereiro:** Department of Physical and Analytical Chemistry, Faculty of Chemistry, University of Oviedo, 33006 Oviedo, Spain; Tel: +34 98510312; E-mail: mrpereiro@uniovi.es

Atta-ur-Rahman, Sibel A. Ozkan & Rida Ahmed (Eds.)

either with direct current (**dc**) or with radiofrequency (**rf**) energy [1]. Continuous powering is typically employed. More recently, pulsed GDs (**PGDs**) have also demonstrated interesting analytical advantages [2].

A GD is generated in a low pressure chamber through which an inert gas (typically argon) continuously flows at a flow rate typically below 0.5 L/min. The chamber contains a grounded anode and a cathode (usually the sample). The application of a voltage between anode and cathode gives rise to the ionization of argon and the generation of a plasma. Then, the argon ions are electrically attracted towards the sample giving rise to the *sputtering process*, so atoms, molecular fragments, electrons and ions are liberated from the sample. The sputtered atoms and molecular species are then excited and ionized through collisions in the plasma (see Fig. **1**). The measurement of the characteristic light emitted by optical emission spectrometry (OES) or the measurement of the ions by mass spectrometry (MS) gives the composition of the sample. Both GD-OES and GD-MS allow multi-elemental analysis directly from the solid sample, including non-metallic elements like hydrogen, carbon, nitrogen, oxygen or sulfur.

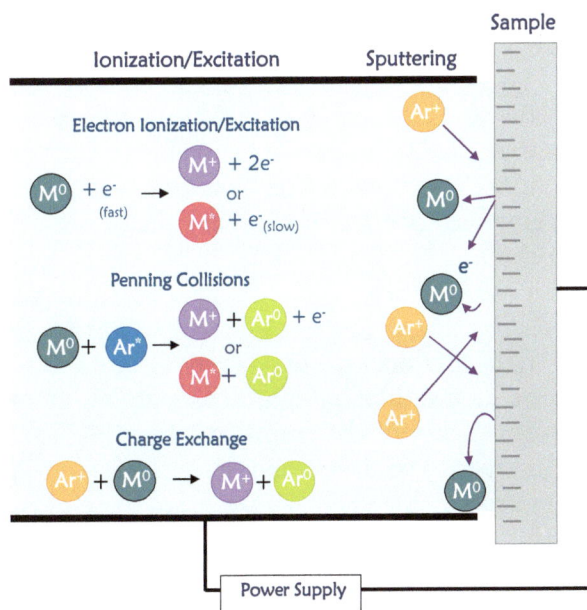

Fig. (1). A scheme of the main processes that take place within a GD: sample sputtering and ionization/excitation of the sputtered atoms. In the Figure are collected the three main ionization/excitation mechanisms: electron ionization/excitation, Penning collisions and charge exchange.

An interesting feature of GD analytical techniques is the possibility to use multi-matrix calibrations. This capability arises from the following two particularities:

(i) sample sputtering and analytes excitation/ionization mechanisms can be considered as rather independent processes taking place separated in time and space, and (ii) as the excitation/ionization of the analytes occur in a plasma where sample constituents are diluted in the plasma gas, "memory effects" from the sample matrix are not very important.

Due to the different particles (electrons, ions, atoms, neutral molecules and photons) present in the plasma, there are different collision processes that can occur, some of them with higher probability compared to others [3, 4]. For instance, it is well known that electron impact is the responsible process of self-maintaining the GD plasma. Nevertheless, in the following lines only the most relevant processes related with the excitation/ionization processes regarding the sputtered atoms from the sample will be described (see Fig. **1**), but still, it is worth mentioning that the same excitation/ionization processes occurring to the discharge gas, also apply to the sputtered atoms. As it was said above, one of the main processes to consider is **electron impact**. This process can lead to a direct ionization by an electron having a higher energy than the ionization energy of the sputtered atom or in a two-step reaction; in this last case, involving the presence of a metastable species that is later ionized. **Penning ionization/excitation** is another important process to consider in analytical GDs. Penning ionization takes place when argon in metastable level collides with an atom in the ground state. Ar has two metastable levels at 11.55 eV and 11.72 eV and, so, elements with an ionization potential below 11.72 eV can be ionized by this process. Although Penning ionization is a more selective mechanism than electron impact, it can still be considered as non-selective, as most of the elements of the periodic table, have an ionization potential below 11.72 eV. Only H, N, O, F, Cl or Br, cannot be ionized by this process (in the case of Br, it has been described that Penning ionization might be possible because the 11.81 eV ionization potential of Br is only slightly higher than the energy of Ar metastable level [5]). Another possibility is when an Ar metastable and a ground state atom collide, giving rise to excitation instead of ionization: differently to Penning ionization, **Penning excitation** is a rather selective process and it requires the energy difference between the excited and ground state level of the atom to be close to the energy of Ar metastable levels (for Penning ionization the excess of energy between the ionization potential and the Ar metastable levels is taken by the ejected electron). The last mechanism to consider is **asymmetric charge transfer**, which consists of an energy (electron) transfer from the analyte atom to the argon ion. This mechanism takes place when the energy difference between the Ar ion and the excited analyte ion is rather close.

Pure argon is, by far, the most commonly used discharge gas. However, other pure gases have been evaluated (*e.g.* helium and krypton) along with gas mixtures (argon and helium, argon and hydrogen, argon and oxygen, *etc.*). Such studies usually have one or several of the following goals: (i) enhance signals of analytical interest, (ii) reduce interferences, and (iii) investigate excitation/ionization mechanisms in the discharge.

BASIC DESIGNS OF GD CHAMBERS

Fig. (**2**) collects three main basic designs of GD chambers used for analytical purposes: hollow cathode, coaxial cathode and flat cathode. The *hollow cathode* (Fig. **2a**) offers high efficiency for the excitation and emission processes and high analytical sensitivity in OES (problems arise in the case of MS detection because of the inherent difficulty of extracting the ions towards the mass analyzer). Disadvantages of the hollow cathode design include the need of sample mechanization before measurement and the fact that in depth profile information cannot be achieved. A proper design to extract ions for MS measurements is the *coaxial cathode* or "pin-type" (Fig. **2b**). In this case, flat samples are not required, but in depth profile analysis is not possible. In 1968, Grimm proposed a discharge chamber with a *flat cathode* (Fig. **2c**) [6].

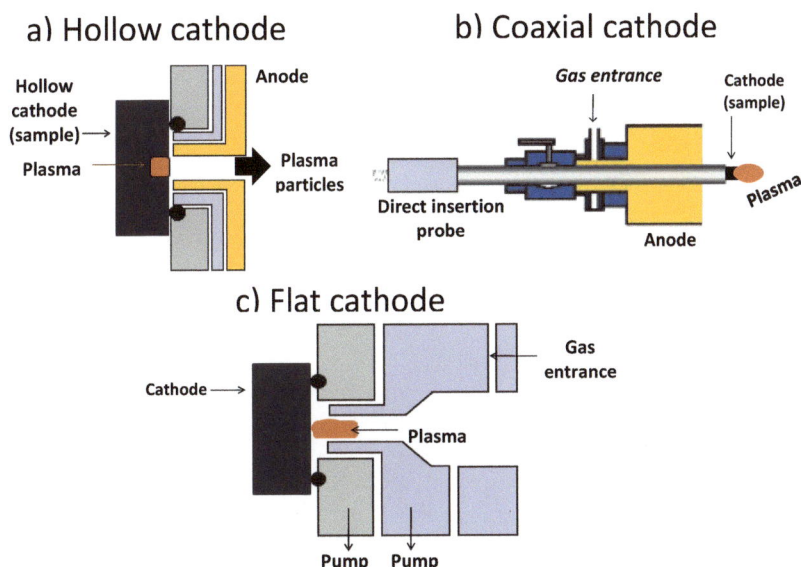

Fig. (2). Basic GD designs for analytical purposes. (**a**) hollow cathode. (**b**) coaxial cathode. (**c**) flat cathode (in the three cases the sample is the cathode).

The Grimm-type source is characterized by a cylindrical open hollow anode (typically 3-4 mm of diameter) by which the sputtering is restricted to a defined area of the sample. The sealing of the GD source is achieved using an O-ring. The sputtering rate under typical operation conditions is quite high (about 1 μm/min). The fast sputtering rate along with the fact that high vacuum inside the discharge chamber is not required, allow for a high sample throughput (just a few minutes per sample). Commercial GD-OES instruments are based on the Grimm source. The Grimm design is also appropriate for MS with slight modifications [7].

The principal advantage of the flat cathode design is that atoms are sputtered atomic layer by atomic layer, allowing to depth profile the specimen (Fig. **3**) with high depth resolution.

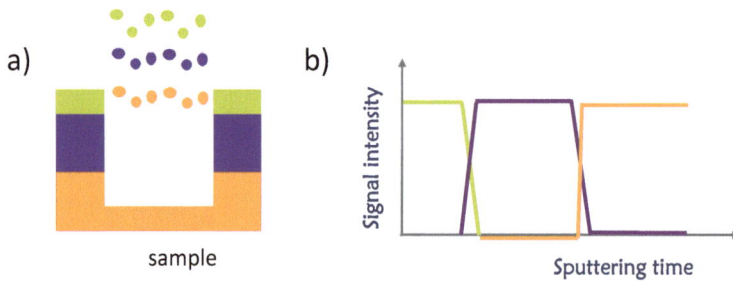

Fig. (3). Depth profile analysis with GD (artist rendition). (**a**) sample sputtered with the GD. (**b**) depth profile obtained.

The generation of an appropriate crater shape in the specimen by the sputtering process is crucial to obtain a well depth-resolved analytical profile. Fig. (**4**) shows a real example of a 3D crater produced in a sample after the sputter process. As can be observed, some material is re-deposited on the crater edges.

Fig. (4). Example of a crater produced in a sample after the sputtering process.

The selection of GD experimental conditions (pressure and electrical parameters) which are able to produce craters with a flat bottom and perpendicular edges is very important for the achievement of good depth resolution, as illustrated in Fig. (**5**). Fig. (**5a**) presents a schematic view of a layered sample; Fig. (**5b**) presents a 2D example of a good crater shape, where the sample has been sputtered layer by layer. By contrast, Fig. (**5c**) presents an example of an undesirable crater (in this case, atoms from different layers are being simultaneously sputtered).

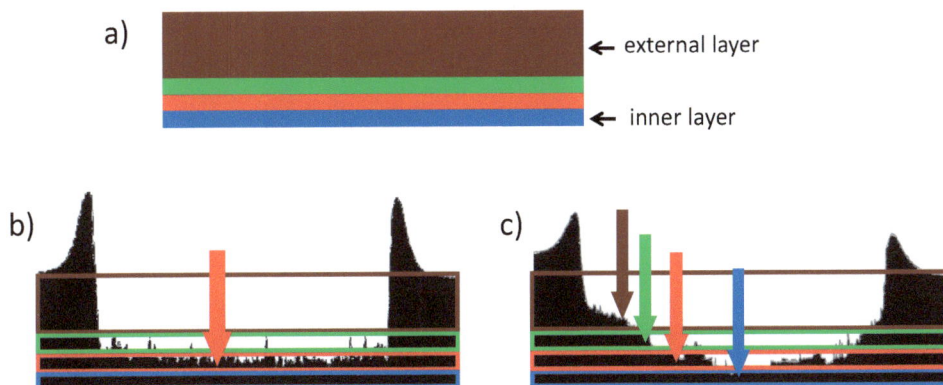

Fig. (5). Importance of obtaining good crater shapes (artist rendition). (**a**) sample. (**b**) correct crater shape. (**c**) Inappropriate crater shape.

NEW INSTRUMENTAL DEVELOPMENTS

Analytical applications of dc-GDs are mostly directed to analysis of conductive materials. The implementation of **rf-GDs** has pushed forward the analyses related with direct depth profiling of non-conductive samples [1]. The sputtering of electrically non-conductive materials is possible because a negative dc self-bias potential is acquired by the surface exposed to the plasma of the non-conductive sample.

GD plasmas can be generated either by applying constant power during the whole analysis time (non-pulsed mode) or by switching it on and off alternately (*i.e.* pulsed glow discharges). The instrumental implementation of **PGDs** (produced either with ms or μs pulses) has brought about interesting analytical advantages in GD spectroscopies. A PGD can be can be considered as a dynamic plasma and, so, preferential ionization/excitation mechanisms along the temporal distribution of power are produced. In particular, three main time domains are distinguished along the pulse duration (see Fig. **6**) [8]. First, the **prepeak region**, that corresponds to the initial moments after plasma ignition; it has been observed that discharge gas species appear earlier as compared to analyte signals in this region, which could be attributed to the production of gas ions before sample sputtering

[9]. After the prepeak, the whole duration of the applied power (pulse) corresponds to the **plateau region** and last, there is the **afterglow or afterpeak region**, which is observed once the applied power is off. It has been described that electron impact and charge-transfer are the preferential mechanisms both in the prepeak and plateau regions, whereas Penning ionization dominates the afterglow region as the population of metastable Ar atoms is increased (as a consequence of recombination between thermalized electrons and argon ions). PGDs have shown interesting advantages compared to their continuous or non-pulsed (dc or rf) counterpart, *e.g.*: (i) enhancement of the excitation/ionization processes (higher instantaneous power can be applied), (ii) less thermal stress on samples since the average power of a PGD is typically smaller than in the non-pulsed GDs, this being of particular interest for the analysis of nonconductors, and (iii) because the different temporal excitation/ionization regimes, are sequentially created along each PGD cycle, time gated detection makes possible to select the time window within the GD pulse period providing highest sensitivity or least interferences. The afterglow is the regime most commonly used for analytical purposes because it usually offers the highest sensitivity. In any case, it should be mentioned that a fast analyzer is required (OES or MS) to follow the pulse profile.

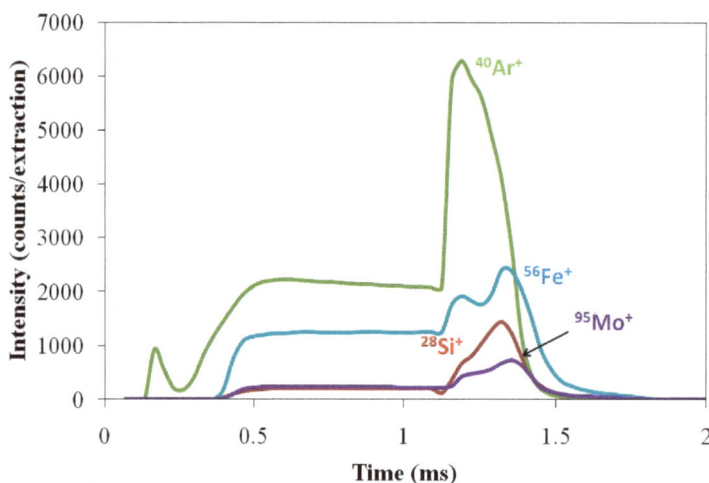

Fig. (6). Typical signal profile obtained with a PGD, indicating the different time regimes within the pulse period.

Other instrumental advancements related with the GD source, though not commercially implemented yet, include for example boosting of the GD with a magnetic field aiming to achieve higher analytical sensitivity [10]. Also, it is of interest to highlight the efforts carried out to improve the lateral resolution of GD-based techniques. The achievable lateral resolution has been historically restricted to the diameter of the sputtered area, typically some millimeters. However, it has been shown that elemental mapping of the sample surface can be obtained with specially designed GD chambers, optimized operating conditions and a proper measuring system that produces information resolved not only in the spectral and temporal dimensions but also in the x and y spatial dimensions [11, 12]. Moreover, it has been proposed the combination of laser ablation (LA) with GD [13]; in this latter case, higher lateral resolution can be obtained than just using GDs.

Regarding detection of the photons or the ions generated, it is worth highlighting two main advances: the implementation of charge-coupled devices (CCDs) in GD-OES and time of flight (TOF) in GD-MS. In the case of GD-OES, most commercial companies provide instruments with polychromators based on the use of photomultiplier tubes (up to almost 60) in a Paschen-Runge configuration. However, a company offers CCDs for GD detection, so allowing the addition of almost an unlimited amount of channels to any given method [14, 15]. The wavelength range in GD-OES instruments can go from 110 nm up to 800 nm. Another interesting feature of some commercial GD-OES instruments is the possibility to directly obtain sample erosion rates in reflective or partially reflective samples with a differential interferometry accessory [16].

In the case of ions, traditional commercial GD-MS systems have been based on sector field (SF) analyzers. More recently, GD-**TOFMS** has been introduced into the market. SFMS offers higher analytical sensitivity (in the low ng/g range) than TOFMS. Nevertheless, monitoring of different ions takes a long time (it is a sequential mass analyzer) not compatible with the fact that the sample is being continuously sputtered; so, depth information can be lost. Steady-state operated-GDs (dc-GDs and rf-GDs) as well as PGDs have been coupled to SFMS. The coupling of a µs PGD to SFMS (using time gated detection of ions) has shown to offer higher signal-to-noise ratios for analytes in comparison to non-pulsed GD [17]. Additionally, PGD-SFMS may offer better depth resolution as compared to the non-pulsed counterpart, thanks to the fact that the short duty cycle permits a more precise control of the sputtering rate. Also, the long power-off time gives the sputtered species time to depart from the crater, while ionized species are no longer accelerated back to the sample, minimizing the possibility of their re-deposition near the crater region [18].

In any case, the coupling of a PGD to a time-gated fast analyzer, like TOFMS, is a most convenient combination [19]. Besides, TOFMS allows to obtain a full mass spectrum quasi-simultaneously, this capability being suitable for multielemental depth profiling, and to discover contaminants or unexpected elements in thin layers.

QUANTIFICATION PROCEDURES

Information directly provided by GD-OES and GD-MS techniques is qualitative. GD-OES is most commonly used for depth profile analysis, while GD-MS traditional routine applications head towards the analysis of traces and ultratraces in high purity metals.

First, there will be briefly described approaches for quantification of traces and ultratraces in homogeneous materials by GD-MS [20, 21]. The simpler approach is called **Ion beam ratio** (IBR), which corresponds to the ratio of the analyte intensity with respect to the intensity of matrix element (corrected by abundances of measured analyte and matrix isotopes). Therefore, in this case, the matrix is taken as internal standard for the correction of signal fluctuations during the measurement due to operational instabilities. This method is only useful when the matrix component keeps constant during the analysis time. As no certified reference materials are required, the information provided is just semiquantitative. Quantitative results require the correction of differences in elemental sensitivities by using certified reference materials (CRMs) similar in composition to the material under study. The **Relative sensitivity factor** (RSF) is defined as the inverse of the sensitivity of a specific element in a specific matrix. Therefore, RSFs require CRMs to construct calibration graphs for the different analytes and, so, they provide more accurate results than the IBR method. This is the most widely employed method for quantification in GD-MS. Another related approach is named **Standard RSF** (typically, RSFs normalized to Fe). The standard RSF values could be therefore determined by analysis of Fe matrix or in the case of other matrices, by division of the RSF value of a given analyte by the RSF value of Fe (Fe was selected because there is a big number of RSF for CRMs based on an iron matrix and because Fe is present in almost every analyzed sample).

Qualitative depth profiles consist of plots of signal intensities of each of the analytes *versus* sample sputtering time. This information can be already useful when comparing differences in similar samples. However, in many cases quantitative information is required. Moreover, it is very risky to compare qualitative depth profiles of different materials, because differences in sputtering rates depending on the matrix and/or ionization/excitation energies for each of the sample components can lead to misinterpretations. **Quantitative (or compositional) depth profiles** show concentration of each analyte *versus* depth. Quantitative depth profiling of traces in high purity materials can be carried out by the IBR and RSF approaches in GD-MS (for example, these approaches have proved to be appropriate for diffusion studies and quantification of trace impurities in silicon solar cells using GD-SFMS [22]). Nonetheless, this is not the case when the sample consists of layers with different matrixes.

Neither the IBR nor the RSF are useful when aiming at depth profile analysis of layered materials. A book, published in 2003, provides a practical guide for GD-OES quantitative depth analysis [23]. Recently, calibration methods in GD-OES have been reviewed [24]. The quite frequent independence of the sputtering and the excitation/ionization processes in GDs allows multi-matrix calibrations based on the emission yield (R) concept (see eq 1):

$$I_{\lambda(E),M} = R_{\lambda(E)} \cdot C_{E,M} \cdot q_M \qquad \textbf{(1)}$$

being $I_{\lambda(E),M}$ the intensity of an emission line of an element E, $C_{E,M}$ the concentration of element E in material M, and q_M the sputter rate of the material M.

The product $C_{E,M} \cdot q_M$ is taken as an independent variable in eq (1) and the intensities are considered to be proportional to it, with the proportionality constant $R_{\lambda(E)}$ to be determined by calibration (slope). Diverse CRMs with the different matrixes of the multilayered sample should be used to construct a calibration plot for a given element (see Fig. **7**).

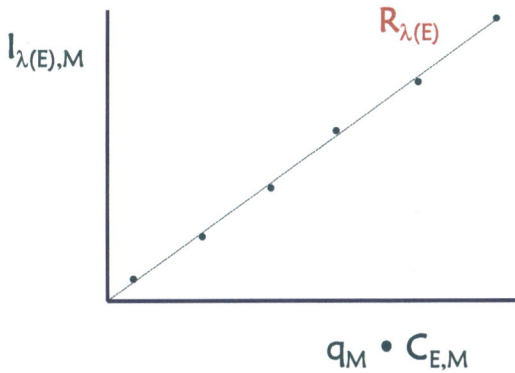

Fig. (7). A typical multi-matrix calibration plot in GD-OES.

The composition of unknown samples is calculated in two steps: (i) the product $C_{E,M} \cdot q_M$ is calculated for each element based on the intensities measured and using the emission yields resulting from calibration, and (ii) the unknown q_M is calculated by normalization to 100%. Concentration in compositional depth profiles can be presented as mass percentage or as atomic percentage. Expressing results in at.% is particularly sensible when chemical bonds are predominant in the analysed material. For example, when analyzing a TiN layer, it would be expected 50% Ti and 50% N expressed in at.%, otherwise the TiN layer would not be stoichiometric. However, care should be taken with data interpretation because most commercial GD software presents data as mass percentage.

The multi-matrix calibration procedure has proved useful with dc-GD-OES, rf-GD-OES [25] and PGD-OES [26]. Furthermore, using a similar concept ("ionization yield" or "absolute sensitivity factor") the methodology can be applied to GD-MS. However, here it is important to note that all the elements in the sample should be simultaneously measured to obtain accurate results, therefore this procedure fits better with TOFMS detection than with SFMS (in TOFMS there are no restrictions in the number of ions quasi-simultaneously measured).

Multi-matrix calibrations for depth profile analysis are typically carried out with homogeneous CRMs. To overcome the lack and/or instability of CRMs containing high concentrations of light elements (*e.g.* hydrogen, nitrogen or oxygen) calibration curves can be obtained using well-characterized coated calibration materials [27]. In this last case, the layer of the calibration material should either be thick enough to produce a measurable crater inside the coating, or the thickness of the coating should be known (in this last case, the sputtering rate can be calculated by measuring the sputtering time taken to reach the interface).

APPLICATIONS

Continuous progresses in materials science are giving rise to new challenges in analytical chemistry. Coatings and thin films are present in an amazing number of real-life applications, from protective coatings for mechanical and corrosion resistance to multilayered materials for electronic, optoelectronic and magnetic applications, coated glasses for buildings, renewable energies, catalysts, sensors, *etc.* Technical demands, with thinner and more complex films being developed to meet market needs, are ever increasing. Layer thicknesses, dopant element concentrations, presence of impurities at interfaces and diffusion processes are among the different types of information required.

GD-OES and GD-MS are accurate and high throughput (analysis in a few minutes) techniques to obtain compositional depth-profile analysis of materials. Initially, depth profile GD applications were focused on the depth profiling of a few micrometers thick conductive coatings. At present, GD-OES and GD-MS are well established techniques for direct solid analysis, and even some ISO norms are available (*e.g.* ISO 16962:2005 "Surface chemical analysis -- Analysis of zinc- and/or aluminium-based metallic coatings by glow-discharge optical-emission spectrometry"). Detection limits achieved with GD-OES are in the order of μg/g and concentrations of ng/g can be reached with GD-MS. The vast majority of latest publications dealing with depth profile analysis by GD-MS have been carried out with TOFMS detection, being almost all elements of the periodic table quasi-simultaneously measured practically atomic layer by atomic layer.

In the 21[rst] century, advances and trends with GD spectroscopies are mainly directed towards the analysis at the nanometric depth scale [28] and even a Round Robin exercise has been conducted for 100 nm CrNi/100 nm Cu multilayers aimed at finding the best conditions for GD-OES depth profile analysis [29]. Impressively, the depth profile analysis of an organic monolayer (thiourea adsorbed on a smooth copper surface) has been also described [30]; in that work it was discussed both the detection of atoms of the organic monolayer and the orientation of the molecules relative to the surface.

It is also worth mentioning that depth profile analysis of very near-surface layers requires a very stable plasma from the beginning of sputtering, which depends on several issues. For instance, it is important to use a high pure plasma gas (at least 99.999%) and to evacuate properly the GD chamber before starting the discharge by removing air introduced during the process of sample exchange. Also, surface contamination (organics, water, *etc.*) not only from the sample but also from the inner GD chamber surfaces can lead to signal spikes at the discharge ignition. To minimize such problems, it is recommended to clean the inner GD walls by sputtering monocrystalline silicon under standard depth profiling conditions prior to the analysis of the specimen [31].

As previously described, GD spectroscopies allow multielemental depth profiling of almost any major, minor and trace element of the periodic table including hydrogen, carbon, oxygen and nitrogen. Examples of depth profile characterization of samples containing light elements are for instance the analysis of anodic alumina and iron oxide films, nitrided austenitic stainless steel, TiN and CrN hard coatings, characterization of thin passive films formed on highly corrosion-resistant stainless steel, studies of decarburization processes in low-carbon steel, among many others [19, 28]. Interesting examples of modern applications of GD-OES include, for example, the depth-profiling of hydrogen in diamond-like films [14] and the study of the degradation of graphite anodes in Li-ion cells [15]. GD-OES and GD-MS are also excellent tools for the characterization of the elemental distribution of thin film solar cells (TFSCs) and recent examples are given in the literature, including amorphous silicon [32], Cu(In,Ga)Se [18], tandem [33], CdTe [34] and perovskite [35] solar cells.

Additionally, the use of a rf powered GD makes possible the fast and easy direct depth analysis of non-conductors. An example of successful application is the quantitative multielemental depth profile of a few nm thin layers on few millimeters thick glass substrates (*e.g.* 18 nm A1/10 nm Nb/10 nm A1//4.8 mm glass substrate, or an industrial glass consisting on 27 nm Si_3N_4/15 nm NbN_x/6 nm Si_3N_4//5.8 mm glass substrate) [25].

GDs have also proved to be convenient for the direct analysis of polymers with comparable or in some cases better performance than established techniques such as secondary ion mass spectrometry (SIMS). SIMS presents some drawbacks, including high cost and long-time analysis. As already indicated, PGDs produce less thermal stress in the specimens than the non-pulsed mode, this being of particular interest for the analysis of fragile and thermally unstable materials like polymers. Using PGD-TOFMS, not only elemental but also molecular information can be obtained, offering a promising alternative tool for identification and/or screening of polymers. Brominated flame retardants (BFRs) are added into polymeric materials, such as plastics and polyurethane foam, to decrease their risk of fire. They are widely used in electric and electronic equipment, construction, textile applications, *etc.* These compounds can leach out into the environment and it is known that some of them present potential adverse health effects. Therefore, there is a need of analytical methods for the rapid and simple detection of BFRs in waste. In this context, promising and convenient capabilities of rf-PGD-TOFMS for fingerprinting of polymers containing BFRs (based on the measurement of characteristic bromine-containing polyatomics) have been demonstrated [36]. Furthermore, rf-PGD-TOFMS has proved useful for identification of polymers just containing carbon, hydrogen and either sulfur or nitrogen. Results showed that some of the polyatomic ions can be eventually considered as fragments coming from the structures of the corresponding monomers constituting each polymer, so making feasible to differentiate rather similar polymers by resorting to the measurement of characteristic fragments [37].

Positive ion detection has yielded high analytical signals for most elements and molecular information capabilities in GD-MS, however some limitations exist. For instance, isobaric interferences restrict the detection sensitivity for some elements, like $^{19}H_3O^+$ in $^{19}F^+$ detection. Samples containing F, Cl, Br or P heteroatoms were measured by GD-MS and, as expected, the analytical sensitivity turned out to be higher with increasing analyte electron affinities when measuring the negative ions [38]. Likewise, the negative ion detection mode has proved useful for identification of polymers [38].

CONCLUDING REMARKS

In recent years, the applications of GD-OES and GD-MS to different samples have increased impressively. This has been possible thanks to three main complementary facts: (i) successful implementation of instrumental advances in commercial GD equipment, (ii) improvement of operation procedures for near-surface layers, and (iii) development of appropriate quantification methods for analysis of complex samples such as multilayers, thin layers, materials containing hydrogen, oxygen, nitrogen, *etc.*

Of course, investigation on plasma basics and instrumental improvements for GD spectroscopies is alive. However, it is fair to say that at present GD-OES and GD-MS have reached a great degree of maturity, being employed both as routine depth profile techniques for quality control issues and as complementary tools for characterization of new developed materials. Research trends in GD-based analytical techniques will most probably be driven by requirements for the characterization of new high-tech products, including nanomaterials.

CONFLICT OF INTEREST

The authors declare no conflict of interest, financial or otherwise.

ACKNOWLEDGEMENTS

Lara Lobo acknowledges financial support through program Juan de la Cierva-Incorporación (IJCI-2015-25801) from the Spanish Ministry of Economy, Industry and Competitiveness.

REFERENCES

[1]   Winchester, M.R.; Payling, R. Radio-frequency glow discharge spectrometry: A critical review. *Spectrochim. Acta B At. Spectrosc.,* **2004,** *59,* 607-666.
        [http://dx.doi.org/10.1016/j.sab.2004.02.013]

[2]   Belenguer, Ph.; Ganciu, M.; Guillot, Ph.; Nelis, Th. Pulsed glow discharges for analytical applications. *Spectrochim. Acta B At. Spectrosc.,* **2009,** *64,* 623-641.
        [http://dx.doi.org/10.1016/j.sab.2009.05.031]

[3]   Bogaerts, A.; Gijbels, R. Fundamental aspects and applications of glow discharge spectrometric techniques. *Spectrochim. Acta B At. Spectrosc.,* **1998,** *53,* 1-42.
        [http://dx.doi.org/10.1016/S0584-8547(97)00122-5]

[4]   Marcus, R.K.; Broekaert, J.A. *Glow discharge plasmas in analytical spectroscopy*; John Wiley & Sons Ltd: Chichester, **2003.**

[5]   González Gago, C.; Bordel, N.; Pereiro, R.; Sanz-Medel, A. Investigation of the afterglow time regime in pulsed radiofrequency glow discharge time of flight mass spectrometry. *J. Mass Spectrom.,* **2011,** *46,* 757-763.
        [http://dx.doi.org/10.1002/jms.1956]

[6]   Grimm, W. Eine neue glimmentladungslampe für die optische emissionsspektralanalyse. *Spectrochim. Acta B At. Spectrosc.,* **1968,** *23,* 443-445.
        [http://dx.doi.org/10.1016/0584-8547(68)80023-0]

[7]   Jakubowski, N.; Dorka, R.; Steers, E.; Tempez, A. Trends in glow discharge spectroscopy. *J. Anal. At. Spectrom.,* **2007,** *22,* 722-735.
        [http://dx.doi.org/10.1039/b705238n]

[8]   Harrison, W.W.; Yang, C.; Oxley, E. Pulsed glow discharge: temporal resolution in analytical spectroscopy. *Anal. Chem.*, **2001**, *73*, 480A-487A.
[http://dx.doi.org/10.1021/ac012502g]

[9]   Pisonero, J.; Valledor, R.; Licciardello, A.; Quirós, C.; Martín, J.I.; Sanz-Medel, A.; Bordel, N. Pulsed rf-GD-TOFMS for depth profile analysis of ultrathin layers using the analyte prepeak region. *Anal. Bioanal. Chem.*, **2012**, *403*, 2437-2448.
[http://dx.doi.org/10.1007/s00216-011-5601-3]

[10]   Vega, P.; Valledor, R.; Pisonero, J.; Bordel, N. An improved analytical performance of magnetically boosted radiofrequency glow discharge. *J. Anal. At. Spectrom.*, **2010**, *27*, 1658-1666.
[http://dx.doi.org/10.1039/c2ja30106g]

[11]   Gamez, G.; Voronov, M.; Ray, S.J.; Hoffmann, V.; Hieftje, G.M.; Michler, J. Surface elemental mapping *via* glow discharge optical emission spectroscopy. *Spectrochim. Acta B At. Spectrosc.*, **2012**, *70*, 1-9.
[http://dx.doi.org/10.1016/j.sab.2012.04.007]

[12]   Kroschk, M.; Usala, J.; Trevor Addesso, T.; Gamez, G. Glow discharge optical emission spectrometry elemental mapping with restrictive anode array masks. *J. Anal. At. Spectrom.*, **2016**, *31*, 163-170.
[http://dx.doi.org/10.1039/C5JA00288E]

[13]   Gonzalez de Vega, C.; Álvarez Llamas, C.; Bordel, N.; Pereiro, R.; Sanz-Medel, A. Analytical potential of a laser ablation – glow discharge – optical emission spectrometry system for the analysis of conducting and insulating materials. *Anal. Chim. Acta*, **2015**, *877*, 33-40.
[http://dx.doi.org/10.1016/j.aca.2015.04.034]

[14]   Takahara, H.; Ishigami, R.; Kodama, K.; Kojyo, A.; Nakamura, T.; Oka, Y. Hydrogen analysis in diamond-like carbon by glow discharge optical emission spectroscopy. *J. Anal. At. Spectrom.*, **2016**, *31*, 940-947.
[http://dx.doi.org/10.1039/C5JA00447K]

[15]   Ghanbari, N.; Waldmann, T.; Kasper, M.; Axmann, P.; Wohlfahrt-Mehrens, M. Inhomogeneous degradation of graphite anodes in Li-ion cells: A postmortem study using glow discharge optical emission spectroscopy (GD-OES). *J. Phys. Chem. C*, **2016**, *120*, 22225-22234.
[http://dx.doi.org/10.1021/acs.jpcc.6b07117]

[16]   Richard, S.; Gaston, J.; Acher, O.; Chapon, P. Glow discharge spectroscopy method and system for measuring *in situ* the etch depth of a sample. United States Patent Application Publication 2017.2017/0045457 A1

[17]   Voronov, M.; Smıd, P.; Hoffmann, V.; Hofmann, Th.; Venzago, C. Microsecond pulsed glow discharge in fast flow Grimm type sources for mass spectrometry. *J. Anal. At. Spectrom.*, **2010**, *25*, 511-518.
[http://dx.doi.org/10.1039/b922551j]

[18]   Churchill, G.; Putyera, K.; Weinstein, V.; Wang, X.; Steers, E.B. New µs-pulsed DC glow discharge assembly on a fast flow high power source for time resolved analysis in high resolution mass spectrometry. *J. Anal. At. Spectrom.*, **2011**, *26*, 2263-2273.
[http://dx.doi.org/10.1039/c1ja10086f]

[19]   Pisonero, J.; Bordel, N.; Gonzalez de Vega, C.; Fernández, B.; Pereiro, R.; Sanz-Medel, A. Critical evaluation of the potential of radiofrequency pulsed glow discharge – time of flight mass spectrometry for depth profile analysis of innovative materials. *Anal. Bioanal. Chem.*, **2013**, *405*, 5655-5662.
[http://dx.doi.org/10.1007/s00216-013-6914-1]

[20]   Hoffmann, V.; Kasik, M.; Robinson, P.K.; Venzago, C. Glow discharge mass spectrometry. *Anal. Bioanal. Chem.*, **2005**, *381*, 173-188.
[http://dx.doi.org/10.1007/s00216-004-2933-2]

[21]   Gusarova, T.; Hofmann, T.; Kipphardt, H.; Venzago, C.; Matschat, R.; Panne, U. Comparison of

different calibration strategies for the analysis of zinc and other pure metals by using the GD-MS instruments VG 9000 and Element GD. *J. Anal. At. Spectrom.,* **2010**, *25*, 314-321.
[http://dx.doi.org/10.1039/b921649a]

[22]   Di Sabatino, M.; Mondanese, C.; Arnberg, L. Depth profile analysis of solar cell silicon by GD-MS. *J. Anal. At. Spectrom.,* **2014**, *29*, 2072-2077.
[http://dx.doi.org/10.1039/C4JA00175C]

[23]   Nelis, Th.; Payling, R. *Glow discharge optical emission spectroscopy: a practical guide; RSC Analytical Spectroscopy Monographs*; The Royal Society of Chemistry: Cambridge, **2003**.

[24]   Weiss, Z. Calibration methods in glow discharge optical emission spectroscopy: a tutorial review. *J. Anal. At. Spectrom.,* **2015**, *30*, 1038-1049.
[http://dx.doi.org/10.1039/C4JA00482E]

[25]   Fernández, B.; Martín, A.; Bordel, N.; Pereiro, R.; Sanz-Medel, A. In-depth profile analysis of thin films deposited on non-conducting glasses by radiofrequency glow discharge – optical emission spectrometry. *Anal. Bioanal. Chem.,* **2006**, *384*, 876-886.
[http://dx.doi.org/10.1007/s00216-005-0123-5]

[26]   Sánchez, P.; Fernández, B.; Menéndez, A.; Orejas, J.; Pereiro, R.; Sanz-Medel, A. Quantitative depth profile analysis of metallic coatings by pulsed radiofrequency glow discharge optical emission spectrometry. *Anal. Chim. Acta,* **2011**, *684*, 47-53.
[http://dx.doi.org/10.1016/j.aca.2010.10.039]

[27]   Malherbe, J.; Fernández, B.; Martinez, H.; Chapon, P.; Panjan, P.; Donard, O.F. In-depth profile analysis of oxide films by radiofrequency glow discharge optical emission spectrometry (rf-GD-OES): possibilities of depth-resolved solid-state speciation. *J. Anal. At. Spectrom.,* **2008**, *23*, 1378-1387.
[http://dx.doi.org/10.1039/b803713b]

[28]   Fernández, B.; Pereiro, R.; Sanz-Medel, A. Glow discharge analysis of nanostructured materials and nanolayers– a review. *Anal. Chim. Acta,* **2010**, *679*, 7-16.
[http://dx.doi.org/10.1016/j.aca.2010.08.031]

[29]   Hodoroaba, V.D.; Hoffmann, V.; Steers, E.B.; Griepentrog, M.; Dück, A.; Beck, U. Round Robin exercise: coated materials for glow discharge spectroscopy. *J. Anal. At. Spectrom.,* **2006**, *21*, 74-81.
[http://dx.doi.org/10.1039/B513426A]

[30]   Shimizu, K.; Payling, R.; Habazaki, H.; Skeldon, P.; Thompson, G.E. Rf-GDOES depth profiling analysis of a monolayer of thiourea adsorbed on copper. *J. Anal. At. Spectrom.,* **2004**, *19*, 692-695.
[http://dx.doi.org/10.1039/b400918p]

[31]   Molchan, S.; Thompson, G.E.; Skeldon, P.; Trigoulet, N.; Chapon, P.; Tempez, A.; Malherbe, J.; Lobo Revilla, L.; Bordel, N.; Belenguer, Ph.; Nelis, T.; Zahri, A.; Therese, L.; Guillot, Ph.; Ganciu, M.; Michler, J.; Hohl, M. The concept of plasma cleaning in glow discharge spectrometry. *J. Anal. At. Spectrom.,* **2009**, *24*, 734-741.
[http://dx.doi.org/10.1039/b818343k]

[32]   Alvarez-Toral, A.; Sanchez, P.; Menéndez, A.; Pereiro, R.; Sanz-Medel, A.; Fernández, B. Depth profile analysis of amorphous silicon thin film solar cells by pulsed radiofrequency glow discharge time of flight mass spectrometry. *J. Am. Soc. Mass Spectrom.,* **2015**, *26*, 305-314.
[http://dx.doi.org/10.1007/s13361-014-1022-9]

[33]   Fernandez, B.; Lobo, L.; Reininghaus, N.; Pereiro, R.; Sanz-Medel, A. Characterization of thin film tandem solar cells by radiofrequency pulsed glow discharge – Time of flight mass spectrometry. *Talanta,* **2017**, *165*, 289-296.
[http://dx.doi.org/10.1016/j.talanta.2016.12.062]

[34]   Kartopu, G.; Tempez, A.; Clayton, A.J.; Barrioz, V.; Irvine, S.J.; Olivero, C.; Chapon, P.; Legendre, S.; Cooper, J. Chemical analysis of $Cd_{1-x}Zn_xS/CdTe$ solar cells by plasma profiling TOFMS. *Materials Research Innovations,* **2014**, *18*, 82-85.
[http://dx.doi.org/10.1179/1433075X14Y.0000000207]

[35] Cojocaru, L.; Uchida, S.; Matsubara, D.; Matsumoto, H.; Ito, K.; Otsu, Y.; Chapon, P.; Nakazaki, J.; Kubo, T.; Segawa, H. Direct confirmation of distribution for Cl⁻ in $CH_3NH_3PbI_{3-x}Cl_x$ layer of perovskite solar cells. *Chem. Lett.,* **2016**, *45*, 884-886.
[http://dx.doi.org/10.1246/cl.160436]

[36] Gonzalez de Vega, C.; Lobo, L.; Fernandez, B.; Bordel, N.; Pereiro, R.; Sanz-Medel, A. Pulsed glow discharge time of flight mass spectrometry for the screening of polymer-based coatings containing brominated flame retardants. *J. Anal. At. Spectrom.,* **2012**, *27*, 318-326.
[http://dx.doi.org/10.1039/C2JA10300A]

[37] Gonzalez de Vega, C.; Fernandez, B.; Bordel, N.; Pereiro, R.; Sanz-Medel, A. Challenging identifications of polymer coatings by radiofrequency pulsed glow discharge-time of flight mass spectrometry. *J. Anal. At. Spectrom.,* **2013**, *28*, 1054-1060.
[http://dx.doi.org/10.1039/c3ja50075f]

[38] Lobo, L.; Fernández, B.; Muñiz, R.; Pereiro, R.; Sanz-Medel, A. Capabilities of radiofrequency pulsed glow discharge-time of flight mass spectrometry for molecular screening in polymeric materials: positive *versus* negative ion mode. *J. Anal. At. Spectrom.,* **2016**, *31*, 212-219.
[http://dx.doi.org/10.1039/C5JA00291E]

CHAPTER 5

Recent Advances and Challenges for Beer Volatile Characterization Based on Gas Chromatographic Techniques

Cátia Martins[1,2], Adelaide Almeida[2] and Sílvia M. Rocha[1]

[1] *Departamento de Química & QOPNA, Universidade de Aveiro, Campus Universitário Santiago, 3810-193 Aveiro, Portugal*

[2] *Departamento de Biologia & CESAM, Universidade de Aveiro, Campus Universitário Santiago, 3810-193 Aveiro, Portugal*

Abstract: Beer is one of the most popular alcoholic beverages worldwide, taste and flavor being the main factors which contribute for consumers' acceptance, and its volatile components represent the major contributors for beer global and peculiar aroma properties. Also, beer volatile components may be used to monitor the impact of raw materials composition, yeast metabolism, beer aging, beer distinction and screening of off-flavors, among others. The analysis of beer volatile fraction is very challenging due to the presence of CO_2, and also a diversity of chemical structures, namely with different polarities, volatilities, and a wide concentration range. Thus, beer volatile analysis requires an effective extraction technique to recover representative information, minimizing modifications or/and loss of components. The different extraction and gas chromatographic techniques that have been used for beer volatile composition will be critically presented, exploring their advantages and drawbacks; wherein solid phase microextraction combined with gas chromatography is the methodology most frequently used, and a special attention will be given to these techniques. Furthermore, over the last decades, significant improvements have occurred on the chromatographs (namely the multidimensional ones), detection systems, columns technology and algorithms that contribute to the reduction of analysis time, making the methods more expeditious and user-friendly. Taking into account the information available in the literature, a comprehensive outline about the volatile components previously determined in beers is included, considering, when existing, data about their detection and quantification limits. This review presents the state-of-the-art for researches who want to study beer volatile characterization.

Keywords: Beer, Beer volatile characterization, Chromatographic techniques, Comprehensive two-dimensional gas chromatography, Decarbonation, Extraction

* **Corresponding Author Sílvia M. Rocha:** Departamento de Química & QOPNA, Universidade de Aveiro, Campus Universitário Santiago, 3810-193 Aveiro, Portugal; Tel: + 351234401524; Fax: + 351234370084; E-mail: smrocha@ua.pt

Atta-ur-Rahman, Sibel A. Ozkan & Rida Ahmed (Eds.)

techniques, One-dimensional gas chromatography, Solid phase microextraction, Volatile composition.

INTRODUCTION

Beer is one of the most popular and widespread alcoholic beverages in the world. More than 1.95 billion liters of beer have been produced per year, whereas an increase of 0.6% (around 10 million of hectoliters) of beer production was registered in 2013 when compared with 2012 [1]. Nevertheless, the beer production trend has been stagnating in Europe [2], as well as in North America; otherwise Asia (particularly China), Africa and South America (namely Brazil) had increased their beer production in 2013, representing attractive markets for Europe' exportation, which increased 15% since 2008 until 2014 [1 - 3]. The actual economic context led to the stagnation of beer consumption, and consequently their production, but other factors also contributed, namely lower per capita consumption, higher taxes, lower purchase of premium brands [4]. In order to counteract this trend, brewing industries have been wagged in the development of innovative products (*e.g.* non-alcoholic and flavored-mixed beers), with improved organoleptic characteristics to attract consumers. Furthermore, the development of microbrewing industries (also known as craft brewing) is an impressive ascending trend; in fact, in Europe, there was *ca.* 13% increase of new microbreweries each year (from 2013 to 2015).

The consumers' acceptance is dependent on the specific organoleptic characteristics of each beer type, which comprises gustative, visual, and aroma perceptions. One of the major contributions for the final quality of beer is its flavor [5]. Indeed, pleasant beer flavor is obtained with a fine and subtle balance between numerous (*ca.* 800 compounds) volatile and non-volatile compounds. Several sources may contribute to their formation, namely: different raw materials [6 - 8], different steps of the brewing process, such as yeast fermentation and pasteurization [9 - 13]; and also along with aging process [14 - 17]. A balanced mixture of volatile and non-volatile compounds comprise the beer flavor, *i.e.* alcohols, aldehydes, carbohydrates, essential oils, fatty acids, ionones, isohumulones (iso-alpha-acids), lactones, methyl esters, mineral ions, nucleotides, organic acids, phenolics, proteins, sulfur-containing volatile compounds, tannins, unsaturated carbonyl compounds, and vicinal diketones [13, 18]. Several of these previous are not key flavor compounds; nevertheless, they give background perception to beer, contributing for the general impression of beer flavor [18].

Volatile organic compounds (VOC) belong to different chemical families, such as acids, aliphatic and aromatic alcohols, esters, carbonyl compounds (aldehydes and ketones), sulfur and terpenic compounds; and they can be present in beer at different concentrations, from ng to mg L^{-1} [13, 19]. The contribution of each VOC to beer aroma is dependent on the ratio between the concentration and sensory threshold (minimum compound concentration needed to be perceived by human nose), it being defined as odor activity value (OAV). The compound will only contribute to individual beer aroma when OAV is greater than 1, otherwise if its level is at least 20% of the OAV, it may contribute to beer overall notes [20, 21]. As VOCs can contribute to beer aroma peculiarities, studies regarding its volatile composition may help brewers to select the most adequate raw materials, the yeast strain, and/or also to monitor the routine product quality [19, 22].

Beer volatile composition has been the subject of several studies within different purposes. A search queried with *beer* and *volatile** and not *non-volatile**, *via* Web of Science™, retrieved a total of 431 original research articles, from 1967 to 2016. After a manual and careful refinement, almost half of the obtained publications were excluded as they are not directly related with the determination of beer volatile composition. The interest on this topic has been increased, especially over the past 14 years, which represents about 67% of all published articles (Fig. **1**).

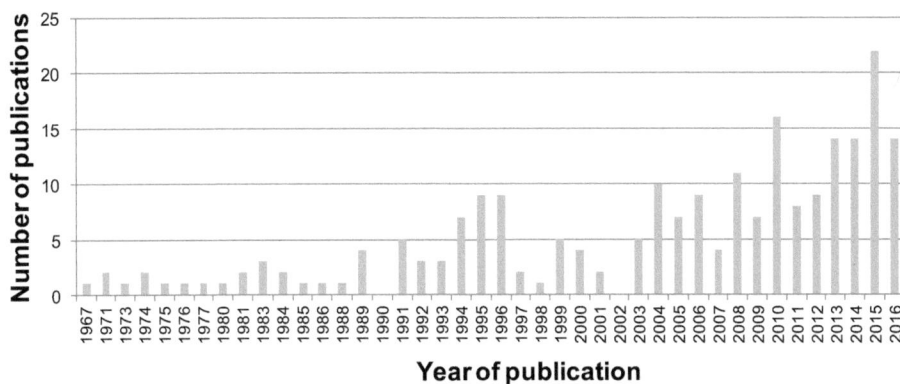

Fig. (1). Literature survey of published original research articles, using a search query with keywords *beer* and *volatile** and not *non-volatile** in topic, from 1967 to 2016 *via* Web of Knowledge, followed a manual refinement.

As different challenges may be exposed regarding the determination of beer volatile composition, it is important to have fast and reliable methods that allow its monitoring. Therefore, the present book chapter intends to contribute with an overview of the state-of-the-art and technical know-how about the extraction and chromatographic techniques, including strategies for analyte quantification and identification that have been used for beer volatile characterization, also giving advantages and drawbacks of the different techniques. Finally, a comprehensive overview about the beer volatile components available in the literature is also included, taking into consideration their detection and quantification limits.

IMPORTANT TOOLS FOR BEER VOLATILE CHARACTERIZATION

After a careful analysis of the literature available, it was observed that three main steps have been currently taken into consideration for beer volatile determination: *i)* decarbonation, *ii)* extraction of volatile components, followed by *iii)* chromatographic and/or mass spectrometry analysis. The presence of CO_2 represents the first challenge to be exceeded once it can interfere in the extraction efficiency and reproducibility. Another important challenge is the establishment of a suitable extraction procedure, once VOCs may present diverse chemical structures and concentration levels. Significant improvements have been made related with sensitivity and specificity, as well as reduction of the sample preparation steps.

Fig. (**2**) systematizes the extraction and chromatographic or mass spectrometry based techniques that have been used for beer volatile determination. Different extraction approaches have been reported, namely, headspace (HS) generation apparatus; solvent based techniques, using microliters (*e.g.* solid drop microextraction – SDME) or milliliters of solvents (*e.g.* liquid-liquid extraction – LLE); sorptive based techniques (solid phase extraction (SPE), stir bar sorptive extraction (SBSE), solid-phase microcolumn extraction (SPMCE), solid phase dynamic extraction (SPDE), or solid phase microextraction (SPME)) and gas diffusion techniques. SPME is a solvent-free technique highly used on the last decade as it allows a simultaneous sample preparation, extraction and concentration. Also, SPME may be combined with analytical tools (*e.g.* gas chromatography), or complemented with organoleptic profiling panels [23]. Gas chromatography has been widely used for beer volatile characterization, namely one dimensional gas chromatography (1D-GC), also comprehensive two-dimensional gas chromatography (GC×GC), or high-performance liquid chromatography (HPLC) were used, which have been combined with several types of detection, including hyphenated techniques and sensorial analysis. GC-MS (gas chromatography-mass spectrometry) and GC-FID (flame ionization detector) are the techniques currently used, due to their easy manipulation and low costs. Also, more advanced mass spectrometry based techniques, such as EESI-Q-ToFMS (Extractive electrospray ionization-quadrupole-time-of-flight) [24], DART –ToFMS (Direct analysis in real-time – time-of-flight) [25] and APCI-MS (Atmospheric pressure chemical ionization-mass spectrometry) [26], were applied as high throughput approaches for beer fingerprinting, that may comprises volatile and non-volatile components.

Beer volatile characterization		
Extraction		**Analysis**
Headspace generation apparatus [19,22,26–39]	**GC**	GC-FID (Gas chromatography-Flame ionization detection) [22,23,28,30,31,33,38,41,49,70,71,75–81]
		GC-ECD (Electron capture detection) [30,33,57,89]
		GC-FPD (Flame photometric detection) [49,87]
Solvent based techniques [21,26,40–50]		GC-PFPD (Pulsed flame photometric detection) [36,88]
		GC-AED (Atomic emission detection) [64]
		GC-Electronic nose [35]
Sorptive based techniques		GC-MS (Mass spectrometry) [5,7,19,23,25–27,29,36,37,39–74]
SPE (Solid phase extraction) [39,51–54]		GC-MS combined with Olfactometry [21,42,44,61]
SPDE (Solid-phase dynamic extraction) [55,56]		1D or 2D-GC combined with MS, FID,PFPD and/or NPD (Nitrogen phosphorus detector) [61]
SBSE (Stir bar sorptive extraction) [7,38,73,77,79,84,90]		GC×GC-ToFMS (Time-of-flight-mass spectrometry) [47,82]
SPMCE (Solid-phase microcolumn extraction) [62,63]	**HPLC**	HPLC-UV (High performance liquid chromatography-spectrophotometric detection) [94]
SPME (Solid-phase microextraction) [5,23,28,29,38,44,46,49–71,77–79,83–89]		HPLC-PAD (Photodiode array detection) [92,93]
		HPLC-APCI (Atmospheric pressure chemical ionization)-MS [93]
Gas diffusion techniques [92–94]	**MS based techniques**	EESI (Extractive electrospray ionization)-Q-ToFMS [24]
		DART (Direct analysis in real time)-ToFMS [25]
		APCI-MS [26]

Fig. (2). Extraction and chromatographic or mass spectrometry based methodologies that have been used for beer volatile determination.

It is important to mention that the majority of these researches are focused on the determination of a set of volatile compounds, usually with specific purposes (*e.g.* to monitor yeasts fermentation performance, detect off-flavors, among others). Thus, some specific apparatus and methodologies (for instance, gas diffusion techniques) have been developed, however their further application by other researchers or even in the breweries is quite limited. As SPME combined with GC is the most frequently used methodologies, a particular attention will be given to these techniques.

Sample Preparation: Decarbonation Procedure

The presence of CO_2 promotes the foam formation during the extraction of volatile analytes, which may affect the extraction efficiency and reproducibility [22, 35, 75], and its impact increased for higher extraction temperatures (50 °C) [35]. Thus, in order to solve this problem and trying to preserve the integrity of the volatile fraction, several procedures have been applied to promote the decarbonation of beer prior to the volatiles' extraction:

- Ultrasonic bath treatment (5 – 20 min) [5, 19, 25, 52, 65, 78, 79, 85, 94];
- Filtration [21, 29, 42, 53, 73, 74, 76, 83, 84, 90, 91];
- Manual shaking (10 s – 1 min) [58, 62, 63];
- Pouring into a beaker [41];
- Diffusion of CO_2 by storage of sample at 4° C (1 h to overnight period) [41, 82].

Ultrasonic bath treatment and filtration were the highly used procedure, however some authors claimed that ultrasonic treatments can degrade analytes due to high internal temperatures [24, 35, 75]; also the application of an ultrasonic treatment can lead to highest loss of ethanol content, along with CO_2 removal. The performance of the ultrasonic treatment was evaluated by American Society of Brewing Chemists (ABSC) concluding that the method Beer-1D (shaken on a rotary shaker for 12 min) promoted a more accurate results expressed as ethanol content than ultrasonic treatment [95]. More recently, ultrasonic procedure was compared with a CO_2 static procedure removal and revealed that the reliable analysis of the beer volatile composition seems to be using the static procedure, where the CO_2 dissolved gas molecules and the volatile analytes are naturally released above the liquid surface and the ultrasonic treatment promoted higher loss of volatile compounds [82]. The mechanical waves of ultrasound cause turbulence in beer, leading to the formation of cavitation bubbles (with CO_2 and the volatile compounds), that disrupt and lead to a faster loss of volatile compounds.

Finally, it is important to point out that some authors reported that the beer decarbonation may not represent a crucial step for targeted analysis of high volatile components or semi-volatile ones [27, 64, 75]. For instance, the beer decarbonation can promote the loss of the very volatile sulfur compounds [27, 64], and not benefit the extraction efficiency or reproducibility of the target analysis of volatile phenols [75], which required the use of extreme conditions of SPME in order to promote the volatility of these compounds (80 °C during 55 min).

Extraction Techniques for Further Gas Chromatographic Analysis

Summary of the Extraction Techniques

Regarding beer volatile determination, it is necessary to perform a proper sample extraction technique prior to instrumental analysis, once GC direct injection is not appropriate. Indeed, the presence of huge quantities of non-volatile analytes in beer can lead to column and/or detector damages [6, 96]. Thus, an extraction step will allow the clean-up (elimination of interfering matrix components) and pre-concentration of samples. Table **1** enumerates the advantages and drawbacks of the extraction techniques that have been used for beer volatile determination, such as: headspace generation apparatus, solvent based techniques, and sorptive based techniques (SPE, SPDE, SBSE, SPMCE, and SPME).

Table 1. Main advantages and drawbacks of different extraction techniques used for beer volatile determination.

	Headspace generation apparatus	Solvent based techniques	Sorptive based techniques				
			SPE	SPDE	SBSE	SPMCE	SPME
Advantages							
Different stationary phases			++	+	+	+	+++
Highly sensitive	++				+++	+++	+++
Small in size		+		++	+++		+++
Direct analysis of sample	++			+	+		+++
Solvent-free	+++			+	+	+++	+++
Possible automation	+++	+		+		++	+++
Friendly user	++	+		++	++		+++
Selectivity		+	+	++	+	+	+++
One-step extraction	++			++			+++
Durability	+++			+	++	+	+
Drawbacks							
Require a specific equipment	+++			++	++		
Use of solvents		+++	++		+		
Time consuming		+++	++		+		
Few trapping materials available			+	+++	+++	+++	
Carryover problems	+++		++	++	+	+	+
Requires concentration step	++	+++	+++		+		

Legend: The + signal means higher degree associated to an advantage or drawbacks.

The extraction of beer volatile compounds can be performed through a headspace sampler device of a GC, called as headspace generation apparatus. This type of sampling/extraction is a solvent-free, simple and fast technique (Table **1**) [32, 33]. Two modes can be used: static or dynamic, being gas-tight syringes or other devices (with different trapping materials and designs) used for this type of sampling. However, the required specific equipment is the main drawback from this technique [97].

Others techniques require the use of organic solvents [21, 40 - 42, 44, 46, 47], whose affinity to the beer components will allow their extraction. Non-polar organic solvents (dielectric constant $\varepsilon < 20$) have been used, namely diethyl ether ($\varepsilon = 4.3$) and dichloromethane ($\varepsilon = 9.1$). Formerly, trichlorofluoromethane ($\varepsilon = 2.3$) was also used, however the use of this solvent is prohibitive once it is one of the solvents that is responsible for the ozone layer destruction [98]. Moreover, a prior solvent evaporation should be performed before the GC injection. This may lead to analytes' loss or degradation, and possible artifacts formation [99], which phenomena can be reduced using a Vigreux column concentration system, that operates under vacuum. This type of technique does not address the actual concerns about health and environment sustainability, thus not following the principles of green chemistry and modern separation science. In fact, waste reduction, through the decrease of toxic reagents amount, and the miniaturization of the sample preparation steps are actual trends in Analytical Chemistry [100, 101]. LLE was used to extract beer volatile components using 80 mL – 2.4 L of organic solvents. Furthermore, these techniques have others drawbacks, such as they are time-consuming (extraction times between 20 min and 24 h), labor-intensive, requiring several steps [102]. SDE (simultaneous distillation–extraction) uses the combination of solvent extraction with distillation, usually performed with a Likens-Nickerson apparatus. The continuous distillation and extraction may promote higher extraction efficiencies, however this technique has been poorly used for beer volatile determination [46]. A recent technique, SDME, requires few microliters of organic solvent which are used for the volatiles extraction. Besides this advantage, it is a very simple technique, which can be performed with only one step. SDME was only applied to extract beer sulfur compounds, either direct or headspace mode [49].

Several sorptive techniques have also been used. SPE is a technique based on partition such as LLE, where analytes (from a liquid sample or solvent extract achieved from a solid sample) are partitioned through a solid sorbent phase. SPE procedure (Fig. **3a**) involves several steps, namely conditioning, sample addition (load of sample onto the solid phase), washing (removal of undesired components), and elution (with appropriate solvent) [103]. This technique requires lower amount of solvents than conventional LLE, which leads to less waste production [104, 105]. After the adsorption of beer components to the hydrophobic solid phase surface, the elution is performed using a non-polar solvent, such as dichloromethane. Then the extract is dried and concentrated for further GC analysis. Few trapping materials were used for beer volatile determination (*i.e.* Strata C18-E column, LiChrolut EN cartridge, Tenax-TA trap, Tenax and carbotrap activated). Due to the multi-steps, this technique is time consuming. If a proper clean-up of the sorbent phase is not performed, carryover problems may occur when re-used. The performance of SPE technique in beer matrix was compared, it was verified good sensitivity for monophenols determination however limitations on quantitative data was observed for analysis of three compounds (comparing with SPME) [51]; and was successfully applied in the analysis of medium-chain fatty acids, being the solvents utilization the main drawback when comparing with SPME and SBSE techniques [39]. Other sorptive based technique that has been applied to extract the beer volatile components is SPDE, which is a recent setting of SPME regarding dynamic extraction of samples (liquid or gaseous). The stationary phase is placed inside a needle (an internal layer forms a tube), which is connected to a gas-tight syringe, and where analytes can be concentrated (Fig. **3b**). A fixed HS volume is pulled in and out (strokes), allowing the dynamic extraction.

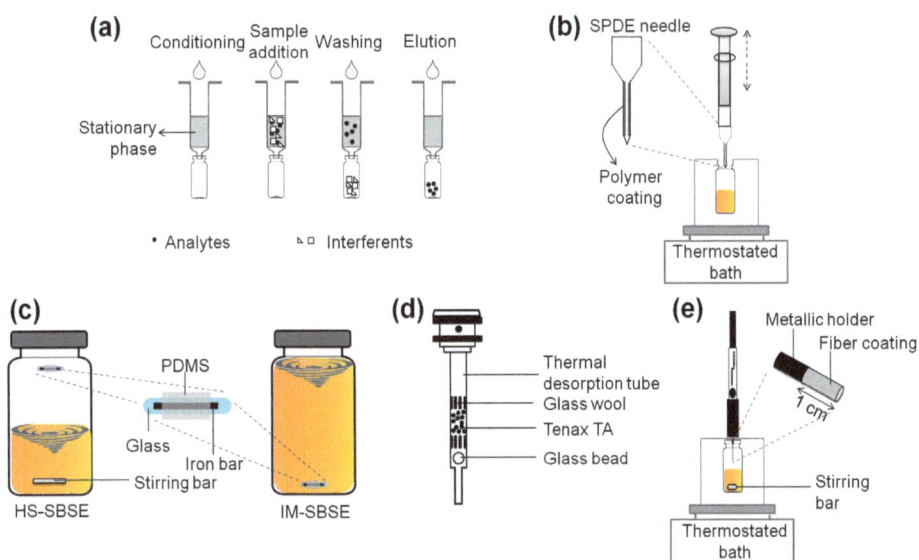

Fig. (3). Schematic illustrations of sorptive based techniques: **a)** SPE using cartridges, **b)** SPDE technique using a SPDE needle and syringe; **c)** SBSE with PDMS stationary phase, in headspace (HS) or immersion (IM) sampling modes; **d)** HS-SPMCE, using a glass microcolumn; **e)** HS-SPME using manual syringe extraction holder.

The robustness of the capillary gives SPDE a great advantage comparing with fragility of SPME fibers, once SPDE devices do not suffer mechanical damages very easily; also the SPDE stationary phase has five to six times the length of SPME stationary phase, which promote higher sorption capacity (Table **1**) [55, 56, 106]. Different polymeric sorbents may be used in SPDE technique, however only polydimethylsiloxane (PDMS) was used to extract the beer volatile components. Nevertheless, there are some drawbacks regarding this technique: the analytes could be retained in the needle wall (carryover problems); some desorption problems can occur due to the coating length, if there is a considerable temperature profile in the GC injector; also it requires a robotic system, once it is not possible to handle it manually [55, 106].

SBSE is other sorptive technique where the extraction of analytes is performed with a magnetic stir bar coated with a sorbent phase (usually a thick film of PDMS or PDMS combined with ethylene glycol (EG), which has more ability to extract non-polar analytes) (Fig. **3c**). Basically, there are two different sampling modes regarding liquid samples: immersion (direct contact of stir bar with aqueous phase) or headspace (stir bar is suspended in the flask top, being the analytes in the headspace allowed to adsorb onto the sorbent, reducing the probabilities of contamination with non-volatile analytes). After the enrichment step, stir bar immersed in the liquid sample is removed and rinsed with distilled water (removal of potential interferents), dipped on a clean paper tissue (removal of water), and then is submitted into a back-extraction process, by thermal desorption GC unit or liquid desorption (using hexane or mixture of dichloromethane/hexane). Stir bar suspended in the flask top (HS mode) may be introduced directly in the GC thermal desorption unit. In order to reuse the stir bar, a proper clean-up is required, which involves several steps (washing with distillated water, drying, followed by the back-extraction process) [100, 107]. The higher volume of PDMS phase used in SBSE (between 24 and 126 µL), comparing with SPME fiber coatings (up to 0.5 µL in a 100 µm film thickness) seems to give an apparent advantage to SBSE, namely higher sensitivity and sorption capacity once the analytes amount is proportional to the thickness of coating [107]. Indeed, this fact was showed through the comparison of fatty acid extraction in beer, where 11 and 5 fatty acids were extracted by SBSE and SPME, respectively [39]. Nevertheless, it is time consuming (*e.g.* total time of extraction time of 160 min when comparing with 35 min for SPME technique [39]), and it seems that only allow to have higher recoveries for major analytes, thus overloaded chromatograms (with co-eluted peaks) can occur, disregarding trace analytes [39, 108]. Another drawback is the poor extraction efficiency of polar analytes (log $K_{O/W}$ < 3) once weak hydrophobic interactions are involved in PDMS extraction (the sorbent phase more commercially used) (Table **1**). Indeed, the analysis of vicinal diketones (considered off-flavors in certain beer types, $K_{O/W}$ of

-1.34 and -0.85 for diacetyl and 2,3-pentanedione, respectively) by SBSE (PDMS) was optimized by Horák *et al*. [57], and better extraction efficiency was obtained for SPME when compared with SBSE extraction.

SPMCE is a large volume HS sampling technique, with similar features to SPME, in which a glass microcolumn, packed with a specific adsorbent, is used to extract the analytes (Fig. **3d**). Basically this microcolumn is temporarily assembled inside a gas-tight syringe, which is then placed in contact with the sample's headspace, and the aspiration of the HS gas occurs. After the extraction procedure, thermal desorption of the trapped analytes is performed into a GC, which need an specific modified inlet or a separated thermal desorption unit connected with the GC injector (Table **1**) [109]. Regarding beer volatile determination, scarce literature is available, nevertheless, Hrivňák *et al*. [62, 63] developed an SPMCE technique, that showed to be more sensitive than the HS generation apparatus technique, decreasing also the equilibrium time among the HS and the liquid sample: 1 – 2 h for HS generation apparatus [19], and only 10 – 15 min to reach the equilibrium time for SPMCE analysis [62, 63].

Solid Phase Microextraction (SPME)

SPME is a rapid and simple sample preparation technique based on sorption of analytes (absorption and/or adsorption) into a stationary phase, allowing their simultaneous sampling, extraction, and pre-concentration. This technique was developed by Pawliszyn, in the beginning of the 1990 decade [110], in order to extract directly analytes from liquid, solid, and gaseous samples. SPME is a non-exhaustive extraction technique once it only extracts a small analyte fraction, which characterize the global composition of analytes in the free form [102]. Several advantages are associated with this technique, namely friendly-user, easy to couple to automation, easy miniaturization, and can be easily coupled with chromatographic instruments, among others [101, 110] (Table **1**). Moreover, SPME has ability to extract analytes with concentration of parts per million (ppm) or even parts per billion (ppb) [110]. Thus, SPME has been currently the extraction technique more commonly used for beer volatile determination (Fig. **2**), which may be the result of the best compromise between advantages and drawbacks, as illustrated in Table **1**. Indeed, volatiles extraction efficiency may be compared with LLE and SPME, which LOD and LOQ of the compounds was significantly higher for LLE than SPME (*e.g.* LOD of decanoic acid was 0.01 [86] and 1270 [50] µg/L, when extraction was performed with HS-SPME and LLE, respectively).

There are commercially available different SPME fiber coatings and geometries, nevertheless a needle, whose base is fused silica coated with a thin layer of a stationary phase, inside a syringe (Fig. **3e**) [110] is the most commonly used. Two different approaches can be used in SPME sampling: immersion or direct extraction (IM-SPME) or headspace extraction (HS-SPME). In the first mode, analytes are released directly from the sample to the fiber stationary phase, once the coated fiber is placed into the sample. In the headspace extraction mode, the analytes are released from the liquid matrix, being transported through air barrier, and the vaporized analytes are sorbed by the fiber stationary phase. Thus, no direct contact between the fiber stationary phase and the sample occurs, which allows the preservation of the SPME fiber stationary phase of damage from matrix interferences, there is a reduction of the matrix and background effect, and also there is an increase of the selectivity through the extraction of volatile and semi-volatile compounds [110]. After fiber exposition, a mass-transfer process takes place promoted by the second law of thermodynamics, and the analyte' amount that is sorbed by the fiber stationary phase can be quantified by the following mathematical equation:

$$n = \frac{C_0 \times V_f \times V_s \times K}{K \times V_f + V_s}$$

where C_0 represents the samples' initial concentration; V_f and V_s are the volume of fiber and sample (or headspace), respectively; K represents $K_{fh} \times K_{hs}$, in which K_{fh} represents the partition coefficient between fiber stationary phase and headspace, and K_{hs} is the partition coefficient between the headspace and the sample.

Several experimental parameters determine the extraction efficiency and reproducibility of SPME technique, namely SPME fiber coating, salt addition (salting-out effect), extraction temperature and time, stirring effect, and sample amount and chemical composition [102, 110]. Moreover, SPME fiber coatings can be reused hundreds of times without any compromise to their physical properties, nevertheless, it is required to have a regular control of their performance (*e.g.* using a chemical standard) and thermal clean-ups between extractions, in order to avoid cross-over contaminations [108].

Different SPME stationary phases are commercially available, regarding polarity, types of analyte interaction (absorption and/or adsorption), and film thickness. In the last years, the beer volatile composition has been characterized through the use of several SPME fiber coatings. The combination of two (PDMS/DVB and CAR/PDMS) or three (DVB/CAR/PDMS) sorbents allows the synergistic effect between adsorption and absorption, leading to higher retention capacity and sensitivity, comparing with fibers with one sorbent (*i.e.* PDMS), that is only based on absorption [110]. As beer matrix is very complex, regarding chemical structures and polarities, beer volatile extraction has been more frequently performed with DVB/CAR/PDMS [25, 29, 73, 74, 84, 90] and PDMS/DVB [70, 76, 79, 81 - 83, 85, 90]. Moreover, PDMS/DVB is commonly used to analyze carbonyl compounds, during aging beer process [76, 81]. As these compounds are present at trace levels, as well as they are highly volatile and reactive, it is required to decrease the interferences from the most volatile compounds present in beer. For this purpose, in order to increase the selectivity and recovery of SPME technique, it can be applied a selective derivatization procedure on the carbonyl moiety [72]. The derivatization agent more used is *O*-(2,3,4,5,6-pentafluorobenzyl)-hydroxylamine (PFBHA), that selectively reacts with carbonyl groups of aldehydes and ketones. In beer volatile determination, two different approaches were used:

- On-fiber derivatization, in which the derivatization agent is loaded on the fiber before the sample's extraction, then the analytes reach the fiber coating, where react directly on the fiber, thus it is allowed the simultaneous derivatization and extraction in the sample headspace [70, 76, 81];
- In-solution derivatization, where the derivatization agent is added to sample matrix, promoting the derivatization reactions, and then is applied the HS-SPME extraction [76].

The combination of a derivatization procedure with SPME can decrease the analysis time, comparing to the use of derivatization followed by LLE, once reduces the number of steps [72, 76]. Nevertheless, these procedure (derivation with SPME) still is time-consuming and need to use solvents and materials for the derivatization procedure, despite the good reproducibility that can offer [111].

SPME extraction efficiency is also conditioned by extraction temperature, once influences the analytes' solubility and their vapor pressure in a liquid sample. This parameter interferes in the diffusion coefficients and Henry's constants of analytes, allowing to reduce the equilibrium time. Two phenomena can be promoted with the extraction temperature increase, namely the analytes' volatilization, and their consequent transference from liquid phase to the headspace; and also the analytes' solubility in the ethanolic/aqueous matrix. Thus, it is important to reach the best compromise between both phenomena to determinate the adequate extraction temperature [110]. Regarding beer volatile profiling determination, the extraction temperature varied between 20 °C until 60 °C (Supplementary material Table **S1**, which presents experimental details from beer volatile profiling studies). Particularly higher extraction temperatures (80 °C) were used to improve the extraction efficiency in phenols determination, which have an important role in the beers aroma profile [75]; and furfural determination, considered a dietary mutagen [71]. However, most of the developed SPME technique for beer volatile determination used lower temperatures, once temperatures higher than 60 °C can lead to artifacts formation, through Maillard reactions [112].

Regarding extraction time, in beer volatile determination, it was applied from 2 to 60 min (Supplementary material Table **S1**), only one research paper mentioned an extraction time of 90 min [78]. The most consensual extraction time was 30 min. Moreover, an additional time can be added, prior to SPME extraction, to allow the analytes' transfer from matrix to the headspace. Thus, the beer samples can be thermostated (20 – 60 °C, according to the extraction temperature) during 5 to 30 min (Supplementary material Table **S1**). The vast majority of the papers mention pre-extraction periods between 5 and 30 min, however in a particular case, sample was keep in the vial for 1 h at 70 °C, previous to SPME extraction, in order to identify and quantify beer volatile components, such as isoamyl acetate, ethyl acetate, and ethyl hexanoate [30].

Salt addition and stirring effect should also be considered as important factors in the extraction efficiency, particularly in aqueous matrices, such as wort and beer. The mass transfer of analytes from sample matrix to headspace can be improved with salt addition through the salting-out effect, where the reduction of analytes solubility in the matrix occurs, while the boundary phase properties are changed [110]. In this particular case, the salt more commonly used was sodium chloride (NaCl) with high purity (> 99%, to avoid introduction of artifacts), with concentrations from 0.2 to 0.350 g/mL (Supplementary material Table **S1**). Firstly, the salting-out effect promotes the increase on extraction efficiency: hydration spheres, from water molecules, are formed around the ionic salt molecules, which leads to less free water molecules to dissolve the analytes, and consequently, analytes transfer to the headspace, and then to SPME fiber coating. When saturation is achieved, the extraction efficiency decreases because the salt ions in solution can participate in electrostatic interactions with analytes, blocking their transfer to the headspace, and consequently to SPME fiber coating [110]. Regarding stirring effect, agitation promotes an increase of extraction efficiency through the reduction of diffusion coefficients of analytes, and also decreases the depletion zone effect that occurs near the fiber as a result of fluid shielding [113]. In beer volatile determination, the agitation was performed with using a magnetic stirring bar, from 250 to 1200 rpm (Supplementary material Table **S1**).

Gas Chromatographic Analysis

The gas chromatography is the most common technique used for beer volatile characterization, more specifically the one dimensional gas chromatography (1D-GC) (Fig. **2**), being also widely used for food products and beverage analysis.

Significant improvements have been occurring driven by the need of analytical tools that can be able to analyze target and untargeted components from complex samples, either from a sensitive and selective point of view. Thus, several developments have been performed, namely the development of new stationary phases of GC columns (fused silica capillary columns have been replaced by packed columns), improvements of chromatographic equipments (*e.g.* development of pneumatics and microfluidic devices) and improvements in the hardware and software (namely development of algorithms for data processing) [114]. Consequently, deeper characterization of samples has been achieved, improved resolution and detection limits, reducing the time of analysis and data processing. Moreover, comprehensive two-dimensional gas chromatography (GC×GC) was developed as a high sensitive and throughput technique, being already applied to determine beer volatile composition [47, 82].

Fig. (**4**) shows a schematic representation of the two configurations of GC systems for beer volatile determination: 1D-GC (Fig. **4a**) and GC×GC-ToFMS (Fig. **4b**).

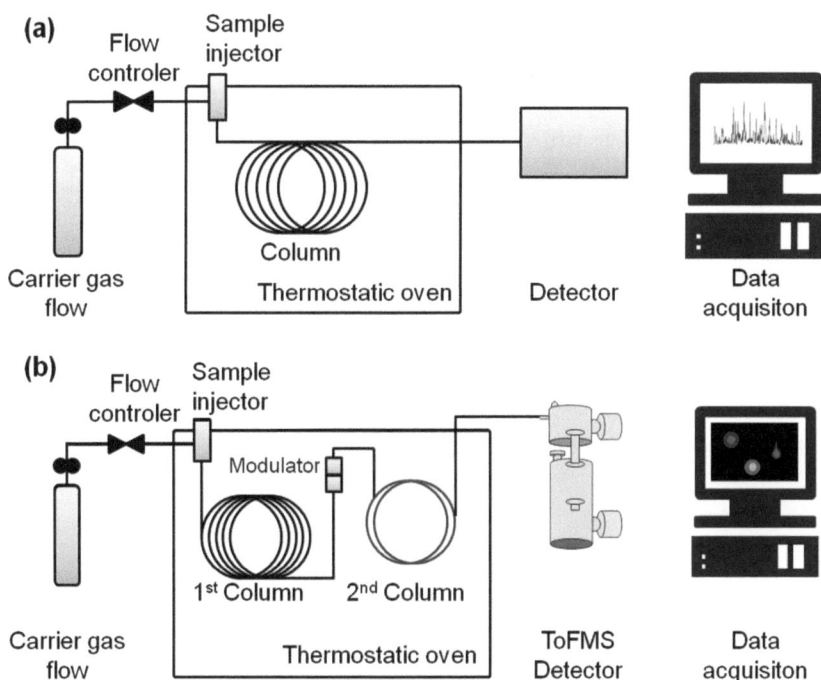

Fig. (4). Schematic illustrations of (**a**) one-dimensional gas chromatographic system (1D-GC); and (**b**) comprehensive two-dimensional (GC×GC) gas chromatographic system coupled with ToFMS (time-of-flight mass spectrometry) detector.

One Dimensional Gas Chromatography

Chromatographic systems allow the separation of mixture' analytes through their partitioning between two phases: one large stationary surface and a mobile phase (*i.e.* inert carrier gas such as helium, nitrogen, or hydrogen), that is in contact with the first one. Dispersion and specific interactions between the stationary phase and the analytes are the main factors that contribute to analytes' elution [115].

A schematic illustration of a 1D-GC is presented in Fig. (**4a**), in which a chromatographic column, which is housed in a temperature programmable oven, is connected with an injector and a detector. In some particular cases, a combination of different detectors may be used [21, 42, 44, 61]. Beer volatile determination has been commonly performed with polar polyethylene glycol columns or non-polar columns, such as those usually composed of 5% phenyl-methylpolysiloxane or equivalents. Several detectors can be coupled to the GC (Fig. **1**), however the most widely used are MS detectors, followed by FID, the first one allows the detection, quantification and identification (based on mass spectrum fragmentation patterns) of analytes within complex mixtures, and promotes high sensitivity. For instance, quantification of hexanoic acid was performed by HS-SPME combined with GC-FID and GC-MS, and data obtained allows to compare LODs, 0.27 [86] and 1.1 [73] µg/L, respectively for GC-MS and GC-FID analysis. It is important to point out, HS-SPME combined with GC-FPD was the methodology that registered the lower LODs, with compounds' concentration in the order of ng/L [49], however only sulfur compounds were analyzed. In fact, this chemical family has been analyzed with a wide range of detectors (FPD, PFPD, AED, and MS), which allow more sensibility through the selective detection of sulfur compounds.

Actually, the robust software algorithms associated to the GC-MS equipment allow an easier processing of the data [116, 117], regarding both qualitative and quantitative purposes. MS data can be acquired and processed through different approaches, namely:

- Full-scan mode (scan using a *m/z* range),
- Single ion monitoring mode (SIM – data acquisition with *m/z* diagnostic ions),
- Ion extraction chromatography mode (IEC – data processing with *m/z* diagnostic ions).

Co-elutions are frequent when full-scan acquisition is performed, particularly in complex samples. SIM and IEC can be used to overcome the former drawback, increasing the sensitivity and specificity, and consequently reducing the co-elution problems and increasing the GC area of specific analyte(s) [108, 118]. Moreover, an in-depth sample analysis can be achievable through IEC, once allows a global profile of the volatile composition (by full-scan acquisition), and simultaneously enables to have improved data of target analytes (selection of specific ions to highlight certain chemical families).

The amount of each analyte is estimated through its chromatographic area, being their quantification achieved mainly through the standard addition method or external standard calibration curves. Isotopically labeled internal standards are the more frequently used when extraction is performed with LLE (*e.g.* [^{2}H$_3$]-ethyl 3-methylbutanoate, 3-[^{2}H]-methylbutanol or 2-[^{2}H]-phenylethanol) [21, 42, 43]. Despite providing highly accurate data, the costs associated with isotopically labeled standards have limited their use to very specific cases. On the other hand, a wide range of other internal standards (IS) have been used depending on the analytes and methodologies. A list of internal standards and respective applications is presented below:

- Alcohol compounds (*i.e.* 4-methyl-1-pentanol, 1-butanol, 3-octanol, 2-heptanol, and 3,4-dimethylphenol) are the IS more used, namely when a general volatile profiling is performed [37, 41, 48, 50, 52, 53, 66, 73, 90, 91];
- *p*-fluorobenzaldehyde, 3-methyl-2-butenal, benzaldehyde-d$_6$, and guaiacol have been used to quantify carbonyl compounds (namely when a derivatization procedure is applied) [29, 72, 74, 76, 81, 90];
- Acidic compounds (*i.e.* heptanoic acid and pentanedecanoic acid) have been used to quantify medium-chain fatty acids [39, 50];
- Sulfur compounds have been quantified through the IS thiophene, ethyl methyl sulphide, 1-propylthioacetate, or isopropyl disulfide [49, 64, 87, 88];
- Methyl 2-methylbutyrate was used to quantify the following analytes: acetaldehyde, ethyl acetate, ethyl propanoate, diacetyl, and ethyl butyrate [37];
- Terpenic compounds were quantified using caryophyllene oxide [69];
- 4-(2-methoxyethyl)phenol was used to quantify volatile phenols [77];
- Vicinal diketones were quantified using 2,3-hexanedione [57];
- And also linalool was used as IS, allowing the quantification of isoamyl acetate, ethyl hexanoate, myrcene and benzaldehyde [60].

Regarding analytes' identification, it must be confirmed through the co-injection of authentic standards. However, standards can be quite expensive, often unattainable in the time available for analysis, and sometimes they are not commercially available. Therefore, the analytes can be tentatively identified using commercial and/or open source mass spectra databases [116, 117]. Another strategy for analytes' identification is the comparison of retention indices (RI) values, improving the identification confidence. Therefore, an *n*-alkane is given a retention index value of 100 times its carbon number. The RI of each analyte can be computed by van den Dool and Kratz equation [119], in which the analyte' retention time is normalized with the adjacent eluting *n*-alkanes' retention times. A *n*-alkanes series (usually ranging from C_6 to C_{20} for volatile and semi-volatile analytes) must be injected in the same GC column, and the RIs values calculated for each analyte may be compared with RIs previously reported in the literature or in open source RI libraries, using GC column similar to the one used experimentally. Indeed, different analytical laboratories can compare RIs once they are independent of some instrumental variables of GCs (*e.g.* length and diameter of column, film thickness, pressure, or carrier gas type and velocity), contrary to retention times of the analytes [115, 120].

The complexity of some matrixes, such as beer, usually implies extended GC runs and overloaded chromatograms (with frequent co-elutions of two or more analytes in some peaks), thus leading to difficulties in MS analytes' identification. Therefore, considerable research has been going towards the development of independent techniques, which allow to improve the resolving power, namely comprehensive two-dimensional gas chromatography [118, 121].

Comprehensive Two-Dimensional Gas Chromatography

The separation efficiency of multidimensional gas chromatography (MDGC) enables an in-depth characterization of complex samples. MDGC implies the multiple sequential separation of a sample, through different mechanisms (*e.g.* use of two or more columns with different characteristics, connected by an interface that allow to preserve the individual analytes' retention) [121, 122]. Comprehensive two dimensional gas chromatography (GC×GC), an alternative MDGC approach, was already used for determination of beer volatile composition [47, 82]. Briefly, analytes are separated on GC×GC system (Fig. **4b**) by orthogonal mechanisms: two GC columns with different coating materials are used, namely a non-polar and long first column (usually 30 m), that separates analytes by volatility; and a polar and short second column (1 – 2 m), that separates analytes by polarity. This set of columns allows a relatively "slow" analytes' separation in ^{1}D column, and then extremely "flash" high-resolution separation in the ^{2}D column. GC columns are connected in series by a modulator, currently by a cryomodulator (use liquid nitrogen for cooling) that accumulates and traps small portions (usually 4 – 8 s) of the eluate from the ^{1}D column by cryofocusing, and then re-injects them into the ^{2}D column. The ^{1}D separation is preserved once each peak is modulated several times. Moreover, the ^{1}D column separation is maintained once the collected fractions are no larger than ¼ of the peak width, thus producing narrow peaks. These peaks are consequently "flash" separated before the subsequent modulation [121, 122]. If the ^{2}D separation of a certain analyte does not finish before the next modulation, the elution time of the analyte exceeds the modulation time, and wrap-around phenomenon occurs. This phenomenon can interfere in the accurate analytes' quantification if there are co-elutions with analytes of interest [121 - 123], thus modulation time should be previously optimized to avoid wrap-around.

Several advantages are covered by GC×GC comparing with 1D-GC, such as improved resolution and peak capacity, also faster run times. The peaks' focusing in the modulator promotes improved detection limits and enhanced mass sensitivity and selectivity [121, 122, 124, 125]. In addition, there is an enhancement of the signal-to-noise ratio for GC×GC [126]. Nevertheless, there are some drawbacks regarding the use of this equipment, namely it requires operational expertise, the instrumentation is complex (which consequently makes complex acquired data), and with high costs of consumables [121].

GC×GC can be combined with a mass spectrometer and time-of-flight analyzer (ToFMS), which can detect analytes in the range of pg (Fig. **4b**). Indeed, the narrow peaks produced by GC×GC (peak width at half height of 0.1 s or less) require a detector with high data acquisition speed (*ca.* hundred full-mass-range spectra per second), such as ToFMS, thus providing sufficient data density [121, 122]. Moreover, ToFMS allows the acquisition of full mass spectra at trace levels and mass spectral continuity, letting a reliable spectra deconvolution of overlapping peaks. Spectral continuity occurs when all the point of the chromatographic peak have the same ion abundance ratios for the different masses in the spectrum [115, 127].

GC×GC produces structured chromatograms due to the orthogonal analytes' separation (Fig. **5**) due to the combination of a non-polar with a polar column which allows that analytes with similar chemical properties display a specific spatial location in the 2D chromatographic space. Thus, chemically-related analytes are easily recognized in the 2D "chemical map", thus simplifying the data analysis through the reduction of analysis time and helping to obtain reliable identifications (particularly if standards are not available) [122, 128]. These structured chromatograms are possibly one of the most important features of GC×GC, comparing with 1D-GC performance. Fig. (**5**) elucidates the principle of the structured chromatogram, where hydrocarbons are the least polar compounds, presenting the lower retention time for the second dimension (^2t_R), and acids, with higher polarity, presenting higher ^2t_R value.

Due to the data complexity, specific softwares have been developed for GC×GC-ToFMS equipments, such as the ChromaTOF® that allows the acquirement, processing, and reporting data, through the True Signal Deconvolution® and automated peak find algorithms. ChromaTOF allows to generate and to visualize the GC×GC chromatogram, either through a contour plot or a 3D plot (example of a 3D plot from a beer is represented in Fig. (**5**) [126]. Thus, data processing and respective analysis makes it easier for the user.

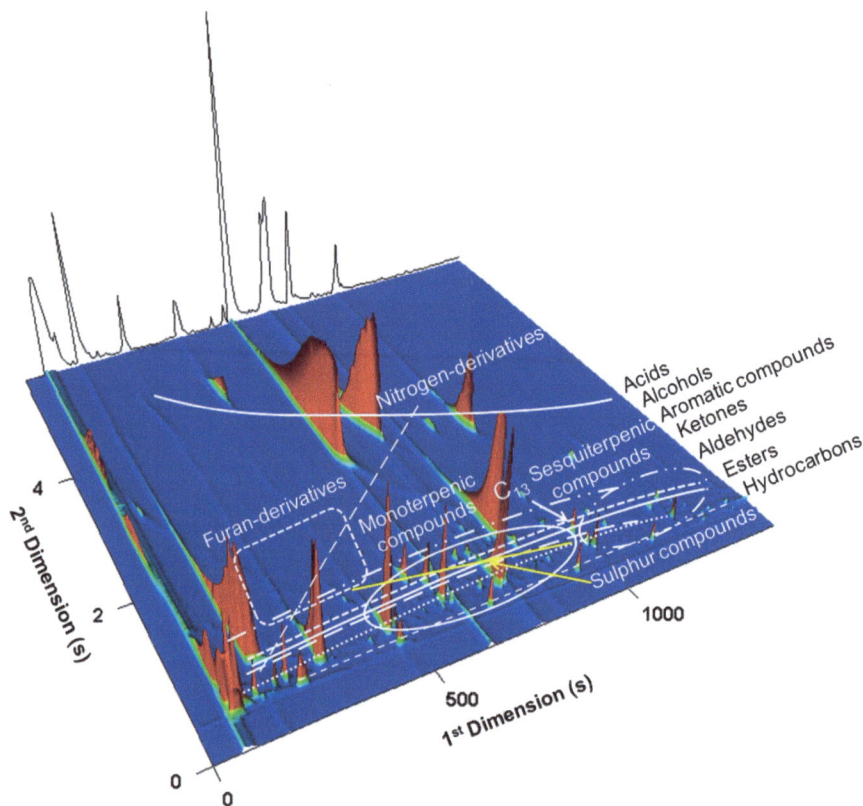

Fig. (5). 3D GC×GC total ion chromatogram plot of one beer highlighting beer volatile chemical families. Bands and clusters formed by structurally related compounds are indicated [82].

Fig. (**6a**) shows a blow-up of a part of GC×GC total ion contour plot presented in Fig. (**5**), in which the utility of the these two dimensional systems can be showed through the separation of analytes that present similar volatility, by their separation according to their polarity. Indeed, ethyl benzoate (peak 2; ^1t_R = 702s, ^2t_R = 1.070 s) and 2-nonen-1-ol (peak 3; ^1t_R = 702 s, ^2t_R = 0.630 s) presented the same ^1t_R (similar volatility), thus co-eluting on the Equity-5 column. However, they exhibited different polarities and, therefore, were separated on the DB-FFAP column, once narrow fractions are cryo-focused and re-injected on the ^{2}D column, allowing the separation of 2 compounds within only half of a second. Comparing the chemical structures of these compounds, the different polarities observed are explained by the presence of aromatic ring that increase the polarity of the molecules (higher ^2t_R), using this non-polar/polar column set [129].

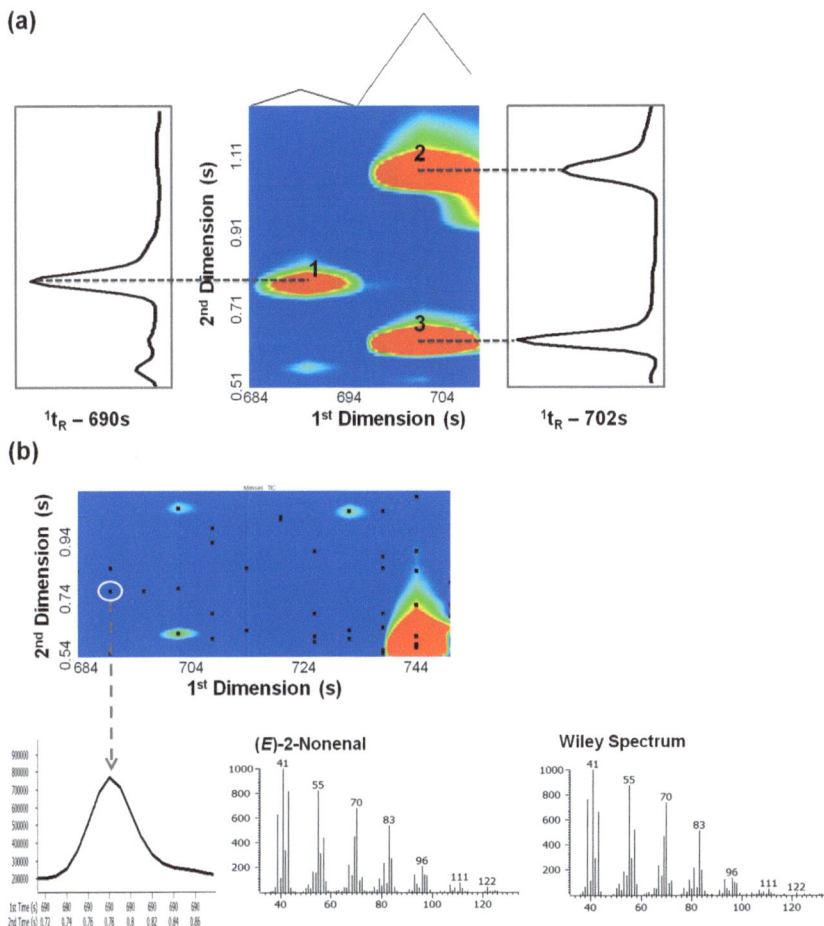

Fig. (6). (a) Blow-up of a part of GC×GC chromatogram contour plot of one beer showing the separation of (1) *(E)*-2-nonenal, (2) ethyl benzoate, and (3) 2-nonen-1-ol. **(b)** The 49 milliseconds-wide *(E)*-2-nonenal (trace beer metabolite) GC×GC peak is easily defined and identified at a mass spectral acquisition of 125 spectra/s and its spectral quality allows its identification by comparison with mass spectrum of commercial databases (as Wiley).

The determination of trace analytes takes additional challenges, which might be overcome using GC×GC-ToFMS systems, where detection limits and spectral quality are particularly determinants. For instance, Fig. (**6b**) reveals that even with narrow peaks achieved for *(E)*-2-nonenal (*ca.* 49 ms), the spectral quality of the acquired spectrum is appropriate to compare it with commercial databases. Thus, the putatively identification of *(E)*-2-nonenal can be achieved through the analysis of the retention times (^1t_R and ^2t_R), the mass spectrum and its similarity with commercial databases, and the calculation of the RI and its comparison with the literature for columns equivalents with those used in the ^1D.

Over the past years, GC×GC-ToFMS has been widely applied in different fields of analysis, namely in food area [118, 125, 128, 130 - 134], providing several advances in research and also new analytes' identification. Regarding the analysis of beer volatile composition, this technique has been poorly explored. Indeed, Inui *et al.* identified 67 hop-derived compounds which revealed to be key compounds that explain the differences of hop aroma characteristics in the analyzed beers (beers produced with different hops) [47]. Martins *et al.* [82] optimized a set of SPME experimental parameters, and then explored the potential of the GC×GC-ToFMS regarding the beer volatile profiling.

EXPLORING THE BEER VOLATILE FRACTION

Several hundreds of analytes have been previously determined in beer samples, which are enumerated in Table **2**. These analytes were grouped according to their chemical family: acids, alcohols, aldehydes, esters, ethers, furan-derivatives, hydrocarbons, ketones, lactones, monoterpenic compounds, norisoprenoids, phenols, pyrans, pyrazines, sulfur compounds, sesquiterpenic compounds, and other nitrogen-containing compounds. This list may be further used as a valuable database, as it includes a wide range of analytes which have been used to studied several issues related to beer raw material, brewing process, aging, contaminants, off-flavors, among others. When available, data about their detection (LOD) and quantification (LOQ) limits are also included. As for several compounds, a wide range of LOD and LOQ values were observed, these systematized data is useful for the readers to select the most convenient methodology, according to the sensitivity requirements.

Table 2. Volatile and semi-volatile analytes previously determined in beers [5, 7, 19, 21, 22, 24–94], including respective CAS number, chemical formula, and when available, limits of detection (LOD) and quantification (LOQ).

Name	CAS	Formula	LOD (µg/L)	LOQ (µg/L)
Acids				
Alifatics				
Methanoic acid	64-18-6	CH_2O_2	–	–
Acetic acid	64-19-7	$C_2H_4O_2$	35.2 [86]	–
Propanoic acid	79-09-4	$C_3H_6O_2$	–	–
Lactic acid	50-21-5	$C_3H_6O_3$	–	–
Butanoic acid	107-92-6	$C_4H_8O_2$	–	–
2-Methylpropanoic acid	79-31-2	$C_4H_8O_2$	–	–
4-Oxopentanoic acid	123-76-2	$C_5H_8O_3$	–	–
3-Methylbutanoic acid	503-74-2	$C_5H_{10}O_2$	–	–
2-Methylbutanoic acid	116-53-0	$C_5H_{10}O_2$	–	–
Pentanoic acid	109-52-4	$C_5H_{10}O_2$	–	–

(Table 2) contd.....

Name	CAS	Formula	LOD (µg/L)	LOQ (µg/L)
2,4-Hexadienoic acid	110-44-1	$C_6H_8O_2$	–	–
(*E*)-2-Methyl-2-pentenoic acid	16957-70-3	$C_6H_{10}O_2$	–	–
2-Methyl-2-ethyl-3-hydroxy propanoic acid	–	$C_6H_{11}O_3$	–	–
Hexanoic acid	142-62-1	$C_6H_{12}O_2$	0.27 [86] 1.1 [73] 610 [50]	– 3.67 [73] –
2,5-Dimethyl-4-hexenoic acid	82898-13-3	$C_8H_{14}O_2$	–	–
Heptanoic acid	111-14-8	$C_7H_{14}O_2$	–	–
2-Methylhexanoic acid	4536-23-6	$C_7H_{14}O_2$	–	–
2-Ethylhexanoic acid	149-57-5	$C_8H_{16}O_2$	–	–
Octanoic acid	124-07-2	$C_8H_{16}O_2$	0.02 [86] 0.32 [73] 810 [50]	– 1.06 [73] –
Nonanoic acid	112-05-0	$C_9H_{18}O_2$	–	–
9-Decenoic acid	14436-32-9	$C_{10}H_{18}O_2$	–	–
Decanoic acid	334-48-5	$C_{10}H_{20}O_2$	0.01 [86] 1270 [50]	– –
Dodecanoic acid	143-07-7	$C_{12}H_{24}O_2$	–	–
Tridecanoic acid	638-53-9	$C_{13}H_{26}O_2$	–	–
Tetradecanoic acid	544-63-8	$C_{14}H_{28}O_2$	–	–
Aromatics				
Benzoic acid	65-85-0	$C_7H_6O_2$	–	–
Phthalic acid	88-99-3	$C_8H_6O_4$	–	–
4-Methylbenzoic acid	99-94-5	$C_8H_8O_2$	–	–
Phenylacetic acid	103-82-2	$C_8H_8O_2$	–	–
4-(1-Methylethyl)-benzoic acid	536-66-3	$C_{10}H_{12}O_2$	–	–
Alcohols				
Alifatics				
Ethanol	64-17-5	C_2H_6O	–	–
Propanol	71-23-8	C_3H_8O	2.1, 7.0 [38] 10.4 [86] 42 [22] 2080, 1830 [26]	– – – –
2-Propanol	67-63-0	C_3H_8O	3.6, 11.3 [38]	
1-Butanol	71-36-3	$C_4H_{10}O$	0.3, 5.0 [38] 130, 280 [26]	–
2-Butanol	78-98-2	$C_4H_{10}O$	1.1, 3.1[38]	–
2-Methyl-2-propanol	75-65-0	$C_4H_{10}O$	0.3, 2.2 [38]	–
Isobutanol	78-83-1	$C_4H_{10}O$	1.87 [86] 1.6, 13.0 [38] 62.41 [73] 71 [22] 130 [37] 130, 320 [26]	– – 208.04 [73] – – –

(Table 2) contd.....

Name	CAS	Formula	LOD (µg/L)	LOQ (µg/L)
2,3-Butanediol	513-85-9	$C_4H_{10}O_2$	–	–
3-Methyl-2-buten-1-ol	556-82-1	$C_5H_{10}O$	–	–
3-Methyl-3-buten-1-ol	763-32-6	$C_5H_{10}O$	–	–
4-Penten-1-ol	821-09-0	$C_5H_{10}O$	–	–
1-Pentanol	71-41-0	$C_5H_{12}O$	0.2, 1.4 [38] 120, 190 [26]	–
2-Methyl-1-butanol	137-32-6	$C_5H_{12}O$	0.1, 1.6 [38]	–
3-Methyl-1-butanol	123-51-3	$C_5H_{12}O$	0.1, 3.0 [38] 0.28 [86] 5.75 [73] 10 [37] 124 [22] 140, 180 [26] 1650 [50]	– – 19.17 [73] – – – –
(Z)-Hex-2-en-1-ol	928-94-9	$C_6H_{12}O$	–	–
Hexanol	111-27-3	$C_6H_{14}O$	0.04 [86] 0.08, 1.1 [38] 0.09 [73]	– – 0.31 [73]
Heptanol	111-70-6	$C_7H_{16}O$	–	–
2-Heptanol	543-49-7	$C_7H_{16}O$	–	–
2-Methyl-1-hexanol	624-22-6	$C_7H_{16}O$	–	–
3-Methyl-1-hexanol	13231-81-7	$C_7H_{16}O$	–	–
1-Octen-3-ol	3391-86-4	$C_8H_{16}O$	–	–
Octanol	111-87-5	$C_8H_{18}O$	–	–
2-Ethyl-1-hexanol	104-76-7	$C_8H_{18}O$	0.08, 1.8 [38]	–
3-Ethyl-1-hexanol	41065-95-6	$C_8H_{18}O$	–	–
2-Octanol	123-96-6	$C_8H_{18}O$	–	–
Isooctanol	26952-21-6	$C_8H_{18}O$	–	–
1-Nonanol	143-08-8	$C_9H_{20}O$	–	–
2-Nonanol	628-99-9	$C_9H_{20}O$	–	–
9-Decen-1-ol	13019-22-2	$C_{10}H_{20}O$	–	–
Decanol	112-30-1	$C_{10}H_{22}O$	–	–
2-Decanol	1120-06-5	$C_{10}H_{22}O$	–	–
1-Undecanol	112-42-5	$C_{11}H_{24}O$	–	–
2-Undecanol	1653-30-1	$C_{11}H_{24}O$	–	–
Dodecanol	112-53-8	$C_{12}H_{26}O_2$	–	–
2-Dodecanol	10203-28-8	$C_{12}H_{26}O$	–	–
2,6,6-Trimethyl-9-undecen-1-ol	–	$C_{14}H_{27}O$	–	–
Tetradecanol	112-72-1	$C_{14}H_{30}O$	–	–
8,10-Hexadecadien-1-ol	–	$C_{16}H_{30}O$	–	–
Aromatics				
Benzenemethanol	100-51-6	C_7H_8O	–	–

(Table 2) contd.....

Name	CAS	Formula	LOD (µg/L)	LOQ (µg/L)
Phenethyl alcohol	60-12-8	$C_8H_{10}O$	0.02 [86] 1.43 [73] 130 [62] 410 [50] 8330, 1560 [26]	– 4.77 [73] 790 [62] – –
Tyrosol	501-94-0	$C_8H_{10}O_2$	–	–
Butylhydroxy toluene	128-37-0	$C_{15}H_{24}O$	–	–
Cyclics				
2,3-Trimethyl cyclopentaneethanol	52363-24-3	$C_{10}H_{20}O$	–	–
1,5,5,8-Tetramethyl-3,7- cycloundecadien-1-ol	118014-38-3	$C_{15}H_{26}O$	–	–
Aldehydes				
Alifatics				
Glyoxal	107-22-2	$C_2H_2O_2$	0.046 [81] 4.5, 67 [76]	0.14 [81] 15, 223 [76]
Acetaldehyde	75-07-0	C_2H_4O	3.1, 5.1 [38] 12 [93] 30 [37] 36 [22] 39 [62]	– 41 [93] – – 240 [62]
2-Propenal	107-02-8	C_3H_4O	0.51 [81] 0.24, 7.96 [76]	1.55 [81] 0.81, 26.54 [76]
Methylglyoxal	78-98-8	$C_3H_4O_2$	0.02 [81] 3.2, 12 [76]	0.061 [81] 10.6, 39 [76]
Propanal	123-38-6	C_3H_6O	0.013 [90] 0.035 [81] 1.2, 1.7 [76]	0.045 [90] 0.107 [81] 4.1, 5.7 [76]
(E)-2-Butenal	123-73-9	C_4H_6O	0.013, 0.69 [76]	0.042, 2.29 [76]
2-Methyl-2-propenal	78-85-3	C_4H_6O	–	–
Butanal	123-72-8	C_4H_8O	0.037 [81] 0.024, 0.04 [76] 0.054 [90]	0.113 [81] 0.081, 0.131 [76] 0.179 [90]
2-Methylpropanal	78-84-2	C_4H_8O	0.038 [81] 0.11 [72] 0.23, 0.28 [76] 1.323 [90] 1.5 [93]	0.116 [81] 0.40 [72] 0.75, 0.93 [76] 4.411 [90] 4.9 [93]
2,4-Pentadienal	764-40-9	C_5H_6O	–	–
(E)-2-Pentenal	1576-87-0	C_5H_8O	0.02, 0.24 [76] 0.061 [81]	0.067, 0.079 [76] 0.186 [81]
2-Methyl-2-butenal	1115-11-3	C_5H_8O	–	–
3-Methyl-2-butenal	107-86-8	C_5H_8O	–	–
Pentanal	110-62-3	$C_5H_{10}O$	0.026, 0.34 [76] 0.031 [90]	0.088, 0.11 [76] 0.102 [90]

(Table 2) contd.....

Name	CAS	Formula	LOD (µg/L)	LOQ (µg/L)
2-Methylbutanal	96-17-3	$C_5H_{10}O$	0.024, 0.038 [76] 0.05 [81] 0.03 [72] 0.494 [90]	0.081, 0.126 [76] 0.15 [81] 0.11 [72] 1.646 [90]
3-Methylbutanal	590-86-3	$C_5H_{10}O$	0.031 [81] 0.041, 0.1 [76] 0.09 [72] 1.426 [90]	0.094 [81] 0.138, 0.33 [76] 0.30 [72] 4.754 [90]
3-Ethoxy-1-propanal	2806-85-1	$C_5H_{10}O_2$	–	–
(E,E)-2,4-Hexadienal	142-83-6	C_6H_8O	0.005 [81]	0.015 [81]
(E)-2-Hexenal	6728-26-3	$C_6H_{10}O$	0.004 [81] 0.011 [90] 0.013, 0.077 [76]	0.012 [81] 0.037 [90] 0.045, 0.26 [76]
Hexanal	66-25-1	$C_6H_{12}O$	0.01, 0.008 [76] 0.043 [81] 0.149 [90]	0.035, 0.028 [76] 0.132 [81] 0.496 [90]
3-Methylpentanal	15877-57-3	$C_6H_{12}O$	–	–
4-Methylpentanal	1119-16-0	$C_6H_{12}O$	–	–
2,4-Heptadienal	4313-03-5	$C_7H_{10}O$	–	–
(E)-2-Heptenal	18829-55-5	$C_7H_{12}O$	0.003 [81] 0.007, 0.021 [76] 0.014 [90]	0.01 [81] 0.025, 0.069 [76] 0.045 [90]
Heptanal	111-71-7	$C_7H_{14}O$	0.028, 0.012 [76] 0.054 [81] 0.082 [90]	0.094, 0.039 [76] 0.164 [81] 0.272 [90]
(E)-2-Octenal	2548-87-0	$C_8H_{14}O$	0.013 [81] 0.015, 0.016 [76]	0.041 [81] 0.051, 0.054 [76]
Octanal	124-13-0	$C_8H_{16}O$	0.035 [81] 0.048, 0.025 [76] 0.075 [90]	0.105 [81] 0.159, 0.083 [76] 0.251 [90]
(E)-2-Nonenal	18829-56-6	$C_9H_{16}O$	0.005 [55] 0.008 [90] 0.009 [81] 0.01 [78] 0.015 [72] 0.042, 0.011 [76] 0.2, 0.5 [38]	0.015 [55] 0.027 [90] 0.027 [81] 0.02 [78] 0.05 [72] 0.142, 0.036 [76] –
(E,E)-2,4-Nonadienal	5910-87-2	$C_9H_{14}O$	0.01 [81]	0.031 [81]
Nonanal	124-19-6	$C_9H_{18}O$	0.024, 0.017 [76] 0.032 [81] 0.252 [90]	0.08, 0.058 [76]; 0.096 [81]; 0.840 [90]

(Table 2) contd.....

Name	CAS	Formula	LOD (μg/L)	LOQ (μg/L)
(E,E)-2,4-Decadienal	25152-84-5	$C_{10}H_{16}O$	0.020 [90] 0.025 [81] 0.039, 0.015 [76]	0.098 [90] 0.074 [81] 0.13, 0.051 [76]
(E)-2-Decenal	3913-81-3	$C_{10}H_{18}O$	0.03 [81]	0.089 [81]
Decanal	112-31-2	$C_{10}H_{20}O$	0.021, 0.005 [76] 0.035 [81] 0.066 [90]	0.071, 0.017 [76] 0.107 [81] 0.332[90]
Undecanal	112-44-7	$C_{11}H_{22}O$	0.003 [81]	0.011 [81]
Dodecanal	112-54-9	$C_{12}H_{24}O$	–	–
Tridecanal	10486-19-8	$C_{13}H_{26}O$	–	–
Tetradecanal	124-25-4	$C_{14}H_{28}O$	–	–
Pentadecanal	2765-11-9	$C_{15}H_{30}O$	–	–
Aromatics				
Benzaldehyde	100-52-7	C_7H_6O	0.001, 0.078 [76] 0.059 [81] 0.201 [90] 0.23, 0.08 [60] 7.2 [38]	0.033, 0.26 [76] 0.18 [81] 0.671 [90] – –
Phenylacetaldehyde	122-78-1	C_8H_8O	0.029 [81] 0.11, 0.2 [76] 0.23, 0.17 [72] 2.327 [90]	0.087 [81] 0.38, 0.66 [76] 0.76, 0.55 [72] 7.758 [90]
2-Methylbenzaldehyde	529-20-4	C_8H_8O	–	–
Cyclics				
2-Cyclopentene-1-propanal	64504-73-0	$C_8H_{12}O$	–	–
Diterpenic compound				
Phytol	150-86-7	$C_{20}H_{40}O$	–	–
Isophytol	505-32-8	$C_{20}H_{40}O$	–	–
Esters				
Alifatics				
Methyl acetate	79-20-9	$C_3H_6O_2$	3 [62]	22 [62]
Ethyl formate	109-94-4	$C_3H_6O_2$	–	–
Ethyl acetate	141-78-6	$C_4H_8O_2$	0.1, 0.6 [38] 0.37 [86] 1 [62] 3.59 [73] 8, 6 [26] 70 [37] 89 [22]	– – 6.8 [62] 11.96 [73] – – –
Methyl lactate	547-64-8	$C_4H_8O_3$	–	–
Ethyl-2-propenoate	140-88-5	$C_5H_8O_2$	–	–
Ethyl pyruvate	617-35-6	$C_5H_8O_3$	0.53, 17 [76]	1.78, 56 [76]

(Table 2) contd.....

Name	CAS	Formula	LOD (µg/L)	LOQ (µg/L)
Ethyl propanoate	105-37-3	$C_5H_{10}O_2$	0.17 [73] 4 [26] 5 [37]	0.58 [73] – –
Propyl acetate	109-60-4	$C_5H_{10}O_2$	–	–
Methyl-2-methyl propanoate	547-63-7	$C_5H_{10}O_2$	–	–
Ethyl lactate	97-64-3	$C_5H_{10}O_3$	16.4 [86]	–
Monoethyl butanedioate	1070-34-4	$C_6H_{10}O_4$	–	–
Butyl acetate	123-86-4	$C_6H_{12}O_2$	–	–
Ethyl 2-methylpropanoate	97-62-1	$C_6H_{12}O_2$	–	–
Ethyl butanoate	105-54-4	$C_6H_{12}O_2$	0.02, 0.5 [38] 0.11 [86] 2 [26] 10 [37]	– – – –
2-Methylpropyl acetate	110-19-0	$C_6H_{12}O_2$	0.01 [86] 0.34 [73] 2, 5 [26]	– 1.12 [73] –
Propyl propionate	106-36-5	$C_6H_{12}O_2$	–	–
Isoamyl formate	110-45-2	$C_6H_{12}O_2$	–	–
Ethyl 4-hydroxybutanoate	–	$C_6H_{12}O_3$	–	–
2-Methylethyl-2-butenoate	55514-48-2	$C_7H_{12}O_2$	–	–
Ethyl pentanoate	539-82-2	$C_7H_{14}O_2$	–	–
3-Methylbutyl acetate	123-92-2	$C_7H_{14}O_2$	0.01, 4.1 [38] 0.01 [86] 0.05 [73] 0.17, 0.07 [60] 65 [22] 72.83 [90] 1030 [50]	– – 0.18 [73] – – 242.77 [90] –
Ethyl 2-methylbutanoate	7452-79-1	$C_7H_{14}O_2$	0.05 [90]	0.15 [90]
Ethyl 3-methylbutanoate	108-64-5	$C_7H_{14}O_2$	0.06, 0.5 [38] 0.08 [90]	– 0.26 [90]
2-Pentyl acetate	626-38-0	$C_7H_{14}O_2$	–	–
Pentyl acetate	628-63-7	$C_7H_{14}O_2$	3, 5 [26]	–
Hexyl formate	629-33-4	$C_7H_{14}O_2$	–	–
Ethyl hexanoate	123-66-0	$C_8H_{16}O_2$	0.01 [73, 86] 0.02 [62] 0.08, 0.4 [38] 0.17, 0.09 [60] 0.2 [22] 2 [26] 380 [50]	0.03 [73] 0.15 [62] – – – – –
(3z)-3-Hexen-1-yl acetate	3681-71-8	$C_8H_{14}O_2$	–	–
Ethyl 4(E)-hexenoate	51368-03-7	$C_8H_{14}O_2$	–	–
2-Hexenyl acetate	2497-18-9	$C_8H_{14}O_2$	–	–
Methyl 4-methylene-hexanoate	73805-48-8	$C_8H_{14}O_2$	–	–
Ethyl-3-hexenoate	2396-83-0	$C_8H_{14}O_2$	–	–

(Table 2) contd.....

Name	CAS	Formula	LOD (μg/L)	LOQ (μg/L)
Diethyl succinate	123-25-1	$C_8H_{14}O_4$	0.08 [73] 0.30 [86] 17.5, 2.1 [38]	0.27 [73] – –
Hexyl acetate	142-92-7	$C_8H_{16}O_2$	0.04 [73]	0.15 [73]
Ethyl 4-methylpentanoate	25415-67-2	$C_8H_{16}O_2$	–	–
Butyl butanoate	109-21-7	$C_8H_{16}O_2$	–	–
2-Methylpropyl butanoate	539-90-2	$C_8H_{16}O_2$	–	–
3-Methylbutyl propanoate	105-68-0	$C_8H_{16}O_2$	–	–
2-Methylpropyl-2-methyl propanoate	97-85-8	$C_8H_{16}O_2$	–	–
Methyl heptanoate	106-73-0	$C_8H_{16}O_2$	–	–
2-Pentanol propanoate	54004-43-2	$C_8H_{16}O_2$	–	–
Pentyl propanate	624-54-4	$C_8H_{16}O_2$	–	–
Pentyl 2-methyl-2-propenoate	2849-98-1	$C_9H_{16}O_2$	–	–
(Z)-3-Hexenyl propanoate	33467-74-2	$C_9H_{16}O_2$	–	–
Ethyl heptanoate	106-30-9	$C_9H_{18}O_2$	0.005 [73]	0.02 [73]
Heptyl acetate	112-06-1	$C_9H_{18}O_2$	–	–
3-Methylbutyl-2-methylpropanoate	2050-01-3	$C_9H_{18}O_2$	–	–
2-Methylbutyl-2-methylpropanoate	2445-69-4	$C_9H_{18}O_2$	–	–
3-Methylbutyl butanoate	106-27-4	$C_9H_{18}O_2$	–	–
2-Methylbutyl butanoate	51115-64-1	$C_9H_{18}O_2$	–	–
Methyl-6-methyl heptanoate	2519-37-1	$C_9H_{18}O_2$	–	–
Methyl-2-methyl-heptanoate	51209-78-0	$C_9H_{18}O_2$	–	–
Pentyl butanoate	540-18-1	$C_9H_{18}O_2$	–	–
Methyl octanoate	111-11-5	$C_9H_{18}O_2$	–	–
Ethyl 5-methylhexanoate	10236-10-9	$C_9H_{18}O_2$	–	–
Ethyl 7-octenoate	35194-38-8	$C_{10}H_{18}O_2$	–	–
3-Hexenyl butanoate	–	$C_{10}H_{18}O_2$	–	–
Ethyl octanoate	106-32-1	$C_{10}H_{20}O_2$	0.01 [86] 0.01, 0.5 [38] 0.04 [62] 0.07 [73] 0.5 [22] 2 [26]	– – 0.28 [62] 0.22 [73] – –
Pentyl pentanoate	2173-56-0	$C_{10}H_{20}O_2$	–	–
Octyl acetate	112-14-1	$C_{10}H_{20}O_2$	–	–
3-Methylbutyl-3-methylbutanoate	659-70-1	$C_{10}H_{20}O_2$	–	–
2-Methylbutyl-3-methylbutanoate	2445-77-4	$C_{10}H_{20}O_2$	–	–
Methyl nonanoate	1731-84-6	$C_{10}H_{20}O_2$	–	–
2-Methylbutyl pentanoate	55590-83-5	$C_{10}H_{20}O_2$	–	–
2-Methylpropyl hexanoate	105-79-3	$C_{10}H_{20}O_2$	–	–
Methyl *trans*-4,9-decadienoate	67140-63-0	$C_{11}H_{18}O_2$	–	–
Methyl *cis*-2-decenoate	–	$C_{11}H_{19}O_2$	–	–

(Table 2) contd.....

Name	CAS	Formula	LOD (µg/L)	LOQ (µg/L)
Methyl-4-decenoate	1191-02-2	$C_{11}H_{20}O_2$	–	–
Ethyl 2-nonenoate	17463-01-3	$C_{11}H_{20}O_2$	–	–
Ethyl nonanoate	123-29-5	$C_{11}H_{22}O_2$	–	–
3-Methylbutyl hexanoate	2198-61-0	$C_{11}H_{22}O_2$	–	–
Methyl decanoate	110-42-9	$C_{11}H_{22}O_2$	–	–
Propyl octanoate	624-13-5	$C_{11}H_{22}O_2$	–	–
Ethyl 4,9-decadienoate	–	$C_{12}H_{20}O_2$		
Ethyl 9-decenoate	67233-91-4	$C_{12}H_{22}O_2$	–	–
Ethyl 4-decenoate	–	$C_{12}H_{22}O_2$	–	–
Ethyl decanoate	110-38-3	$C_{12}H_{24}O_2$	0.01 [73, 86] 0.1 [62] 0.1, 0.7 [38]	0.05 [73] 0.66 [62] –
Hexyl hexanoate	6378-65-0	$C_{12}H_{24}O_2$	–	–
Butyl octanoate	589-75-3	$C_{12}H_{24}O_2$	–	–
3-Hydroxy-2,4,4-trimethylpentyl 2-methyl-propanoate	74367-34-3	$C_{12}H_{24}O_3$	–	–
Ethyl undecanoate	627-90-7	$C_{13}H_{26}O_2$	–	–
Ethyl 3-methyldecanoate	73105-70-1	$C_{13}H_{26}O_2$	–	–
3-Methylbutyl octanoate	2035-99-6	$C_{13}H_{26}O_2$	–	–
2-Methylbutyl octanoate	67121-39-5	$C_{13}H_{26}O_2$	–	–
Ethyl dodecanoate	106-33-2	$C_{14}H_{28}O_2$	0.03 [73]	0.09 [73]
Ethyl tridecanoate	28267-29-0	$C_{15}H_{30}O_2$	–	–
Propyl dodecanoate	3681-78-5	$C_{15}H_{30}O_2$	–	–
2,2,4-Trimethyl-1,3-pentanediol diisobutanoate	6846-50-0	$C_{16}H_{30}O_2$	–	–
Ethyl tetradecanoate	124-06-1	$C_{16}H_{32}O_2$	–	–
Tetradecyl acetate	638-59-5	$C_{16}H_{32}O_2$	–	–
Ethyl hexadecanoate	628-97-7	$C_{18}H_{36}O_2$	–	–
Ethyl docosanoate	5908-87-2	$C_{24}H_{48}O_2$	–	–
Aromatics				
Ethyl nicotinate	614-18-6	$C_8H_9NO_2$	4.81 [90]	16.04 [90]
Phenylethyl formate	104-62-1	$C_9H_{10}O_2$	–	–
Ethyl benzoate	93-89-0	$C_9H_{10}O_2$	–	–
Methyl 3-phenylpropenoate	103-26-4	$C_{10}H_{10}O_2$	–	–
Phenethyl acetate	103-45-7	$C_{10}H_{12}O_2$	0.07 [73] 1.0, 0.3 [38] 13.40 [90] 177, 47 [26] 540 [50]	0.22 [73] – 44.68 [90] – –
Ethyl-2-phenylethanoate	101-97-3	$C_{10}H_{12}O_2$	0.07 [90]	0.25 [90]
Ethyl 3-phenyl-2-propenoate	103-36-6	$C_{11}H_{12}O_2$	–	–
Ethyl 3-phenylpropanoate	2021-28-5	$C_{11}H_{14}O_2$	–	–
Phenylethyl butanoate	103-52-6	$C_{12}H_{16}O_2$	–	–
2-Phenylethyl 2-methylpropanoate	103-48-0	$C_{12}H_{16}O_2$	–	–

(Table 2) contd.....

Name	CAS	Formula	LOD (µg/L)	LOQ (µg/L)
2-Phenylethyl hexanoate	6290-37-5	$C_{14}H_{20}O_2$	–	–
Diethyl benzene-1,2-dicarboxylate	7487-65-2	$C_{16}H_{22}O_7$	–	–
Cyclics				
Vinyl cyclohexanecarboxylate	4840-76-0	$C_9H_{14}O_2$	–	–
Decyl cyclobutanecarboxylate	–	$C_{15}H_{28}O_2$	–	–
Ethers				
Methoxyethane	540-67-0	C_3H_8O	–	–
1-Methoxybutane	628-28-4	$C_5H_{12}O$	–	–
1,1-Diethoxyethane	105-57-7	$C_6H_{14}O_2$	–	–
2-Furfuryl ethyl ether	6270-56-0	$C_7H_{10}O_2$	0.20 [90]	0.66 [90]
1,1-Diethoxybutane	3658-95-5	$C_8H_{18}O_2$	–	–
Difurfuryl ether	4437-22-3	$C_{10}H_{10}O_3$	–	–
1-Methoxy-4-(4-methyl-4-pentenyl)benzene	65912-66-5	$C_{13}H_{18}O$	–	–
Furan-derivatives compounds				
Furan	110-00-9	C_4H_4O	–	–
Furfural	98-01-1	$C_5H_4O_2$	0.66, 6 [76] 1.54 [81] 1.60 [72] 2.791 [90] 5.4 [93]	2.2, 20 [76] 4.68 [81] 5.20 [72] 9.304 [90] 18 [93]
2-Furoic acid	88-14-2	$C_5H_4O_3$	–	–
2-Methylfuran	534-22-5	C_5H_6O	–	–
3-Methylfuran	930-27-8	C_5H_6O	–	–
Furfuryl alcohol	98-00-0	$C_5H_6O_2$	–	–
2-Methyltetrahydrofuran-3-one	3188-00-9	$C_5H_8O_2$	–	–
2-Vinylfuran	1487-18-9	C_6H_6O	–	–
5-Methylfurfural	620-02-0	$C_6H_6O_2$	0.059, 0.86 [76] 0.259 [90] 1.92 [81]	0.198, 2.85 [76] 0.864 [90] 5.82 [81]
Acetylfuran	1192-62-7	$C_6H_6O_2$	0.35, 87 [76] 1.86 [90]	1.16, 289 [76] 6.20 [90]
5-Hydroxymethylfurfural	67-47-0	$C_6H_6O_3$	3.44 [81] 310, 20000 [76]	10.4 [81] 1000, 66000 [76]
2-Ethylfuran	3208-16-0	C_6H_8O	–	–
2,5-Dimethylfuran	625-86-5	C_6H_8O	–	–
Furaneol	3658-77-3	$C_6H_8O_3$	–	–
Tetrahydro-2,5-dimethylfuran	1003-38-9	$C_6H_{12}O$	–	–
1-(2-Furanyl) propane-1-one	3194-15-8	$C_7H_8O_2$	–	–
4,5-Dimethyl-furane-2-carbaldehyde	52480-43-0	$C_7H_8O_2$	–	–
Furfuryl acetate	623-17-6	$C_7H_8O_3$	–	–
2-Butylfuran	4466-24-4	$C_8H_{12}O$	–	–

(Table 2) contd.....

Name	CAS	Formula	LOD (µg/L)	LOQ (µg/L)
2-Pentylfuran	3777-69-3	$C_9H_{14}O$	–	–
3-(4-Methyl-3-pentenyl)-furan	539-52-6	$C_{10}H_{14}O$	–	–
Hexahydro-1,1-dimethyl-4-methylene-4H-cyclopenta[c]furan	19901-95-2	$C_{10}H_{16}O$	–	–
2-Heptylfuran	3777-71-7	$C_{11}H_{18}O$	–	–
Furfuryl octanoate	39252-03-4	$C_{13}H_{20}O_3$	–	–
Hydrocarbons				
Alifatics				
1,3-Butadiene	106-99-0	C_4H_6	–	–
1,3-Pentadiene	504-60-9	C_5H_8	–	–
2-Methyl-1,3-butadiene	78-79-5	C_5H_8	–	–
1,4-Pentadiene	591-93-5	C_5H_8	–	–
3-Methyl-1,4-pentadiene	1115-08-8	C_6H_{10}	–	–
4-Methyl-1,3-pentadiene	926-56-7	C_6H_{10}	–	–
2,4-Hexadiene	592-46-1	C_6H_{10}	–	–
2-Methyl-1,3-pentadiene	1118-58-7	C_6H_{10}	–	–
1,3,5-Hexadiene	2235-12-3	C_6H_8	–	–
Heptane	142-82-5	C_7H_{16}	–	–
2-Methyl-2-hexene	2738-19-4	C_7H_{14}	–	–
1,3,6-Heptatriene	1002-27-3	C_7H_{10}	–	–
1,3,5-Heptatriene	2196-23-8	C_7H_{10}	–	–
2,2-Dimethylhexane	590-73-8	C_8H_{18}	–	–
2,2,4-Trimethylpentane	540-84-1	C_8H_{18}	–	–
2,4-Dimethyl-1-heptene	19549-87-2	C_9H_{18}	–	–
2,6-Dimethyl-2,6-octadiene	2792-39-4	$C_{10}H_{18}$	–	–
1-Undecene	821-95-4	$C_{11}H_{22}$	–	–
Dodecane	112-40-3	$C_{12}H_{26}$	–	–
2,4-Dimethyl-1-decene	55170-80-4	$C_{12}H_{24}$	–	–
Tetradecane	629-59-4	$C_{14}H_{30}$	–	–
2-Butyl-1-decene	51655-65-3	$C_{14}H_{28}$	–	–
3-Tetradecene	–	$C_{14}H_{28}$	–	–
5-Tetradecene	–	$C_{14}H_{28}$	–	–
Pentadecane	629-62-9	$C_{15}H_{32}$	–	–
Heptadecane	629-78-7	$C_{17}H_{36}$	–	–
2,6,10-Trimethylpentadecane	3892-00-0	$C_{18}H_{38}$	–	–
10-Methylnonadecane	56862-62-5	$C_{20}H_{42}$	–	–
Aromatics				
Benzene	71-43-2	C_6H_6	–	–
Toluene	108-88-3	C_7H_8	–	–
Ethenylbenzene	100-42-5	C_8H_8	–	–

(Table 2) contd.....

Name	CAS	Formula	LOD (µg/L)	LOQ (µg/L)
Ethylbenzene	100-41-4	C_8H_{10}	–	–
o-Xylene	95-47-6	C_8H_{10}	–	–
1-Propynylbenzene	673-32-5	C_9H_8	–	–
2-Propenylbenzene	300-57-2	C_9H_{10}	–	–
p-Cymene	99-87-6	$C_{10}H_{14}$	–	–
(1-Methylethyl)-naphthalene	29253-36-9	$C_{13}H_{14}$	–	–
(1-Pentylhexyl)-benzene	4537-14-8	$C_{17}H_{28}$	–	–
(1-Butylheptyl)-benzene	4537-15-9	$C_{17}H_{28}$	–	–
(1-Propyloctyl)-benzene	4536-86-1	$C_{17}H_{28}$	–	–
Cyclics				
1,3-Cyclopentadiene	542-92-7	C_5H_6	–	–
1,3-Cyclohexadiene	592-57-4	C_6H_8	–	–
Methyl-1,3-cyclopentadiene	96-39-9	C_6H_8		
1,3-Cycloheptadiene	4054-38-0	C_7H_{10}	–	–
1-Methyl-1,4-cyclohexadiene	4313-57-9	C_7H_{10}	–	–
3-Methylene-cyclohexene	1888-90-0	C_7H_{10}	–	–
1,2-Cyclooctadiene	7124-40-5	C_8H_{12}	–	–
Pentyl cyclopropane	2511-91-3	C_8H_{16}	–	–
1,3,5-Trimethylcyclohexane	1839-63-0	C_9H_{18}	–	–
Cyclodecane	293-96-9	$C_{10}H_{20}$	–	–
1,1-Dimethylethyl cyclohexane	3178-22-1	$C_{10}H_{20}$	–	–
1,5-Dimethyl-1,5-cyclooctadiene	3760-14-3	$C_{10}H_{16}$	–	–
1,2,3,4,5-Pentamethyl-1,3-cyclopentadiene	4045-44-7	$C_{10}H_{16}$	–	–
Cycloundecene	13151-60-5	$C_{11}H_{20}$	–	–
Octyl cyclopropane	1472-09-9	$C_{11}H_{22}$	–	–
5-Butyl-6-hexyloctahydro-1H-Indene	55044-36-5	$C_{19}H_{36}$	–	–
Ketones				
Alifatics				
Acetone	67-64-1	C_3H_6O	1.1 [62]	7 [62]
Hydroxyacetone	116-09-6	$C_3H_6O_2$	4.2, 220 [76]	14, 733 [76]
1,3-Dihydroxy-2-propanone	96-26-4	$C_3H_6O_3$	–	–
3-Buten-2-one	78-94-4	C_4H_6O	–	–
2,3-Butanedione	431-03-8	$C_4H_6O_2$	0.041 [81] 3 [92] 3.8 [94] 10, 393 [76] 25 [37]	0.124 [81] 10 [92] 12.6 [94] 34, 1310 [76] –
2-Butanone	78-93-3	C_4H_8O	0.024 [76]	0.08 [76]
Acetoin	513-86-0	$C_4H_8O_2$	14 [76]	46 [76]
4-Hydroxy-2-butanone	590-90-9	$C_4H_8O_2$	–	–

(Table 2) contd.....

Name	CAS	Formula	LOD (µg/L)	LOQ (µg/L)
2,3-Pentanedione	600-14-6	$C_5H_8O_2$	0.26, 30 [76] 4 [92] 4.6 [94]	0.86, 101 [76] 12 [92] 15.2 [94]
2-Pentanone	107-87-9	$C_5H_{10}O$	0.024, 2.4 [76]	0.079, 8.1 [76]
3-Pentanone	96-22-0	$C_5H_{10}O$	–	–
3-Methyl-2-butanone	563-80-4	$C_5H_{10}O$	–	–
1-Hydroxy-3-methyl-2- butanone	36960-22-2	$C_5H_{10}O_2$	–	–
3-Hydroxy-2-pentanone	3142-66-3	$C_5H_{10}O_2$	–	–
2-Hexanone	591-78-6	$C_6H_{12}O$	0.003, 1.4 [76] 0.009 [81]	0.011, 4.8 [76] 0.028 [81]
3-Hexanone	589-38-8	$C_6H_{12}O$	–	–
Methyl isobutyl ketone	108-10-1	$C_6H_{12}O$	0.04, 0.15 [76] 2.00 [90]	0.13, 0.49 [76] 6.67 [90]
2-Methyl-3-pentanone	565-69-5	$C_6H_{12}O$	–	–
3-Heptanone	106-35-4	$C_7H_{14}O$	–	–
2-Heptanone	110-43-0	$C_7H_{14}O$	–	–
4-Methyl-2-hexanone	105-42-0	$C_7H_{14}O$	–	–
2-Methyl-2-hepten-6-one	110-93-0	$C_8H_{14}O$	–	–
1-Octen-3-one	4312-99-6	$C_8H_{14}O$	–	–
2-Octanone	111-13-7	$C_8H_{16}O$	0.01, 0.018 [76] 0.021 [81]	0.011, 4.8 [76] 0.028 [81]
2-Methyl heptane-4-one	626-33-5	$C_8H_{16}O$	–	–
3-Octanone	106-68-3	$C_8H_{16}O$	–	–
6-Methyl-2-heptanone	928-68-7	$C_8H_{16}O$	–	–
3-Methyl-4-heptanone	15726-15-5	$C_8H_{16}O$	–	–
2-Nonanone	821-55-6	$C_9H_{18}O$	0.009 [81] 0.255, 0.046 [76]	0.026 [81] 0.849, 0.152 [76]
2-Decanone	693-54-9	$C_{10}H_{20}O$	0.025 [81] 0.05, 0.015 [76]	0.076 [81] 0.167, 0.05 [76]
2-Methyl-3-nonanone	5445-31-8	$C_{10}H_{20}O$	–	–
2-Undecanone	112-12-9	$C_{11}H_{22}O$	0.019 [81]	0.058 [81]
3-Tridecanone	1534-26-5	$C_{13}H_{26}O$	–	–
10-Nonadecanone	504-57-4	$C_{19}H_{38}O$	–	–
Aromatics				
1-Phenylethanone	98-86-2	C_8H_8O	–	–
2-Aminoacetophenone	613-89-8	C_8H_9NO	–	–
Benzylacetone	2550-26-7	$C_{10}H_{12}O$	–	–
3-Methyl-1-phenyl-1-butanone	582-62-7	$C_{11}H_{14}O$	–	–
Cyclics				
Cyclopentanone	120-92-3	C_5H_8O	–	–
2-Cyclopenten-1,4-dione	930-60-9	$C_5H_4O_2$	–	–

(Table 2) contd.....

Name	CAS	Formula	LOD (µg/L)	LOQ (µg/L)
2-Cyclohexenone	930-68-7	C_6H_8O	–	–
p-Benzoquinone	106-51-4	$C_6H_4O_2$	–	–
1-Cyclohexyl-2-methyl-2-propen-1-one	25183-82-8	$C_{10}H_{16}O$	–	–
Dihydrojasmone	1128-08-1	$C_{11}H_{18}O$	–	–
Lactones				
γ-Butyrolactone	96-48-0	$C_4H_6O_2$	–	–
5,5-Dimethyl-2(5H)-furanone	20019-64-1	$C_6H_8O_2$	–	–
3-Hydroxy-4,5-dimethyl-2(5H)-furanone	28664-35-9	$C_6H_8O_3$	–	–
γ-Hexalactone	695-06-7	$C_6H_{10}O_2$	–	–
δ-Octalactone	698-76-0	$C_8H_{14}O_2$	–	–
γ-Octalactone	104-50-7	$C_8H_{14}O_2$	–	–
γ-Nonalactone	104-61-0	$C_9H_{16}O_2$	0.19 [73] 4.86 [90]	0.65 [73] 14.86 [90]
γ-Decalactone	706-14-9	$C_{10}H_{18}O_2$	–	–
δ-Decalactone	705-86-2	$C_{10}H_{18}O_2$	–	–
Dihydroactinidiolide	17092-92-1	$C_{11}H_{16}O_2$	–	–
δ-Dodecalactone	713-95-1	$C_{12}H_{22}O_2$	–	–
Monoterpenic compounds				
Hydrocarbon-type				
Myrcene	123-35-3	$C_{10}H_{16}$	0.21, 0.11 [60]	–
Limonene	138-86-3	$C_{10}H_{16}$	0.02 [73] 0.07, 0.7 [38]	0.05 [73] –
Ocimene	13877-91-3	$C_{10}H_{16}$	–	–
Terpinene	99-86-5	$C_{10}H_{16}$	–	–
Camphene	79-92-5	$C_{10}H_{16}$	–	–
2-Carene	554-61-0	$C_{10}H_{16}$	–	–
p-Mentha-2,4(8)-diene	586-63-0	$C_{10}H_{16}$	–	–
(1S)-1,7,7-Trimethylnorbornan-2-one	464-17-5	$C_{10}H_{16}$	–	–
β-Pinene	127-91-3	$C_{10}H_{16}$	–	–
(-)-*trans*-Pinane	33626-25-4	$C_{10}H_{18}$	–	–
2,7-Dimethyl-1,6-octadiene	40195-09-3	$C_{10}H_{18}$	–	–
Oxygen-containing type				
Verbenone	1196-01-6	$C_{10}H_{14}O$	–	–
p-Cymen-8-ol	1197-01-9	$C_{10}H_{14}O$	–	–
3-Nopinenone / pinocarvone	30460-92-5	$C_{10}H_{14}O$	–	–
Citral	5392-40-5	$C_{10}H_{16}O$	–	–
B-Thujone	471-15-8	$C_{10}H_{16}O$	–	–
Camphor	76-22-2	$C_{10}H_{16}O$	–	–
Linalool	78-70-6	$C_{10}H_{18}O$	0.003 [73] 0.01 [86] 0.06, 0.9 [38]	0.01 [73] – –

(Table 2) contd.....

Name	CAS	Formula	LOD (µg/L)	LOQ (µg/L)
α-Terpineol	98-55-5	$C_{10}H_{18}O$	0.06 [73]	0.18 [73]
Nerol	106-25-2	$C_{10}H_{18}O$	0.02 [73]	0.07 [73]
Geraniol	106-24-1	$C_{10}H_{18}O$	5.8, 0.8 [38]	–
4-Terpineol	20126-76-5	$C_{10}H_{18}O$	–	–
trans-Shisool	22451-48-5	$C_{10}H_{18}O$	–	–
Fenchyl alcohol	1632-73-1	$C_{10}H_{18}O$	–	–
Myrcenol	543-39-5	$C_{10}H_{18}O$	–	–
β-Terpineol	138-87-4	$C_{10}H_{18}O$	–	–
1-Borneol	507-70-0	$C_{10}H_{18}O$	–	–
p-Menth-8(10)-en-9-ol	5502-99-8	$C_{10}H_{18}O$	–	–
8-Hydroxylinalool	64142-78-5	$C_{10}H_{18}O_2$	–	–
Linalool oxide	60047-17-8	$C_{10}H_{18}O_2$	–	–
Rose oxide	16409-43-1	$C_{10}H_{18}O_2$	–	–
7-Hydroxy-α-terpineol	–	$C_{10}H_{18}O_2$	–	–
Citronellol	106-22-9	$C_{10}H_{20}O$	0.01 [73]	0.05 [73]
Linalool tetrahydride	78-69-3	$C_{10}H_{22}O$	–	–
Menthol	1490-04-6	$C_{10}H_{20}O$	–	–
Methyl *(E)*-geranate	2349-14-6	$C_{11}H_{18}O_2$	–	–
Methyl nerolate	1862-61-9	$C_{11}H_{18}O_2$	–	–
Isobornyl acetate	125-12-2	$C_{12}H_{20}O_2$	–	–
Geranyl acetate	105-87-3	$C_{12}H_{20}O_2$	–	–
Ethyl geranate	32659-21-5	$C_{12}H_{20}O_2$	–	–
Neryl acetate	141-12-8	$C_{12}H_{20}O_2$	–	–
8-Acetoxylinalool	–	$C_{12}H_{20}O_3$	–	–
Citronellol acetate	150-84-5	$C_{12}H_{22}O_2$	–	–
Citronellyl propionate	141-14-0	$C_{13}H_{24}O_2$	–	–
Neryl butyrate	999-40-6	$C_{14}H_{24}O_2$	–	–
Norisoprenoids				
β-Damascenone	23726-93-4	$C_{13}H_{18}O$	0.01 [73] 7.54 [90]	0.03 [73] 25.15 [90]
3-Hydroxy-α-damascone	–	$C_{13}H_{19}O_2$	–	–
α-Ionone	127-41-3	$C_{13}H_{20}O$	–	–
β-Ionone	14901-07-6	$C_{13}H_{20}O$	0.01 [73]	0.03 [73]
β-Ionone epoxide	23267-57-4	$C_{13}H_{20}O_2$	–	–
Geranyl acetone	689-67-8	$C_{13}H_{22}O$	–	–
Phenols				
Phenol	108-95-2	C_6H_6O	–	–
Salicylaldehyde	90-02-8	$C_7H_6O_2$	0.38 [77]	1.27 [77]
p-Cresol (4-methylphenol)	106-44-5	C_7H_8O	–	–
m-Cresol (3-methylphenol)	108-39-4	C_7H_8O	–	–

(Table 2) contd.....

Name	CAS	Formula	LOD (µg/L)	LOQ (µg/L)
2-Methoxyphenol	90-05-1	$C_7H_8O_2$	0.78 [77]	2.61 [77]
4-Vinylphenol	2628-17-3	C_8H_8O	0.01 [75] 0.50 [59] 0.76 [73]	0.03 [75] 1.24 [59] 2.53 [73]
Vanillin	121-33-5	$C_8H_8O_3$	6.59 [77]	21.96 [77]
4-Ethylphenol	123-07-9	$C_8H_{10}O$	0.02 [75] 0.08 [73] 0.10 [59] 0.34 [77]	0.06 [75] 0.26 [73] 0.25 [59] 1.12 [77]
2-Methoxy-4-vinylphenol	7786-61-0	$C_9H_{10}O_2$	0.01 [75] 1.39 [59] 1.98 [77] 5.84 [73]	0.03 [75] 2.77 [59] 6.59 [77] 19.45 [73]
Methyl vanillate	3943-74-6	$C_9H_{10}O_4$	3.10 [77]	10.34 [77]
4-Ethylguaiacol	2785-89-9	$C_9H_{12}O_2$	0.01 [75] 0.03 [73] 0.24 [59] 0.51 [77]	0.02 [75] 0.1 [73] 0.47 [59] 1.69 [77]
Eugenol	97-53-0	$C_{10}H_{12}O_2$	–	–
Thymol	89-83-8	$C_{10}H_{14}O$	0.17 [77]	0.56 [77]
Pyrazines				
2-Methylpyrazine	109-08-0	$C_5H_6N_2$	–	–
2,6-Dimethylpyrazine	108-50-9	$C_6H_8N_2$	–	–
2,3-Dimethylpyrazine	5910-89-4	$C_6H_8N_2$	–	–
2-Ethyl-3-methylpyrazine	15707-23-0	$C_7H_{10}N_2$	–	–
Sesquiterpenic compounds				
Hydrocarbon-type				
Cadalene	483-78-3	$C_{15}H_{18}$	–	–
α-Calacorene	21391-99-1	$C_{15}H_{20}$	–	–
3,8-Triene Cadala-1(10)	–	$C_{15}H_{22}$	–	–
trans-Calamenene	73209-42-4	$C_{15}H_{22}$	–	–
Calamenene	483-77-2	$C_{15}H_{22}$	–	–
Humulene	6753-98-6	$C_{15}H_{24}$	–	–
(+)-Valencene	4630-07-3	$C_{15}H_{24}$	–	–
(-)-α-Neoclovene	4545-68-0	$C_{15}H_{24}$	–	–
Caryophyllene	87-44-5	$C_{15}H_{24}$	–	–
β-Cadinene	523-47-7	$C_{15}H_{24}$	–	–
α-Farnesene	502-61-4	$C_{15}H_{24}$	–	–
2-Isopropyl-5-methyl-9-methylene-byciclo-1-decene (4.4.0)	150320-52-8	$C_{15}H_{24}$	–	–
τ-Cadinene	–	$C_{15}H_{24}$	–	–
γ-Muurolene	30021-74-0	$C_{15}H_{24}$	–	–
α-Amorphene	483-75-0	$C_{15}H_{24}$	–	–
β-Selinene	17066-67-0	$C_{15}H_{24}$	–	–

(Table 2) contd.....

Name	CAS	Formula	LOD (µg/L)	LOQ (µg/L)
α-Selinene	473-13-2	$C_{15}H_{24}$	–	–
Epizonarene	41702-63-0	$C_{15}H_{24}$	–	–
γ-Cadinene	39029-41-9	$C_{15}H_{24}$	–	–
delta-Cadinene	483-76-1	$C_{15}H_{24}$	–	–
β-Copaene	18252-44-3	$C_{15}H_{24}$	–	–
γ-Amorphene	6980-46-7	$C_{15}H_{24}$	–	–
(*trans-*)Cadina-1,4-diene	38758-02-0	$C_{15}H_{24}$	–	–
Farnesene	–	$C_{15}H_{24}$	–	–
Germacrene	–	$C_{15}H_{24}$	–	–
Oxygen-containing type				
Caryophyllene oxide	1139-30-6	$C_{15}H_{24}O$	–	–
trans-Z-α-Bisabolene epoxide	–	$C_{15}H_{24}O$	–	–
cis-α-Bisabolene epoxide	–	$C_{15}H_{24}O$	–	–
Humulene epoxide I	–	$C_{15}H_{24}O$	–	–
Humulene epoxide II	19888-34-7	$C_{15}H_{24}O$	–	–
Humulene epoxide III	21624-36-2	$C_{15}H_{24}O$	–	–
Humulenol II	–	$C_{15}H_{24}O$	–	–
Caryophylladienol	104010-36-8	$C_{15}H_{24}O$	–	–
β-Eudesmol	473-15-4	$C_{15}H_{26}O$	–	–
Nerolidol	7212-44-4	$C_{15}H_{26}O$	–	–
Germacren-4-ol	72120-50-4	$C_{15}H_{26}O$	–	–
Globulol	51371-47-2	$C_{15}H_{26}O$	–	–
Caryophyllenyl alcohol	–	$C_{15}H_{26}O$	–	–
α-Bisabolol	515-69-5	$C_{15}H_{26}O$	–	–
α-Cadinol	481-34-5	$C_{15}H_{26}O$	–	–
Farnesol	4602-84-0	$C_{15}H_{26}O$	–	–
τ-Cadinol	5937-11-1	$C_{15}H_{26}O$	–	–
Caryolan-1-ol	472-97-9	$C_{15}H_{26}O$	–	–
Humulol	28446-26-6	$C_{15}H_{26}O$	–	–
3,7,11-Trimethyl-6,10- dodecadien-1-ol	27745-36-4	$C_{15}H_{28}O$	–	–
Sulfur compounds				
Hydrogen sulfide	7783-06-4	H_2S	0.8 [36]	–
Carbon disulfide	75-15-0	CS_2	0.010-0.060 [88]	–
Methanethiol	74-93-1	CH_4S	0.1 [36]	–
Ethylene sulfide	420-12-2	C_2H_4S	–	–
Ethanethiol	75-08-1	C_2H_6S	0.2 [36]	–
Dimethyl sulphide	75-18-3	C_2H_6S	0.0019, 0.0305, 0.2081 [49] 0.09 [27] 0.2 [64] 2 [36]	– 0.23 [27] – –

(Table 2) contd.....

Name	CAS	Formula	LOD (µg/L)	LOQ (µg/L)
Dimethyl disulfide	624-92-0	$C_2H_6S_2$	0.0018, 0.0434, 0.0103 [49] 0.010-0.060 [88], 0.010 [64]	– – –
Dimethyl trisulfide	3658-80-8	$C_2H_6S_3$	0.019 [90]	0.063 [90]
Dimethyl sulfoxide	67-68-5	C_2H_6OS	–	
Thiazole	288-47-1	C_3H_3NS		
Methyl thioacetate	1534-08-3	C_3H_6OS	–	–
1-Propanethiol	107-03-9	C_3H_8S	–	–
Ethyl thioacetate	625-60-5	C_4H_8OS	0.010-0.060 [88]	–
3-(Methylthio)propanal	3268-49-3	C_4H_8OS	0.046 [81] 0.087, 0.24 [76] 0.259 [90] 0.30 [72]	0.139 [81] 0.29, 0.8 [76]; 0.862 [90]; 1.00 [72]
Diethyl sulfide	352-93-2	$C_4H_{10}S$	0.0015, 0.0252, 0.0088 [49]	–
Methyl propyl sulfide	3877-15-4	$C_4H_{10}S$	0.025 [64]	
3-Methyl thiopropanol	505-10-2	$C_4H_{10}OS$	–	–
1-(2-Thiazoly)-ethanone	24295-03-2	C_5H_5NOS	–	–
3-Methylthiophene	616-44-4	C_5H_6S	0.001-0.005 [88]	
3-Methyl-2-butene-1-thiol	5287-45-6	$C_5H_{10}S$	0.010-0.060 [88]	–
Dipropyl disulfide	629-19-6	$C_6H_{14}S_2$	0.0002, 0.0107, 0.0005 [49]	–
Dipropyl trisulfide	6028-61-1	$C_6H_{14}S_3$	0.0015, 0.0571, 0.0029 [49]	–
Ethyl thiooctanoate	–	$C_{10}H_{20}OS$	–	–
4-n-Hexylthiane,S,S-dioxide	70928-52-8	$C_{11}H_{22}O_2S$	–	–
Others				
2,3-Dihydro-3,5-dihydroxy-6-methyl-4(H)-pyran-4-one (DDMP)	28564-83-2	$C_6H_8O_4$	–	–
Tetrahydro-4-hydroxy-4-methyl-2(H)-pyran-2-one	674-26-0	$C_6H_{10}O_3$	–	–
γ-Butyrolactam	616-45-5	C_4H_7NO	–	–
Pyrrolidine	123-75-1	C_4H_9N	–	–
Pyrrole-2-carboxaldehyde	1003-29-8	C_5H_5NO	–	–
4-Hydroxypyridine	626-64-2	C_5H_5NO	–	–
2-Acetylpyrrole	1072-83-9	C_6H_7NO	5.5, 1178 [76]	18.2, 3927 [76]
2-Acetyl-1-pyrroline	85213-22-5	C_6H_9NO	–	–
1-Methyl-2-nitrobenzene	88-72-2	$C_7H_7NO_2$	–	–
Benzyl nitrile	140-29-4	C_8H_7N	–	–
Ethyl-3-pyridinecarboxylate	614-18-6	$C_8H_9NO_2$	–	–
2-Ethylacridine	55751-83-2	$C_{15}H_{13}N$	–	–

Alcohols (*e.g.* phenethyl alcohol, 3-methyl 1-butanol, and isobutanol) and esters (phenethyl acetate, ethyl caprylate, ethyl caproate, isoamyl acetate, ethyl butyrate), as the beer major components, have been widely studied and quantified due to their potential impact on aroma and flavor properties [11, 135]. Moreover, yeasts produce acidic compounds (being octanoic acid, the metabolite more reported), which are intermediate metabolites of phospholipid biosynthesis and important substrates for ethyl esters formation. These metabolites may have negative impact on beer, regarding both taste and foam stability [11, 50].

The detection of off-flavors has been also widely studied, namely the determination of volatile phenols and sulfur compounds. Volatile phenols may lead to aroma defects (*i.e.* phenolic, clove-like, horsy, or smoky), depending on their concentration. These compounds may be formed through the decarboxylation of phenolic compounds during wort boiling or by enzymatic reactions from yeasts [136 - 138]. Sulfur compounds content has also been widely studied, once they have low flavor thresholds and can contribute with unpleasant notes (*e.g.* rotten eggs, cooked cabbage), if they are present at concentrations above their sensorial perception limit (*e.g.* 0.004 – 0.005 mg/L for hydrogen sulfide, 0.050 mg/L for dimethyl sulfide, and 1.2 mg/L for 3-methylthio-1-propanol). Different sources can contribute to their formation, namely from raw materials (malts), from brewing process (during malt kilning, wort boiling or yeast fermentation), and from possible contaminations of spoilage microorganisms [27, 36, 49, 64, 88].

Raw materials, namely hops, can contribute to the presence of hydrocarbons, monoterpenic and sesquiterpenic compounds, and some norisoprenoids [139]. Furthermore, yeasts can biotransformate hop aroma terpenoids during fermentation [140 - 142]. Few data have been reported for terpenic compounds' quantification, however linalool is vastly reported on beer volatile composition, it being considered an analytical marker for beer hoppy aroma [69].

Beer aging is a complex phenomenon, which involves different types of reactions. Several researches have been dedicated to this phenomenon, due to its impact on beer volatile composition. Carbonyl compounds (*i.e.* aldehydes, ketones, and furan-derivatives) have been the analytes more studied related with beer aging, once there are reported changes in their content along this process, which may impact on flavor stability (due to their low flavor thresholds) [14 - 17, 72, 76, 90]. They can be originated from raw materials, alcoholic fermentation or from chemical families, such as Maillard reactions, lipid oxidation, Strecker degradation, etc. A particular attention has been given to *(E)*-2-nonenal, once it may be considered an off-flavor (contributes with cardboard notes, odor threshold of 0.15 mg/L), and may be perceived in aged beers [55, 78]. Moreover, the presence of Maillard reactions products (*e.g.* furan-derivatives, pyrans, lactones, pyrazines) may derive from the degradation of analytes along storage of malted cereals or along beer aging [13] (furfural being the main studied analyte).

CONCLUDING REMARKS AND FUTURE TRENDS

Beer volatile determination may combine three main steps: *i)* decarbonation, *ii)* extraction, and *iii)* chromatographic and/or mass spectrometry analysis. The beer volatiles' extraction efficiency and reproducibility may be compromised if a not proper decarbonation is applied (CO_2 promotes the formation of foam). Therefore, ultrasonic bath treatment and filtration have been widely applied to promote beer decarbonation, prior to volatiles' extraction. Nevertheless, for some targeted analysis, namely for less volatile components, this procedure may not represent a crucial step.

After beer decarbonation, a suitable sample preparation technique should be applied in order to achieve a reliable characterization of the volatile and semi-volatile analytes. Headspace extraction has gained particular interest, more specifically, the microextraction techniques. Actually, SPME is the most studied and applied technique to promote the extraction of beer volatile compounds (reported in about 45% of the articles cited in this book chapter), mainly because it combines sampling, extraction and concentration in one step, is also a solvent-free technique, with high selectivity and sensitivity, and easily integrated in an analytical platform. There are under development new fiber coatings for SPME, which can be more selective, sensitive and robust, which opens new future applications on beer matrix, depending on the desired strategy: target or non-target analytes, only in one analysis. Also the reduction of the SPME fibers size is under development, which can allow direct sample introduction into portable mass spectrometers. This can potentially be an asset for brewing industries, allowing a real-time, on-line monitoring of the process. Moreover, the use of auto-samplers for this technique can enhance reliability and reproducibility.

Different GC systems have been used in the analysis of beer volatile composition, being the samples' complexity and the analytes' concentration the main factors that should be considered on the choice of the most suitable equipment. In fact, 1D-GC is the equipment highly used once it is an inexpensive option, with low costs of consumables. However, for complex samples, as beer, using 1D-GC equipments co-elution problems can occur, which may compromise reliable identification and quantification, especially for trace analytes. Moreover, part of the volatile fraction of beer is characterized by the secondary metabolites from raw materials (namely those of vegetable origin, such as hops or cereals). Therefore, it is possible to have analytes with similar moieties once they are formed through the same biosynthetic pathways [128]. Thus, GC×GC is required for a reliable in-depth characterization and profiling of the volatile composition of beer, with combination of an adequate sample preparation technique. Several advantages can be covered by GC×GC, when comparing with conventional 1D-GC, especially the presence of the structured 2D chromatographic space which is a useful tool in the analytes' identification. This feature is used as a complement with the analysis of mass spectra fragmentation pattern and determination of retention indices values, for the achievement of reliable analytes' identification (particularly if standards are not available). GC×GC-ToFMS showed to be valuable for comprehensive characterization of the volatile composition of beer, whose fingerprinting may be dependent on the used raw materials, on the transformations that are promoted along brewing process, and also during beer storage. Indeed, the detection of hundreds of analytes may help in the characterization of key odorants and elucidate their formation; they can be explored in order to differentiate samples (*e.g.* comparison of different industrial treatments, raw materials' geographical origin, among others), also can be used to monitor food authenticity, quality or safety. The analysis of complex samples by GC×GC-ToFMS generates unequalled information regarding their volatile composition. However, there is a challenging problem concerning the extraction of relevant and useful information, once the obtained high-dimensional data is extremely complex [128]. Indeed, the processing algorithms that are available for data treatment are not efficient, thus there is still need for developments on more effective and mainstream GC×GC-ToFMS data processing methods. Also, the development of more efficient pre-processing of data will allow the diminution of some uncertainty on analytes' peak detection and respective area. Nevertheless, there have been significant improvements on the GC systems, with increased resolution, sensitivity, and also new stationary phases of GC columns. Finally, it is important to point out that in the near future, it is expected that developments will be on fastest and miniaturized solutions that allow real-time monitoring of the brewing process and that can be routinely used in the labs.

SUPPORTIVE/SUPPLEMENTARY MATERIAL

Table **S1**. General methodology used in each study of beer volatile profiling (extraction and detection/analysis technique), and the specificities of each sample preparation procedure.

CONFLICT OF INTEREST

The authors declare no conflict of interest, financial or otherwise.

ACKNOWLEDGEMENTS

The authors would like to thank FCT/MEC for the financial support to the QOPNA research Unit (FCT UID/QUI/00062/2013) and CESAM (project PEst-C/MAR/LA0017/2013), through national funds and where applicable co-financed by the FEDER, within the PT2020 Partnership Agreement. C. Martins thanks FCT/MEC for the PhD grant (SFRH/BD/77988/2011) through the program POPH/FSE.

FCT
Fundação para a Ciência e a Tecnologia
MINISTÉRIO DA CIÊNCIA, TECNOLOGIA E ENSINO SUPERIOR

COMPETE

QREN
QUADRO DE REFERÊNCIA ESTRATÉGICO NACIONAL

UNIÃO EUROPEIA
Fundo Europeu de Desenvolvimento Regional

PoPH
PROGRAMA OPERACIONAL **POTENCIAL HUMANO**

UNIÃO EUROPEIA
Fundo Social Europeu

REFERENCES

[1]   *The Barth Report 2013-2014,* Available at: http://www.barthhaasgroup.com/images/pdfs/reports/2014/BarthReport_2013-2014.pdf

[2]   Walle, M. Van de, Ed. *Beer Statistics: 2015 Edition*; Brussels, **2015**.

[3]   *Kirin Beer University Report,* **2015**.

[4]   Berkhout, B.; Bertling, L.; Bleeker, Y.; de Wit, W.; Kruis, G.; Stokkel, R.; Theuws, R. *The Contribution Made by Beer to the European Economy*; Amsterdam, **2013**.

[5]   Pinho, O.; Ferreira, I.M.; Santos, L.H. Method optimization by solid-phase microextraction in combination with gas chromatography with mass spectrometry for analysis of beer volatile fraction. *J. Chromatogr. A,* **2006**, *1121*(2), 145-153.
[http://dx.doi.org/10.1016/j.chroma.2006.04.013] [PMID: 16687150]

[6]   Kobayashi, M.; Shimizu, H.; Shioya, S. Beer volatile compounds and their application to low-malt beer fermentation. *J. Biosci. Bioeng.,* **2008**, *106*(4), 317-323.
[http://dx.doi.org/10.1263/jbb.106.317] [PMID: 19000606]

[7]   Tsuji, H.; Mizuno, A. Volatile compounds and the changes in their concentration levels during storage in beers containing varying malt concentrations. *J. Food Sci.,* **2010**, *75*(1), C79-C84.
[http://dx.doi.org/10.1111/j.1750-3841.2009.01428.x] [PMID: 20492154]

[8]   Gonçalves, J.L.; Figueira, J.A.; Rodrigues, F.P.; Ornelas, L.P.; Branco, R.N.; Silva, C.L.; Câmara, J.S. A powerful methodological approach combining headspace solid phase microextraction, mass spectrometry and multivariate analysis for profiling the volatile metabolomic pattern of beer starting raw materials. *Food Chem.,* **2014**, *160*, 266-280.
[http://dx.doi.org/10.1016/j.foodchem.2014.03.065] [PMID: 24799238]

[9]   Brányik, T.; Vicente, A.A.; Dostálek, P.; Teixeira, J.A. A Review of Flavour Formation in Continuous Beer Fermentations. *J. Inst. Brew.,* **2008**, *114*(1), 3-13.
[http://dx.doi.org/10.1002/j.2050-0416.2008.tb00299.x]

[10]   Brown, A.K.; Hammond, J.R. Flavour Control in Small-Scale Beer Fermentations. *Food Bioprod. Process.,* **2003**, *81*(1), 40-49.
[http://dx.doi.org/10.1205/096030803765208652]

[11]   Lodolo, E.J.; Kock, J.L.; Axcell, B.C.; Brooks, M. The yeast *Saccharomyces cerevisiae-* the main character in beer brewing. *FEMS Yeast Res.,* **2008**, *8*(7), 1018-1036.
[http://dx.doi.org/10.1111/j.1567-1364.2008.00433.x] [PMID: 18795959]

[12]　Pires, E.J.; Teixeira, J.A.; Brányik, T.; Vicente, A.A. Yeast: the soul of beer's aroma--a review of flavour-active esters and higher alcohols produced by the brewing yeast. *Appl. Microbiol. Biotechnol.,* **2014**, *98*(5), 1937-1949.
[http://dx.doi.org/10.1007/s00253-013-5470-0] [PMID: 24384752]

[13]　Eβlinger, H.M., Ed. *Handbook of Brewing,* **2009**,

[14]　Vanderhaegen, B.; Neven, H.; Coghe, S.; Verstrepen, K.J.; Verachtert, H.; Derdelinckx, G. Evolution of chemical and sensory properties during aging of top-fermented beer. *J. Agric. Food Chem.,* **2003**, *51*(23), 6782-6790.
[http://dx.doi.org/10.1021/jf034631z] [PMID: 14582975]

[15]　Vanderhaegen, B.; Neven, H.; Verachtert, H.; Derdelinckx, G. The Chemistry of Beer Aging – a Critical Review. *Food Chem.,* **2006**, *95*(3), 357-381.
[http://dx.doi.org/10.1016/j.foodchem.2005.01.006]

[16]　Vanderhaegen, B.; Delvaux, F.; Daenen, L.; Verachtert, H.; Delvaux, F.R. Aging Characteristics of Different Beer Types. *Food Chem.,* **2007**, *103*(2), 404-412.
[http://dx.doi.org/10.1016/j.foodchem.2006.07.062]

[17]　Malfliet, S.; Van Opstaele, F.; De Clippeleer, J.; Syryn, E.; Goiris, K.; De Cooman, L.; Aerts, G. Flavour Instability of Pale Lager Beers: Determination of Analytical Markers in Relation to Sensory Ageing. *J. Inst. Brew.,* **2008**, *114*(2), 180-192.
[http://dx.doi.org/10.1002/j.2050-0416.2008.tb00324.x]

[18]　Sohrabvandi, S.; Mortazavian, A.M.; Rezaei, K. Advanced Analytical Methods for the Analysis of Chemical and Microbiological Properties of Beer. *J. Food Drug Anal.,* **2011**, *19*(2), 202-222.

[19]　Vera, L.; Aceña, L.; Guasch, J.; Boqué, R.; Mestres, M.; Busto, O. Characterization and classification of the aroma of beer samples by means of an MS e-nose and chemometric tools. *Anal. Bioanal. Chem.,* **2011**, *399*(6), 2073-2081.
[http://dx.doi.org/10.1007/s00216-010-4343-y] [PMID: 21061001]

[20]　Martí, M.P.; Boqué, R.; Busto, O.; Guasch, J. Electronic Noses in the Quality Control of Alcoholic Beverages. *Trends Analyt. Chem.,* **2005**, *24*(1), 57-66.
[http://dx.doi.org/10.1016/j.trac.2004.09.006]

[21]　Fritsch, H.T.; Schieberle, P. Identification based on quantitative measurements and aroma recombination of the character impact odorants in a Bavarian Pilsner-type beer. *J. Agric. Food Chem.,* **2005**, *53*(19), 7544-7551.
[http://dx.doi.org/10.1021/jf051167k] [PMID: 16159184]

[22]　Huimin, L.; Hongjun, L.; Xiuhua, L.; Bing, C. Analysis of Volatile Flavor Compounds in Top Fermented Wheat Beer by Headspace Sampling-Gas Chromatography. *Int. J. Agric. Biol. Eng.,* **2012**, *5*(2), 67-75.

[23]　Ghasemi-Varnamkhasti, M.; Mohtasebi, S.S.; Rodriguez-Mendez, M.L.; Lozano, J.; Razavi, S.H.; Ahmadi, H. Potential Application of Electronic Nose Technology in Brewery. *Trends Food Sci. Technol.,* **2011**, *22*(4), 165-174.
[http://dx.doi.org/10.1016/j.tifs.2010.12.005]

[24]　Zhu, L.; Hu, Z.; Gamez, G.; Law, W.S.; Chen, H.; Yang, S.; Chingin, K.; Balabin, R.M.; Wang, R.; Zhang, T.; Zenobi, R. Simultaneous sampling of volatile and non-volatile analytes in beer for fast fingerprinting by extractive electrospray ionization mass spectrometry. *Anal. Bioanal. Chem.,* **2010**, *398*(1), 405-413.
[http://dx.doi.org/10.1007/s00216-010-3945-8] [PMID: 20644917]

[25]　Cajka, T.; Riddellova, K.; Tomaniova, M.; Hajslova, J. Recognition of beer brand based on multivariate analysis of volatile fingerprint. *J. Chromatogr. A,* **2010**, *1217*(25), 4195-4203.
[http://dx.doi.org/10.1016/j.chroma.2009.12.049] [PMID: 20074737]

[26]　Ashraf, N.; Linforth, R.S.; Bealin-Kelly, F.; Smart, K.; Taylor, A.J. Rapid Analysis of Selected Beer

Volatiles by Atmospheric Pressure Chemical Ionisation-Mass Spectrometry. *Int. J. Mass Spectrom.,* **2010**, *294*(1), 47-53.
[http://dx.doi.org/10.1016/j.ijms.2010.05.007]

[27] Stafisso, A.; Marconi, O.; Perretti, G.; Fantozzi, P. Determination of Dimethyl Sulfide in Brewery Samples by Headspace Gas Chromatography Mass Spectrometry (HS-GC/MS). *Ital. J. Food Sci.,* **2011**, *23*, 19-28.

[28] Kobayashi, F.; Odake, S. Quality Evaluation of Unfiltered Beer as Affected by Inactivated Yeast Using Two-Stage System of Low Pressure Carbon Dioxide Microbubbles. *Food Bioprocess Technol.,* **2015**, *8*(8), 1690-1698.
[http://dx.doi.org/10.1007/s11947-015-1530-z]

[29] Saison, D.; De Schutter, D.P.; Vanbeneden, N.; Daenen, L.; Delvaux, F.; Delvaux, F.R. Decrease of aged beer aroma by the reducing activity of brewing yeast. *J. Agric. Food Chem.,* **2010**, *58*(5), 3107-3115.
[http://dx.doi.org/10.1021/jf9037387] [PMID: 20143776]

[30] da Silva, G.C.; da Silva, A.A.; da Silva, L.S.; Godoy, R.L.; Nogueira, L.C.; Quitério, S.L.; Raices, R.S. Method development by GC-ECD and HS-SPME-GC-MS for beer volatile analysis. *Food Chem.,* **2015**, *167*, 71-77.
[http://dx.doi.org/10.1016/j.foodchem.2014.06.033] [PMID: 25148961]

[31] Tian, J. Determination of Several Flavours in Beer with Headspace Sampling–gas Chromatography. *Food Chem.,* **2010**, *123*(4), 1318-1321.
[http://dx.doi.org/10.1016/j.foodchem.2010.06.013]

[32] Tian, J. Application of Static Headspace Gas Chromatography for Determination of Acetaldehyde in Beer. *J. Food Compos. Anal.,* **2010**, *23*(5), 475-479.
[http://dx.doi.org/10.1016/j.jfca.2010.02.002]

[33] Tian, J.; Yu, J.; Chen, X.; Zhang, W. Determination and Quantitative Analysis of Acetoin in Beer with Headspace Sampling-Gas Chromatography. *Food Chem.,* **2009**, *112*(4), 1079-1083.
[http://dx.doi.org/10.1016/j.foodchem.2008.06.044]

[34] Hiralal, L.; Olaniran, A.O.; Pillay, B. Aroma-active ester profile of ale beer produced under different fermentation and nutritional conditions. *J. Biosci. Bioeng.,* **2014**, *117*(1), 57-64.
[http://dx.doi.org/10.1016/j.jbiosc.2013.06.002] [PMID: 23845914]

[35] Ragazzo-Sanchez, J.A.; Chalier, P.; Chevalier-Lucia, D.; Calderon-Santoyo, M.; Ghommidh, C. Off-Flavours Detection in Alcoholic Beverages by Electronic Nose Coupled to GC. *Sens. Actuators B Chem.,* **2009**, *140*(1), 29-34.
[http://dx.doi.org/10.1016/j.snb.2009.02.061]

[36] Li, H.; Jia, S.; Zhang, W. Rapid Determination of Low-Level Sulfur Compounds in Beer by Headspace Gas Chromatography with a Pulsed Flame Photometric Detector. *J. Am. Soc. Brew. Chem.,* **2008**, *66*(3), 188-191.

[37] Zapata, J.; Mateo-Vivaracho, L.; Lopez, R.; Ferreira, V. Automated and quantitative headspace in-tube extraction for the accurate determination of highly volatile compounds from wines and beers. *J. Chromatogr. A,* **2012**, *1230*, 1-7.
[http://dx.doi.org/10.1016/j.chroma.2012.01.037] [PMID: 22340891]

[38] Laaks, J.; Jochmann, M.A.; Schilling, B.; Molt, K.; Schmidt, T.C. In-Tube Extraction-GC-MS as a High-Capacity Enrichment Technique for the Analysis of Alcoholic Beverages. *J. Agric. Food Chem.,* **2014**, *62*, 3081-3091.
[http://dx.doi.org/10.1021/jf405832u] [PMID: 24579867]

[39] Horák, T.; Culík, J.; Čejka, P.; Jurková, M.; Kellner, V.; Dvorák, J.; Hasková, D. Analysis of free fatty acids in beer: comparison of solid-phase extraction, solid-phase microextraction, and stir bar sorptive extraction. *J. Agric. Food Chem.,* **2009**, *57*(23), 11081-11085.
[http://dx.doi.org/10.1021/jf9028305] [PMID: 19904941]

[40]   Harayama, K.; Hayase, F.; Kato, H. New Method for Analyzing the Volatiles in Beer. *Biosci. Biotechnol. Biochem. J.,* **1994**, *58*(12), 2246-2247.
[http://dx.doi.org/10.1271/bbb.58.2246]

[41]   Wei, A.; Mura, K.; Shibamoto, T. Antioxidative activity of volatile chemicals extracted from beer. *J. Agric. Food Chem.,* **2001**, *49*(8), 4097-4101.
[http://dx.doi.org/10.1021/jf010325e] [PMID: 11513716]

[42]   Langos, D.; Granvogl, M.; Schieberle, P. Characterization of the key aroma compounds in two bavarian wheat beers by means of the sensomics approach. *J. Agric. Food Chem.,* **2013**, *61*(47), 11303-11311.
[http://dx.doi.org/10.1021/jf403912j] [PMID: 24219571]

[43]   Schieberle, P. Primary Odorants of Pale Lager Beer. *Zeitschrift für Leb. Und-Forshcung,* **1991**, *193*, 558-565.

[44]   Thompson-Witrick, K.A.; Rouseff, R.L.; Cadawallader, K.R.; Duncan, S.E.; Eigel, W.N.; Tanko, J.M.; O'Keefe, S.F. Comparison of two extraction techniques, solid-phase microextraction *versus* continuous liquid-liquid extraction/solvent-assisted flavor evaporation, for the analysis of flavor compounds in gueuze lambic beer. *J. Food Sci.,* **2015**, *80*(3), C571-C576.
[http://dx.doi.org/10.1111/1750-3841.12795] [PMID: 25675965]

[45]   Horák, T.; Kellner, V.; Jurková, M.; Pavel, Č. Using Faster Gas Chromatography Analyses in Brewing Analytics. *J. Inst. Brew.,* **2009**, *115*(3), 214-219.
[http://dx.doi.org/10.1002/j.2050-0416.2009.tb00371.x]

[46]   Šmogrovičová, D.; Dömény, Z. Beer Volatile by-Product Formation at Different Fermentation Temperature Using Immobilised Yeasts. *Process Biochem.,* **1999**, *34*(8), 785-794.
[http://dx.doi.org/10.1016/S0032-9592(98)00154-X]

[47]   Inui, T.; Tsuchiya, F.; Ishimaru, M.; Oka, K.; Komura, H. Different beers with different hops. Relevant compounds for their aroma characteristics. *J. Agric. Food Chem.,* **2013**, *61*(20), 4758-4764.
[http://dx.doi.org/10.1021/jf3053737] [PMID: 23627300]

[48]   Lyu, J.; Nam, P-W.; Lee, S-J.; Lee, K-G. Volatile Compounds Isolated from Rice Beers Brewed with Three Medicinal Plants. *J. Inst. Brew.,* **2013**, *119*(4), 271-279.
[http://dx.doi.org/10.1002/jib.98]

[49]   Xiao, Q.; Yu, C.; Xing, J.; Hu, B. Comparison of headspace and direct single-drop microextraction and headspace solid-phase microextraction for the measurement of volatile sulfur compounds in beer and beverage by gas chromatography with flame photometric detection. *J. Chromatogr. A,* **2006**, *1125*(1), 133-137.
[http://dx.doi.org/10.1016/j.chroma.2006.06.096] [PMID: 16859693]

[50]   Takahashi, K.; Goto-Yamamoto, N. Simple method for the simultaneous quantification of medium-chain fatty acids and ethyl hexanoate in alcoholic beverages by gas chromatography-flame ionization detector: development of a direct injection method. *J. Chromatogr. A,* **2011**, *1218*(43), 7850-7856.
[http://dx.doi.org/10.1016/j.chroma.2011.08.074] [PMID: 21925662]

[51]   Sterckx, F.L.; Missiaen, J.; Saison, D.; Delvaux, F.R. Contribution of monophenols to beer flavour based on flavour thresholds, interactions and recombination experiments. *Food Chem.,* **2011**, *126*(4), 1679-1685.
[http://dx.doi.org/10.1016/j.foodchem.2010.12.055] [PMID: 25213944]

[52]   Rodrigues, J.A.; Barros, A.S.; Carvalho, B.; Brandão, T.; Gil, A.M.; Ferreira, A.C. Evaluation of beer deterioration by gas chromatography-mass spectrometry/multivariate analysis: a rapid tool for assessing beer composition. *J. Chromatogr. A,* **2011**, *1218*(7), 990-996.
[http://dx.doi.org/10.1016/j.chroma.2010.12.088] [PMID: 21227435]

[53]   Hoff, S.; Lund, M.N.; Petersen, M.A.; Frank, W.; Andersen, M.L. Storage stability of pasteurized non-filtered beer. *J. Inst. Brew.,* **2013**, *119*(3), 172-181.

[54]   Olaniran, A.O.; Maharaj, Y.R.; Pillay, B. Effects of fermentation temperature on the composition of beer volatile compounds, organoleptic quality and spent yeast density. *Electron. J. Biotechnol.,* **2011**, *14*(2), 1-10.
[http://dx.doi.org/10.2225/vol14-issue2-fulltext-5]

[55]   Svoboda, Z.; Mikulíková, R.; Běláková, S.; Benešová, K.; Márová, I.; Nesvadba, Z. Optimization of modern analytical SPME and SPDE methods for determination of trans-2-nonenal in barley, malt and beer. *Chromatographia,* **2011**, *73*(S1), S157-S161.
[http://dx.doi.org/10.1007/s10337-011-1958-x]

[56]   Castro, L.F.; Ross, C.F.; Vixie, K.R. Optimization of a solid phase dynamic extraction (SPDE) method for beer volatile profiling. *Food Anal. Methods,* **2015**, *8*, 2115-2124.
[http://dx.doi.org/10.1007/s12161-015-0104-z]

[57]   Horák, T.; Čulík, J.; Jurková, M.; Čejka, P.; Kellner, V. Application of some modern sample preparation procedures for quantitative determination of vicinal diketones in beer. *Kvasný průmysl,* **2009**, *55*(3), 66-72.

[58]   Horák, T.; Kellner, V.; Jurková, M.; Pavel, Č.; Hašková, D.; Dvořák, J. Analysis of selected esters in beer: comparison of solid-phase microextraction and stir bar sorptive extraction. *J. Inst. Brew.,* **2010**, *116*(1), 81-85.
[http://dx.doi.org/10.1002/j.2050-0416.2010.tb00402.x]

[59]   Zhou, Q.; Qian, Y.; Qian, M.C. Analysis of volatile phenols in alcoholic beverage by ethylene glycol-polydimethylsiloxane based stir bar sorptive extraction and gas chromatography-mass spectrometry. *J. Chromatogr. A,* **2015**, *1390*, 22-27.
[http://dx.doi.org/10.1016/j.chroma.2015.02.064] [PMID: 25766496]

[60]   Castro, L.F.; Ross, C.F. Determination of flavour compounds in beer using stir-bar sorptive extraction and solid-phase microextraction. *J. Inst. Brew.,* **2015**, *121*(2), 197-203.
[http://dx.doi.org/10.1002/jib.219]

[61]   Sasamoto, K.; Ochiai, N. Selectable one-dimensional or two-dimensional gas chromatography-mass spectrometry with simultaneous olfactometry or element-specific detection. *J. Chromatogr. A,* **2010**, *1217*(17), 2903-2910.
[http://dx.doi.org/10.1016/j.chroma.2010.02.045] [PMID: 20299024]

[62]   Hrivňák, J.; Smogrovičová, D.; Nádaský, P.; Lakatošová, J. Determination of beer aroma compounds using headspace solid-phase microcolumn extraction. *Talanta,* **2010**, *83*(1), 294-296.
[http://dx.doi.org/10.1016/j.talanta.2010.08.041] [PMID: 21035679]

[63]   Hrivňák, J.; Smogrovičová, D.; Lakatošová, J.; Nádaský, P. Technical note – analysis of beer aroma compounds by solid-phase microcolumn extraction. *J. Inst. Brew.,* **2010**, *116*(2), 167-169.
[http://dx.doi.org/10.1002/j.2050-0416.2010.tb00413.x]

[64]   Campillo, N.; Peñalver, R.; López-García, I.; Hernández-Córdoba, M. Headspace solid-phase microextraction for the determination of volatile organic sulphur and selenium compounds in beers, wines and spirits using gas chromatography and atomic emission detection. *J. Chromatogr. A,* **2009**, *1216*(39), 6735-6740.
[http://dx.doi.org/10.1016/j.chroma.2009.08.019] [PMID: 19700163]

[65]   Jiao, J.; Ding, N.; Shi, T.; Chai, X.; Cong, P.; Zhu, Z. Study of chromatographic fingerprint of the flavor in beer by HS-SPME-GC. *Anal. Lett.,* **2011**, *44*(4), 648-655.
[http://dx.doi.org/10.1080/00032711003783044]

[66]   Charry-Parra, G.; Dejesus-Echevarria, M.; Perez, F.J. Beer volatile analysis: optimization of HS/SPME coupled to GC/MS/FID. *J. Food Sci.,* **2011**, *76*(2), C205-C211.
[http://dx.doi.org/10.1111/j.1750-3841.2010.01979.x] [PMID: 21535736]

[67]   Lei, H.; Zhao, H.; Yu, Z.; Zhao, M. Effects of wort gravity and nitrogen level on fermentation performance of brewer's yeast and the formation of flavor volatiles. *Appl. Biochem. Biotechnol.,* **2012**,

166(6), 1562-1574.
[http://dx.doi.org/10.1007/s12010-012-9560-8] [PMID: 22281783]

[68]   Takoi, K.; Itoga, Y.; Koie, K.; Kosugi, T.; Katayama, Y.; Nakayama, Y.; Watari, J. The contribution of geraniol metabolism to the citrus flavour of beer: synergy of geraniol and β-citronellol under coexistence with excess linalool. *J. Inst. Brew.*, **2010**, *116*(3), 251-260.
[http://dx.doi.org/10.1002/j.2050-0416.2010.tb00428.x]

[69]   Van Opstaele, F.; De Rouck, G.; De Clippeleer, J.; Aerts, G.; De Cooman, L. Analytical and sensory assessment of hoppy aroma and bitterness of conventionally hopped and advanced hopped pilsner beers. *Cerevisia (Gedrukt)*, **2011**, *36*(2), 47-59.
[http://dx.doi.org/10.1016/j.cervis.2011.04.001]

[70]   Rossi, S.; Sileoni, V.; Perretti, G.; Marconi, O. Characterization of the volatile profiles of beer using headspace solid-phase microextraction and gas chromatography-mass spectrometry. *J. Sci. Food Agric.*, **2014**, *94*(5), 919-928.
[http://dx.doi.org/10.1002/jsfa.6336] [PMID: 23929274]

[71]   Tsai, S-W.; Kao, K-Y. Determination of furfural in beers, vinegars and infant formulas by solid-phase microextraction and gas chromatography/mass spectrometry. *Int. J. Environ. Anal. Chem.*, **2012**, *92*(1), 76-84.
[http://dx.doi.org/10.1080/03067319.2010.496050]

[72]   Carrillo, G.; Bravo, A.; Zufall, C. Application of factorial designs to study factors involved in the determination of aldehydes present in beer by on-fiber derivatization in combination with gas chromatography and mass spectrometry. *J. Agric. Food Chem.*, **2011**, *59*(9), 4403-4411.
[http://dx.doi.org/10.1021/jf200167h] [PMID: 21456621]

[73]   Rodriguez-Bencomo, J.J.; Muñoz-González, C.; Martín-Álvarez, P.J.; Lázaro, E.; Mancebo, R.; Castañé, X.; Pozo-Bayón, M.A. Optimization of a HS-SPME-GC-MS procedure for beer volatile profiling using response surface methodology: application to follow aroma stability of beers under different storage conditions. *Food Anal. Methods*, **2012**, *5*(6), 1386-1397.
[http://dx.doi.org/10.1007/s12161-012-9390-x]

[74]   Riu-Aumatell, M.; Miró, P.; Serra-Cayuela, A.; Buxaderas, S.; López-Tamames, E. Assessment of the aroma profiles of low-alcohol beers using HS-SPME–GC-MS. *Food Res. Int.*, **2014**, *57*, 196-202.
[http://dx.doi.org/10.1016/j.foodres.2014.01.016]

[75]   Pizarro, C.; Pérez-del-Notario, N.; González-Sáiz, J.M. Optimisation of a simple and reliable method based on headspace solid-phase microextraction for the determination of volatile phenols in beer. *J. Chromatogr. A*, **2010**, *1217*(39), 6013-6021.
[http://dx.doi.org/10.1016/j.chroma.2010.07.021] [PMID: 20728896]

[76]   Saison, D.; De Schutter, D.P.; Delvaux, F.; Delvaux, F.R. Determination of carbonyl compounds in beer by derivatisation and headspace solid-phase microextraction in combination with gas chromatography and mass spectrometry. *J. Chromatogr. A*, **2009**, *1216*(26), 5061-5068.
[http://dx.doi.org/10.1016/j.chroma.2009.04.077] [PMID: 19450805]

[77]   Sterckx, F.L.; Saison, D.; Delvaux, F.R. Determination of volatile monophenols in beer using acetylation and headspace solid-phase microextraction in combination with gas chromatography and mass spectrometry. *Anal. Chim. Acta*, **2010**, *676*(1-2), 53-59.
[http://dx.doi.org/10.1016/j.aca.2010.07.043] [PMID: 20800742]

[78]   Scherer, R.; Wagner, R.; Kowalski, C.H.; Godoy, H.T. (E)-2-nonenal determination in brazilian beers using headspace solid-phase microextraction and gas chromatographic coupled mass spectrometry (HS-SPME-GC-MS). *Food Sci. Technol. (Campinas)*, **2010**, *30*(1), 161-165.
[http://dx.doi.org/10.1590/S0101-20612010000500024]

[79]   da Silva, G.A.; Maretto, D.A.; Bolini, H.M.; Teófilo, R.F.; Augusto, F.; Poppi, R.J. Correlation of quantitative sensorial descriptors and chromatographic signals of beer using multivariate calibration strategies. *Food Chem.*, **2012**, *134*(3), 1673-1681.

[http://dx.doi.org/10.1016/j.foodchem.2012.03.080] [PMID: 25005998]

[80] Biazon, C.L.; Brambilla, R.; Rigacci, A.; Pizzolato, T.M.; Dos Santos, J.H. Combining silica-based adsorbents and SPME fibers in the extraction of the volatiles of beer: an exploratory study. *Anal. Bioanal. Chem.,* **2009**, *394*(2), 549-556.
[http://dx.doi.org/10.1007/s00216-009-2695-y] [PMID: 19283367]

[81] Moreira, N.; Meireles, S.; Brandão, T.; de Pinho, P.G. Optimization of the HS-SPME-GC-IT/MS method using a central composite design for volatile carbonyl compounds determination in beers. *Talanta,* **2013**, *117*, 523-531.
[http://dx.doi.org/10.1016/j.talanta.2013.09.027] [PMID: 24209376]

[82] Martins, C.; Brandão, T.; Almeida, A.; Rocha, S.M. Insights on beer volatile profile: Optimization of solid-phase microextraction procedure taking advantage of the comprehensive two-dimensional gas chromatography structured separation. *J. Sep. Sci.,* **2015**, *38*(12), 2140-2148.
[http://dx.doi.org/10.1002/jssc.201401388] [PMID: 25907307]

[83] Dresel, M.; Praet, T.; Van Opstaele, F.; Van Holle, A.; Naudts, D.; De Keukeleire, D.; De Cooman, L.; Aerts, G. Comparison of the analytical profiles of volatiles in single-hopped worts and beers as a function of the hop variety. *BrewingScience,* **2015**, *68*(1–2), 8-28.

[84] Rendall, R.; Reis, M.S.; Pereira, A.C.; Pestana, C.; Pereira, V.; Marques, J.C. Chemometric analysis of the volatile fraction evolution of portuguese beer under shelf storage conditions. *Chemom. Intell. Lab. Syst.,* **2015**, *142*, 131-142.
[http://dx.doi.org/10.1016/j.chemolab.2015.01.015]

[85] da Silva, G.A.; Augusto, F.; Poppi, R.J. Exploratory analysis of the volatile profile of beers by HS–SPME–GC. *Food Chem.,* **2008**, *111*(4), 1057-1063.
[http://dx.doi.org/10.1016/j.foodchem.2008.05.022]

[86] Liu, M.; Zeng, Z.; Xiong, B. Preparation of novel solid-phase microextraction fibers by sol-gel technology for headspace solid-phase microextraction-gas chromatographic analysis of aroma compounds in beer. *J. Chromatogr. A,* **2005**, *1065*(2), 287-299.
[http://dx.doi.org/10.1016/j.chroma.2004.12.073] [PMID: 15782975]

[87] Scarlata, C.J.; Ebeler, S.E. Headspace solid-phase microextraction for the analysis of dimethyl sulfide in beer. *J. Agric. Food Chem.,* **1999**, *47*(7), 2505-2508.
[http://dx.doi.org/10.1021/jf990298g] [PMID: 10552517]

[88] Hill, P.G.; Smith, R.M. Determination of sulphur compounds in beer using headspace solid-phase microextraction and gas chromatographic analysis with pulsed flame photometric detection. *J. Chromatogr. A,* **2000**, *872*(1-2), 203-213.
[http://dx.doi.org/10.1016/S0021-9673(99)01307-2] [PMID: 10749498]

[89] Jeleń, H.H.; Dąbrowska, A.; Klensporf, D.; Nawrocki, J.; Wąsowicz, E. Determination of C3-C10 aliphatic aldehydes using PFBHA derivatization and solid phase microextraction (SPME). Application to the analysis of beer. *Chem. Anal.,* **2004**, *869*, 869-880.

[90] Saison, D.; De Schutter, D.P.; Delvaux, F.; Delvaux, F.R. Optimisation of a complete method for the analysis of volatiles involved in the flavour stability of beer by solid-phase microextraction in combination with gas chromatography and mass spectrometry. *J. Chromatogr. A,* **2008**, *1190*(1-2), 342-349.
[http://dx.doi.org/10.1016/j.chroma.2008.03.015] [PMID: 18378248]

[91] Andrés-Iglesias, C.; Blanco, C.A.; García-Serna, J.; Pando, V.; Montero, O. Volatile compound profiling in commercial lager regular beers and derived alcohol-free beers after dealcoholization by vacuum distillation. *Food Anal. Methods,* **2016**, *9*(11), 3230-3241.
[http://dx.doi.org/10.1007/s12161-016-0513-7]

[92] Pacheco, J.G.; Valente, I.M.; Gonçalves, L.M.; Magalhães, P.J.; Rodrigues, J.A.; Barros, A.A. Development of a membraneless extraction module for the extraction of volatile compounds: application in the chromatographic analysis of vicinal diketones in beer. *Talanta,* **2010**, *81*(1-2), 372-

376.
[http://dx.doi.org/10.1016/j.talanta.2009.12.011] [PMID: 20188933]

[93] Gonçalves, L.M.; Magalhães, P.J.; Valente, I.M.; Pacheco, J.G.; Dostálek, P.; Sýkora, D.; Rodrigues, J.A.; Barros, A.A. Analysis of aldehydes in beer by gas-diffusion microextraction: characterization by high-performance liquid chromatography-diode-array detection-atmospheric pressure chemical ionization-mass spectrometry. *J. Chromatogr. A,* **2010**, *1217*(24), 3717-3722.
[http://dx.doi.org/10.1016/j.chroma.2010.04.002] [PMID: 20451914]

[94] Pacheco, J.G.; Valente, I.M.; Gonçalves, L.M.; Rodrigues, J.A.; Barros, A.A. Gas-diffusion microextraction. *J. Sep. Sci.,* **2010**, *33*(20), 3207-3212.
[http://dx.doi.org/10.1002/jssc.201000351] [PMID: 20954176]

[95] Siebert, K.J.; Lynn, P.Y. Comparison of methods for degassing beer for analysis. *J. Am. Soc. Brew. Chem.,* **2007**, *65*(4), 229-231.

[96] Andrés-Iglesias, C.; Montero, O.; Sancho, D.; Blanco, C.A. New trends in beer flavour compound analysis. *J. Sci. Food Agric.,* **2015**, *95*(8), 1571-1576.
[http://dx.doi.org/10.1002/jsfa.6905] [PMID: 25205443]

[97] Soria, A.C.; García-Sarrió, M.J.; Sanz, M.L. Volatile sampling by headspace techniques. *Trends Analyt. Chem.,* **2015**, *71*, 85-99.
[http://dx.doi.org/10.1016/j.trac.2015.04.015]

[98] *Trichlorofluoromethane,* [Mar 13, 2017]; Available at: https://toxnet.nlm.nih.gov/cgi-bin/sis/search/a?dbs+hsdb:@term+@DOCNO+138

[99] Mamede, M.E.; Pastore, G.M. Study of methods for the extraction of volatile compounds from fermented grape must. *Food Chem.,* **2006**, *96*(4), 586-590.
[http://dx.doi.org/10.1016/j.foodchem.2005.03.013]

[100] Camino-Sánchez, F.J.; Rodríguez-Gómez, R.; Zafra-Gómez, A.; Santos-Fandila, A.; Vílchez, J.L. Stir bar sorptive extraction: recent applications, limitations and future trends. *Talanta,* **2014**, *130*, 388-399.
[http://dx.doi.org/10.1016/j.talanta.2014.07.022] [PMID: 25159426]

[101] Xu, C.H.; Chen, G.S.; Xiong, Z.H.; Fan, Y.X.; Wang, X.C.; Liu, Y. Applications of solid-phase microextraction in food analysis. *Trends Analyt. Chem.,* **2016**, *80*, 12-29.
[http://dx.doi.org/10.1016/j.trac.2016.02.022]

[102] Vas, G.; Vékey, K. Solid-phase microextraction: a powerful sample preparation tool prior to mass spectrometric analysis. *J. Mass Spectrom.,* **2004**, *39*(3), 233-254.
[http://dx.doi.org/10.1002/jms.606] [PMID: 15039931]

[103] Andrade-Eiroa, A.; Canle, M.; Leroy-Cancellieri, V.; Cerdà, V. Solid-phase extraction of organic compounds: a critical review. Part Ii. *Trends Analyt. Chem.,* **2016**, *80*, 655-667.
[http://dx.doi.org/10.1016/j.trac.2015.08.014]

[104] Andrade-Eiroa, A.; Canle, M.; Leroy-Cancellieri, V.; Cerdà, V. Solid-phase extraction of organic compounds: a critical review (part I). *Trends Analyt. Chem.,* **2016**, *80*, 641-654.
[http://dx.doi.org/10.1016/j.trac.2015.08.015]

[105] Buszewski, B.; Szultka, M. Past, present, and future of solid phase extraction: a review. *Crit. Rev. Anal. Chem.,* **2012**, *42*(3), 198-213.
[http://dx.doi.org/10.1080/07373937.2011.645413]

[106] Castro, R.; Natera, R.; Durán, E.; García-Barroso, C. Application of solid phase extraction techniques to analyse volatile compounds in wines and other enological products. *Eur. Food Res. Technol.,* **2008**, *228*, 1-18.
[http://dx.doi.org/10.1007/s00217-008-0900-4]

[107] Nogueira, J.M. Novel sorption-based methodologies for static microextraction analysis: a review on SBSE and related techniques. *Anal. Chim. Acta,* **2012**, *757*, 1-10.
[http://dx.doi.org/10.1016/j.aca.2012.10.033] [PMID: 23206390]

[108] Petronilho, S.; Coimbra, M.A.; Rocha, S.M. A critical review on extraction techniques and gas chromatography based determination of grapevine derived sesquiterpenes. *Anal. Chim. Acta,* **2014**, *846*, 8-35.
[http://dx.doi.org/10.1016/j.aca.2014.05.049] [PMID: 25220138]

[109] Tölgyessy, P.; Vrana, B.; Hrivňák, J. Large volume headspace analysis using solid-phase microcolumn extraction. *Chromatographia,* **2007**, *66*(9), 815-817.
[http://dx.doi.org/10.1365/s10337-007-0385-5]

[110] Pawliszyn, J., Ed. *Handbook of solid phase microextraction*; Chemical Industry Press, **2009**.

[111] Vesely, P.; Lusk, L.; Basarova, G.; Seabrooks, J.; Ryder, D. Analysis of aldehydes in beer using solid-phase microextraction with on-fiber derivatization and gas chromatography/mass spectrometry. *J. Agric. Food Chem.,* **2003**, *51*(24), 6941-6944.
[http://dx.doi.org/10.1021/jf034410t] [PMID: 14611150]

[112] Villamiel, M.; del Castillo, M.D.; Corzo, N. Browning Reactions. In: *Food Biochemistry and Food Processing; Hui, Y. H*; Nollet, L.M.; Paliyath, G.; Simpson, B.K., Eds.; Wiley-Blackwell: Nip, W., **2006**; pp. 71-100.
[http://dx.doi.org/10.1002/9780470277577.ch4]

[113] Lord, H.; Pawliszyn, J. Evolution of solid-phase microextraction technology. *J. Chromatogr. A,* **2000**, *885*(1-2), 153-193.
[http://dx.doi.org/10.1016/S0021-9673(00)00535-5] [PMID: 10941672]

[114] Wong, J.W.; Hayward, D.G.; Zhang, K. Gas chromatography-mass spectrometry techniques for multiresidue pesticide analysis in agricultural commodities. In: *Comprehensive Analytical Chemistry*; Ferrer, I.; Thurman, E.M., Eds.; Elsevier B.V.: Oxford, **2013**; Vol. 61, pp. 3-22.
[http://dx.doi.org/10.1016/B978-0-444-62623-3.00001-0]

[115] Grob, R.L.; Barry, E.F., Eds. *Modern Practice of Gas Chromatography,* 4th ed; John Wiley & Sons, Inc.: New Jersey, United States of America, **2004**.
[http://dx.doi.org/10.1002/0471651141]

[116] Dettmer, K.; Aronov, P.A.; Hammock, B.D. Mass spectrometry-based metabolomics. *Mass Spectrom. Rev.,* **2007**, *26*(1), 51-78.
[http://dx.doi.org/10.1002/mas.20108] [PMID: 16921475]

[117] Milman, B.L. General principles of identification by mass spectrometry. *Trends Analyt. Chem.,* **2015**, *69*, 24-33.
[http://dx.doi.org/10.1016/j.trac.2014.12.009]

[118] Tranchida, P.Q.; Dugo, P.; Dugo, G.; Mondello, L. Comprehensive two-dimensional chromatography in food analysis. *J. Chromatogr. A,* **2004**, *1054*(1-2), 3-16.
[http://dx.doi.org/10.1016/S0021-9673(04)01301-9] [PMID: 15553126]

[119] van Den Dool, H.; Dec, Kratz, P. A generalization of the retention index system including linear temperature programmed gas—liquid partition chromatography. *J. Chromatogr. A,* **1963**, *11*, 463-471.
[http://dx.doi.org/10.1016/S0021-9673(01)80947-X]

[120] Babushok, V.I. Chromatographic retention indices in identification of chemical compounds. *Trends Analyt. Chem.,* **2015**, *69*, 98-104.
[http://dx.doi.org/10.1016/j.trac.2015.04.001]

[121] Mondello, L.; Tranchida, P.Q.; Dugo, P.; Dugo, G. Comprehensive two-dimensional gas chromatography-mass spectrometry: a review. *Mass Spectrom. Rev.,* **2008**, *27*(2), 101-124.
[http://dx.doi.org/10.1002/mas.20158] [PMID: 18240151]

[122] Marriott, P.J.; Haglund, P.; Ong, R.C.; Schmarr, H.G.; Bieri, S. A review of environmental toxicant analysis by using multidimensional gas chromatography and comprehensive GC. *Clin. Chim. Acta,* **2003**, *328*(1-2), 1-19.
[http://dx.doi.org/10.1016/S0009-8981(02)00382-0] [PMID: 12559594]

[123] Marriott, P.J.; Massil, T.; Hügel, H. Molecular structure retention relationships in comprehensive two-dimensional gas chromatography. *J. Sep. Sci.,* **2004**, *27*(15-16), 1273-1284.
[http://dx.doi.org/10.1002/jssc.200401917] [PMID: 15587276]

[124] Seeley, J.V.; Seeley, S.K. Multidimensional gas chromatography: fundamental advances and new applications. *Anal. Chem.,* **2013**, *85*(2), 557-578.
[http://dx.doi.org/10.1021/ac303195u] [PMID: 23137217]

[125] Tranchida, P.Q.; Purcaro, G.; Maimone, M.; Mondello, L. Impact of comprehensive two-dimensional gas chromatography with mass spectrometry on food analysis. *J. Sep. Sci.,* **2016**, *39*(1), 149-161.
[http://dx.doi.org/10.1002/jssc.201500379] [PMID: 26179510]

[126] Dallüge, J.; Beens, J.; Brinkman, U.A. Comprehensive two-dimensional gas chromatography: a powerful and versatile analytical tool. *J. Chromatogr. A,* **2003**, *1000*(1-2), 69-108.
[http://dx.doi.org/10.1016/S0021-9673(03)00242-5] [PMID: 12877167]

[127] Górecki, T.; Panić, O.; Oldridge, N. Recent advances in comprehensive two-dimensional gas chromatography (GC×GC). *J. Liq. Chromatogr. Relat. Technol.,* **2006**, *29*(7–8), 1077-1104.
[http://dx.doi.org/10.1080/10826070600574762]

[128] Cordero, C.; Kiefl, J.; Schieberle, P.; Reichenbach, S.E.; Bicchi, C. Comprehensive two-dimensional gas chromatography and food sensory properties: potential and challenges. *Anal. Bioanal. Chem.,* **2015**, *407*(1), 169-191.
[http://dx.doi.org/10.1007/s00216-014-8248-z] [PMID: 25354891]

[129] Tran, T.C.; Logan, G.A.; Grosjean, E.; Harynuk, J.; Ryan, D.; Marriott, P. Comparison of Column Phase Configurations for Comprehensive Two Dimensional Gas Chromatographic Analysis of Crude Oil and Bitumen. *Org. Geochem.,* **2006**, *37*(9), 1190-1194.
[http://dx.doi.org/10.1016/j.orggeochem.2006.05.006]

[130] Silva, I.; Coimbra, M.A.; Barros, A.S.; Marriott, P.J.; Rocha, S.M. Can volatile organic compounds be markers of sea salt? *Food Chem.,* **2015**, *169*, 102-113.
[http://dx.doi.org/10.1016/j.foodchem.2014.07.120] [PMID: 25236204]

[131] Perestrelo, R.; Petronilho, S.; Câmara, J.S.; Rocha, S.M. Comprehensive two-dimensional gas chromatography with time-of-flight mass spectrometry combined with solid phase microextraction as a powerful tool for quantification of ethyl carbamate in fortified wines. The case study of Madeira wine. *J. Chromatogr. A,* **2010**, *1217*(20), 3441-3445.
[http://dx.doi.org/10.1016/j.chroma.2010.03.027] [PMID: 20388567]

[132] Jalali, H.T.; Petronilho, S.; Villaverde, J.J.; Coimbra, M.A.; Domingues, M.R.; Ebrahimian, Z.J.; Silvestre, A.J.; Rocha, S.M. Assessment of the sesquiterpenic profile of *Ferula gummosa* oleo-gu--resin (*Galbanum*) from Iran. Contributes to its valuation as a potential source of sesquiterpenic compounds. *Ind. Crops Prod.,* **2013**, *44*, 185-191.
[http://dx.doi.org/10.1016/j.indcrop.2012.10.031]

[133] Salvador, Â.C.; Rudnitskaya, A.; Silvestre, A.J.; Rocha, S.M. Metabolomic-based strategy for fingerprinting of *Sambucus nigra* L. berry volatile terpenoids and norisoprenoids: influence of ripening and cultivar. *J. Agric. Food Chem.,* **2016**, *64*(26), 5428-5438.
[http://dx.doi.org/10.1021/acs.jafc.6b00984] [PMID: 27348582]

[134] Herrero, M.; Ibáñez, E.; Cifuentes, A.; Bernal, J. Multidimensional chromatography in food analysis. *J. Chromatogr. A,* **2009**, *1216*(43), 7110-7129.
[http://dx.doi.org/10.1016/j.chroma.2009.08.014] [PMID: 19699483]

[135] Gamero, A.; Ferreira, V.; Pretorius, I.S.; Querol, A. Wine, beer and cider: unravelling the aroma profile. In: *Molecular mechanisms in yeast carbon metabolism*; Piškur, J.; Compagno, C., Eds.; Springer Berlin Heidelberg, **2014**; pp. 261-297.
[http://dx.doi.org/10.1007/978-3-662-45782-5_10]

[136] Vanbeneden, N.; Delvaux, F.; Delvaux, F.R. Determination of hydroxycinnamic acids and volatile

phenols in wort and beer by isocratic high-performance liquid chromatography using electrochemical detection. *J. Chromatogr. A,* **2006**, *1136*(2), 237-242.
[http://dx.doi.org/10.1016/j.chroma.2006.11.001] [PMID: 17109870]

[137] Vanbeneden, N.; Gils, F.; Delvaux, F.; Delvaux, F.R. Formation of 4-vinyl and 4-ethyl derivatives from hydroxycinnamic acids: occurrence of volatile phenolic flavour compounds in beer and distribution of pad1-activity among brewing yeasts. *Food Chem.,* **2008**, *107*(1), 221-230.
[http://dx.doi.org/10.1016/j.foodchem.2007.08.008]

[138] Chatonnet, P.; Dubourdieu, D.; Boidron, J.; Lavigne, V. Synthesis of volatile phenols by *saccharomyces cerevisiae* in wines. *J. Sci. Food Agric.,* **1993**, *62*(2), 191-202.
[http://dx.doi.org/10.1002/jsfa.2740620213]

[139] Almaguer, C.; Schönberger, C.; Gastl, M.; Arendt, E.K.; Becker, T. *Humulus lupulus* - a story that begs to be told. A review. *J. Inst. Brew.,* **2014**, *120*(4), 289-314.

[140] Takoi, K.; Koie, K.; Itoga, Y.; Katayama, Y.; Shimase, M.; Nakayama, Y.; Watari, J. Biotransformation of hop-derived monoterpene alcohols by lager yeast and their contribution to the flavor of hopped beer. *J. Agric. Food Chem.,* **2010**, *58*(8), 5050-5058.
[http://dx.doi.org/10.1021/jf1000524] [PMID: 20364865]

[141] King, A.J.; Dickinson, J.R. Biotransformation of hop aroma terpenoids by ale and lager yeasts. *FEMS Yeast Res.,* **2003**, *3*(1), 53-62.
[http://dx.doi.org/10.1111/j.1567-1364.2003.tb00138.x] [PMID: 12702246]

[142] King, A.; Richard Dickinson, J. Biotransformation of monoterpene alcohols by *Saccharomyces cerevisiae, Torulaspora delbrueckii* and *Kluyveromyces lactis. Yeast,* **2000**, *16*(6), 499-506.
[http://dx.doi.org/10.1002/(SICI)1097-0061(200004)16:6<499::AID-YEA548>3.0.CO;2-E] [PMID: 10790686]

SUPPLEMENTARY MATERIAL

Table S1. General methodology used in each study of beer volatile profiling (extraction and detection/analysis technique), and the specificities of each sample preparation procedure.

Extraction Technique	Detection/Analysis Technique	Sample Preparation			Ref.
		Decarbonation Procedure	**Extraction Device[p]**	**Experimental Conditions[q]**	
Headspace generation apparatus	HS-GC-FID[i]	–	Headspace sampler device	–	[45]
		Filtration		Sample was thermostated at 60 °C, 16 min	[29]
	HS-MS e-nose[j]	Ultrasonication (15 min)	Headspace sampler device	Sample was thermostated at 65°C, 1 h	[19]
	DHS-GC-MS[k]	–	Small cartridge filled with tenax® and carbotrap® activated charcoal	Sample was assessed by enclosing it in a polyacetate bag and pumping air from the bag through a small cartridge (50 mL min[-1]) for 30 min	[54]
	HS-GC-MS	–	Entech 7200	Sample was heated in a water bath for 15 min at 50 °C	[28]

(Table S1) contd.....

Extraction Technique	Detection/Analysis Technique	Sample Preparation			Ref.
		Decarbonation Procedure	**Extraction Device[p]**	**Experimental Conditions[q]**	
LLE[a]	GC-FID	–	Likens–Nickerson apparatus	–	[46]
	GC-MS	–	–	Trichlorofluoromethane extraction (20 min), headspace trapping method with Tenax TA	[40]
	GC-MS	–	–	Diethyl ether (150 mL) continuous extraction was conducted over a 24h period at 45 °C. Then, extract was separated using SAFE	[44]
	GC-MS	Poured into a beaker, and placed at 3 °C, for 1 h	–	Dichloromethane extraction (50 °C, 6h), then dried with anhydrous sodium sulphate	[41]
	HRGC-MS	–	–	Diethyl ether extraction (1 L), treatment with sodium bicarbonate, then separation by column chromatography on silica gel	[43]
	HRGC-MS	Filtration with paper filter	–	Diethyl ether extraction (2,4 L) and dried with anhydrous sodium sulphate	[21, 42]
	GC×GC-ToFMS[l]	–	–	Dichloromethane extraction (room temp., 60 min, 100 rpm), then left to separate into two phases (15 min), and dried with anhydrous sodium sulphate	[47]
SAFE[b]	GC-MS	–	SAFE apparatus	SAFE extraction (100 mL beer), then distillate was stirred with dichloromethane (50 mL, 1 h). Then, it was stored in a freezer compartment with sodium sulphate for 12 h. It was concentrated to 1 mL in a water bath and condensed to 0.4 mL	[48]

(Table S1) contd.....

Extraction Technique	Detection/Analysis Technique	Sample Preparation			Ref.
		Decarbonation Procedure	**Extraction Device[p]**	**Experimental Conditions[q]**	
HS-SPDE[c]	GC-MS	–	SPDE needles (50 µm PDMS)	10 mL of sample was incubated 15 min at 60 °C, NaCl addition (1 g in 20 mL vial). Needle temperature was 55 °C. For extraction, 55 extraction cycles were used, 250 rpm, extraction volume 1 mL, fill/eject speed 50 µL/s, and 1 mL of desorption volume	[56]
SPE[d]	GC-MS/MS	Ultrasonic bath (10 min), filtration	LiChrolut EN (500 mg) cartridge	Sample passed through the SPE cartridge bed at a speed lower than 2 mL min⁻¹. Analytes were eluted with dichloromethane (6 mL), and dried by adding anhydrous sodium sulphate	[52]
	GC-MS	Stirring with 1-octanol (0.01% (5 min), filtration through funnel filter	Tenax-TA with mesh size 60/80 and density of 0.37 g/mL	Sample was equilibrated to $30 \pm 1\,°C$ for 5 min, and then purged with nitrogen (75 mL/min) for 15 min	[53]
HS-SPMCE[e]	GC-FID	Vigorously shaken (1 min)	Microcolumn was packed with 60–80 mesh Tenax TA	Sample was thermostated at 22 °C, 10-15 min, then headspace of 10 mL was aspirated through the microcolumn at a flow rate of 2–3 mL min⁻¹	[62, 63]
SBSE[f]	TD-GC-MS[m]	–	Twister (10 × 0.5 mm)	Extraction: at room temp., 60 min, stirring (750 rpm)	[7]
IM-SPME[g]	DART-ToFMS[n]	Ultrasonic bath (5 min, 5 °C)	SPME syringe (50/30 µm DVB/CAR/PDMS)	Sample was thermostated at 30 °C, 2 min, stirring (500 rpm), then SPME fibre was exposed to the sample headspace during 15 min	[25]

(Table S1) contd.....

Extraction Technique	Detection/Analysis Technique	Sample Preparation			Ref.
		Decarbonation Procedure	**Extraction Device[p]**	**Experimental Conditions[q]**	
HS-SPME[h]	GC-MS	–	SPME syringe (65 µm PDMS/DVB)	On-fiber derivatization with *O*-PFBHA Sample was thermostated at 50 °C, 26 min to allow the derivatization, then the SPME fibre was exposed to the sample headspace during 2 min	[70]
HS-SPME	GC-MS	–	SPME syringe (65 µm PDMS/DVB)	Sample was thermostated at 40 °C, then SPME fibre was exposed to the sample headspace during 30 min	[83]
HS-SPME	GC-MS	Filtration	SPME syringe (100 µm PDMS)	Sample was thermostated at 60 °C, NaCl addition (2 g), then SPME fibre was exposed to the sample headspace during 45 min	[83]
HS-SPME	GC-MS	Filtration	SPME syringe (100 µm PDMS)	SPME fibre was exposed to the sample headspace during 45 min at 30 °C	[91]
HS-SPME	GC-MS	Filtration	SPME syringe (50/30 µm DVB/CAR/PDMS)	Sample was thermostated at 44.8 °C, 10 min, NaCl addition (0.300 g mL^{-1}), then SPME fibre was exposed to the sample headspace during 46.8 min	[73]
HS-SPME	GC-MS	Filtration	SPME syringe (50/30 µm DVB/CAR/PDMS)	Sample was thermostated at 45 °C, 20 min, NaCl addition (0.350 g mL^{-1}), then the SPME fibre was exposed to the sample headspace during 40 min	[74]
HS-SPME	GC-MS	Filtration	SPME syringe (50/30 µm DVB/CAR/PDMS)	Sample was thermostated at 40 °C, 10 min, stirring (250 rpm), NaCl addition (350 g L^{-1}), then SPME fibre was exposed to the sample headspace during 30 min	[29]

(Table S1) contd.....

Extraction Technique	Detection/Analysis Technique	Sample Preparation			Ref.
		Decarbonation Procedure	**Extraction Device[p]**	**Experimental Conditions[q]**	
HS-SPME	GC-MS	Filtration	SPME syringe (65 μm PDMS/DVB)	On-fiber derivatization with *O*-PFBHA Sample was thermostated at 45 °C, 10 min, stirring (250 rpm), NaCl addition (0.350 g mL^{-1}), then SPME fibre was exposed to the sample headspace during 30 min	[76]
HS-SPME	GC-MS	Filtration	SPME syringe (65 μm PDMS/DVB)	In-solution derivatization with *O*-PFBHA Sample was thermostated at 60 °C, 10 min, stirring (500 rpm), then SPME fibre was exposed to the sample headspace during 40 min, 250 rpm	[76]
HS-SPME	GC-MS	Filtration	SPME syringe (65 μm PDMS/DVB)	On-fiber derivatization with *O*-PFBHA Sample was thermostated at 45 °C, 10 min, stirring (500 rpm), NaCl addition (0.350 g mL^{-1}), then SPME fibre was exposed to the sample headspace during 30 min, 250 rpm	[90]
HS-SPME	GC-MS	Filtration	SPME syringe (50/30 μm DVB/CAR/PDMS)	Sample was thermostated at 40 °C, 10 min., stirring (500 rpm), NaCl addition (3.5 g), then SPME fibre was exposed to the sample headspace during 30 min, stirring (250 rpm)	[84]
HS-SPME	GC-MS	Filtration	SPME syringe (50/30 μm DVB/CAR/PDMS)	Sample was thermostated at 45 °C, 10 min, stirring (500 rpm), NaCl addition (0.350 g mL^{-1}), then SPME fibre was exposed to the sample headspace at 40 °C, during 30 min, 250 rpm	[90]

(Table S1) contd.....

Extraction Technique	Detection/Analysis Technique	Sample Preparation			Ref.
		Decarbonation Procedure	Extraction Device[p]	Experimental Conditions[q]	
HS-SPME	GC-MS	Ultrasonic bath (20 min at 20 °C)	SPME syringe (75 μm CAR/PDMS)	Sample was thermostated at 20 °C, 30 min, NaCl addition (2g), then SPME fibre was exposed to the sample headspace during 30 min	[5]
HS-SPME	GC-MS	Ultrasonic bath (15 min, 5°C)	SPME syringe (65 μm PDMS/DVB)	Sample was thermostated at 50 °C, 5 min, 1200 rpm, NaCl addition (0.270 g mL^{-1}) then SPME fibre was exposed to the sample headspace during 30 min	[85]
HS-SPME	GC-MS	Ultrasonic bath (15 min)	SPME syringe (65 μm PDMS/DVB)	Sample was thermostated at 50 °C, 5 min, NaCl addition (0.300 g mL^{-1}), then SPME fibre was exposed to the sample headspace during 30 min	[79]
HS-SPME	GC–IT-MS[o]	–	SPME syringe (65 μm PDMS/DVB)	On-fiber derivatization with *O*-PFBHA Sample was thermostated at 45 °C, 7 min, then the SPME fibre was exposed to the sample headspace during 20 min	[81]
HS-SPME	GC-ToFMS	Ultrasonic bath (5min, 5 °C)	SPME syringe (50/30 μm DVB/CAR/PDMS)	Sample was thermostated at 30 °C, 5 min, stirring (500 rpm), then SPME fibre was exposed to the sample headspace during 5 min	[25]
HS-SPME	DART-ToFMS	Ultrasonic bath (5min, 5 °C)	SPME syringe (50/30 μm DVB/CAR/PDMS)	Sample was thermostated at 30 °C, 5 min, stirring (500 rpm), then SPME fibre was exposed to the sample headspace during 15 min	[25]
HS-SPME	GC×GC-ToFMS	Overnight at 4 °C	SPME syringe (65 μm PDMS/DVB)	Sample was thermostated at 40 °C, 10 min, NaCl addition (2g), 400 rpm, then SPME fibre was exposed to the sample headspace during 30 min	[82]

[a] Liquid-liquid extraction; [b] Solvent-assisted flavour evaporation; [c] Solid phase dynamic extraction; [d] Solid

phase extraction; [e] Headspace-solid-phase microcolumn extraction; [f] Stir bar sorptive extraction; [g] Immersion-solid phase microextraction; [h] Headspace-solid phase microextraction; [i] Headspace-gas chromatography-flame ionization detector; [j] Electronic nose coupled headspace-mass spectrometry; [k] Dynamic headspace-gas chromatography-mass spectrometry; [l] Comprehensive two-dimensional gas chromatography coupled with a time-of-flight mass spectrometer; [m] Direct analysis in real time coupled with a time-of-flight mass spectrometer; [n] Gas chromatography-ion trap mass spectrometry; [o] Thermal desorption-gas chromatography-mass spectrometry; [p] DVB: divinylbenzene; CAR: carboxen; PDMS: polydimethylsiloxane; [q] O-PFBHA: O-(2,3,4,5,6-pentafluorobenzyl)hydroxylamine.

CHAPTER 6

Recent Advances in Nano Photodynamic Therapy

Ufana Riaz[1,*], S.M. Ashraf[1,±] and Juraj Bujdak[2,3]

[1] Materials Research Laboratory, Department of Chemistry, Jamia Millia Islamia, New Delhi-110025, India

[2] Comenius University in Bratislava, Faculty of Natural Sciences, Department of Physical and Theoretical Chemistry, 842 15 Bratislava, Slovakia

[3] Institute of Inorganic Chemistry, Slovak Academy of Sciences, Bratislava, Slovakia

Abstract: Nanomaterials are being investigated for a vast variety of biomedical applications such as drug delivery, biosensors, tissue engineering, bioimaging *etc*. Most of the approaches utilize nano-scale biomaterials for developing unique functionalities required by fabricating biomedical devices. During the past few decades, photodynamic therapy (PDT) has been developed as an upcoming therapy for the treatment of various life threatening diseases such as cancer and some common bacterial infections. It is based on the photochemical reactions between light and tumour tissue generated using photosensitizing agents. The present chapter focuses on the various materials developed as photodynamic therapeutic agents and their future prospects.

Keywords: Bioimaging, Dendrimers, Nanomaterials, Photochemical reaction, Photodynamic therapy, Polymers, Polymeric nanoparticles, Quantum dots, Singlet oxygen.

INTRODUCTION

With latest advancements in the diagnosis and treatment of life threatening diseases, the demand for alternative therapeutic methods that are non-invasive have led to intense research radiation therapy [1, 2], ultrasonic therapy [3, 4], and in particular photodynamic therapy (PDT) [5 - 8]. Photodynamic therapy (PDT) deals with excitation of photosensitizers (PSs) using light of appropriate wavelength, that causes the generation of cytotoxic products such as singlet oxygen which leads to irreversible destruction of the target cells/tissues [9]. Therefore, PDT is extensively used to treat various types of cancers, macular

[*] **Corresponding Author Ufana Riaz:** Materials Research Laboratory, Department of Chemistry, Jamia Millia Islamia, New Delhi-110025, India; Tel: +91(011) 26981717,26980337; Fax: +91(011) 26980229; E-mail: ufana2002@yahoo.co.in
[±] now retired

Atta-ur-Rahman, Sibel A. Ozkan & Rida Ahmed (Eds.)

degeneration and microbial infections [10]. The major type of ROS generated in PDT is a reactive form of O_2 -singlet oxygen ($^1O_2$) which is known to cause the destruction of target cells through various processes such as apoptosis, necrosis or auto-phagocytosis. The preferred types of photosensitizers are those which can be excited in the near-infrared range (635–760 nm), since the penetration of light into biological tissues is efficient in this region. The general characteristics of an ideal photosensitizer are:

- Chemical stability and functionality under the various conditions to be applied in PDT
- Low self-aggregation tendency to avoid reduced photoactivity
- Good solubility in water
- Photostability
- High molar extinction coefficient
- High yields of triplet state formation and in the activation of molecular oxygen
- Target specificity

Although many photosensitizers have been designed and investigated, only few of them have made a market place. This is due to the fact that most of the photosensitizers developed till date show poor selectivity in terms of target tissue and low singlet oxygen generation due to their tendency to undergo aggregation in aqueous media. This eventually leads to reduction in their photoactivity and the tendency to produce singlet oxygen [11 - 13]. Moreover, most photosensitizers are excited by UV light, which damages even the healthy cells/tissues. Therefore, the development of PDT photosensitizers with minimum aggregation tendency, and higher photoactivity in the near infrared range is still a big challenge [14, 15].

The earliest material that has been reported to be used as a photosensitizer (PS) was acridine, reported in 1900 and was used to kill paramecia. It was also used for the treatment of skin cancer during 1903 [16]. Porphyrin based photosensitizers have shown potential clinical applications because of their high quantum yields and preferential retention in cancer tissues [17]. Some of the most common sensitizers used are porphyrin and its derivatives, chlorins, and phthalocyanines, all of which exhibit various photophysical properties pertaining to the mode of action as well as light activation.

MECHANISM OF PHOTODYNAMIC THERAPY

The mechanism of photodynamic response is basically governed by the processes of absorption of light by the photosensitizer followed by energy transfer. The light source needed for photodynamic therapy must be similar to the wavelength of light required to generate singlet oxygen in a photosensitizer. The requirements of an ideal photosensitizer are specificity towards target cells, physico-chemical stability, and activation at wavelengths of tissue penetration while efficiency of photodynamic process depends on the concentration of the photosensitizer in the target tissue, maximum activation in minimal light dosage and the presence of oxygen.

Singlet Oxygen Generation in PDT

Photosensitizers employed in PDT can transfer their energy from triplet excited state to neighboring oxygen molecules upon activation by light of suitable wavelength (Fig. **1**). As light is absorbed by the photosensitizer, one of these electrons is promoted to a higher energy level, keeping its spin retained in the first excited singlet state. This state is short-lived and loses its energy by emitting fluorescent light or other, non-radiant relaxation pathways. The singlet state in the photosensitization process acts as a transition state of the triplet state, which is achieved *via* intersystem crossing. In the triplet state molecules contain two electrons with parallel spins [18, 19]. This triplet state is the photoactive state, which can generate ROS *via* two mechanisms: electron or energy transfer (Fig. **1**). The latter reaction between molecular oxygen and the photosensitizer in the triplet state produces singlet oxygen, which is the main cytotoxic species. The electron transfer mechanism leads to the formation of reactive radical species. Although both of the reactions occur simultaneously, the ratio between the two processes depends on the type of photosensitizer and the concentrations of photosensitizer molecules and oxygen in the substrate [20, 21]. Since singlet oxygen has a very short lifetime, the photosensitizer is optimally placed inside or in the vicinity of the target tissues/cells [22 - 25].

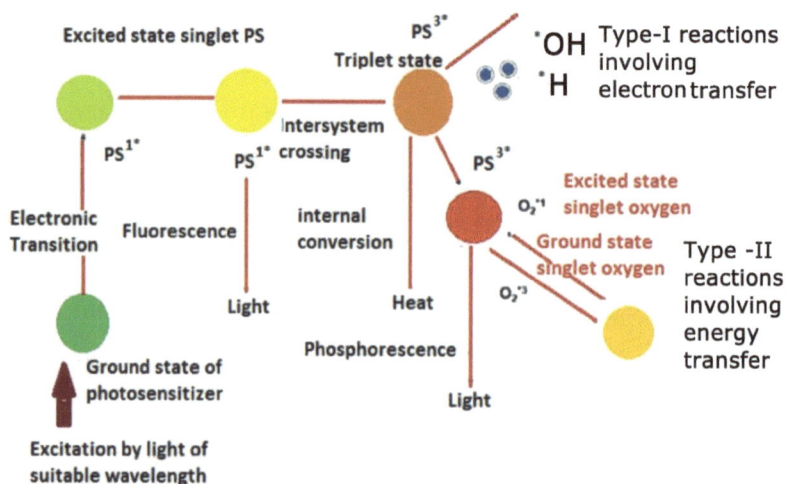

Fig. (1). Singlet oxygen generation in photosensitizers.

Therapeutic Action

An extensive area of research in PDT deals with the mode of action of the photosensitizer in the presence of biological specimens, signal transduction pathways (calcium expression levels, tyrosine kinase expression, transcription factors), the effect on the cell cycle, and the cause of cell death. The spatial distribution of light within the target tissue is therefore crucial. Light is either scattered or absorbed upon entering a cell/tissue, and therefore both processes are dependent on the type of target tissue and also on the wavelength of light chosen for the PDT. In order to study the therapeutic action of PDT, cell genotype and the PDT dose are monitored. Apoptosis or necrosis mechanisms follow pathways that are either intrinsic (mitochondria mediated) or extrinsic (death receptor mediated). In the extrinsic pathway, cell surface receptors are stimulated through the activation of initiator caspase-8 [26] while in the intrinsic pathway, the reaction occurs by the disruption of mitochondrial function. The perturbation in the mitochondrial process causes the release of cytochrome c [27] which binds to Apaf-1 and in the presence of dATP undergoes oligomerization to form an apoptosome. The apoptosome activates the initiator caspase-9, which also leads to the activation of caspase-3, caspase-6 and caspase-7. The cells then undergo shrinkage and eventually break the cytoplasm and nucleus into apoptotic bodies. Inter-nucleosomal sites digest DNA-forming fragments, while the apoptotic cells as well as apoptotic bodies are engulfed by phagocytes and cleared [28 - 31]. It can therefore be concluded that upon PDT treatment, there is the creation of an acute stress in the cellular metabolism which results in apoptosis either through the mitochondrial pathway involving caspases or through the release of cytochrome c.

TYPES OF PHOTOSENSITIZERS

As discussed earlier, a photosensitizer (PS) should exhibit specific features such as high singlet oxygen yields, and long-lived triplet state. It should also show specificity towards target cells/tissues along with efficient killing of the diseased cells/tissues causing minimal damage to the normal tissues around the area of infection/malignancy.

Phenothiazines [32], phthalocyanines [33], and porphyrins [34] are among the highly investigated photosensitizers characterized by the presence of quaternary nitrogen atoms [35, 36]. The nature and number of ionic substituents has a profound influence on the kinetics and the extent of photosensitizer binding in biological cells/tissues. The degree of hydrophobicity plays a major role in determining the type of binding, which can be controlled by varying the number of hydrophobic hydrocarbon chains and hydrophylic ionic groups. As phenothiazines, porphyrins, and phthalocyanines have been shown to induce the inactivation of bacteria, fungi, and pathogenic protozoa, they are considered a prime choice for photosensitizers for PDT [37, 38].

Porphyrins, Phthalocyanines and their Derivatives

Porphyrins and their substituted derivatives comprised of thioethers, esters, and amino groups have been widely investigated Fig. (2), and tuning of their photophysical properties has been carried by selecting appropriate macrocycle substituents [39 - 49]. The design of water-soluble porphyrins *via* alkylation, carboxylation, or sulfonation of N-pyridyl-substituted compounds has also been reported [50 - 55]. Haematoporphyrin derivatives such as Photofrin and Photoheme, are popularly known as first generation photosensitizers [56 - 58]. Chlorins and bacteriochlorins of the porphyrin family are also well documented in literature having porphyrin cores with saturation present at one or two double bonds. Texaphyrins consist of five coplanar nitrogen- atoms while its atomic periphery consists of a 22 π-electron system, that is known to be responsible for the red-shift in absorption spectrum [59, 60].

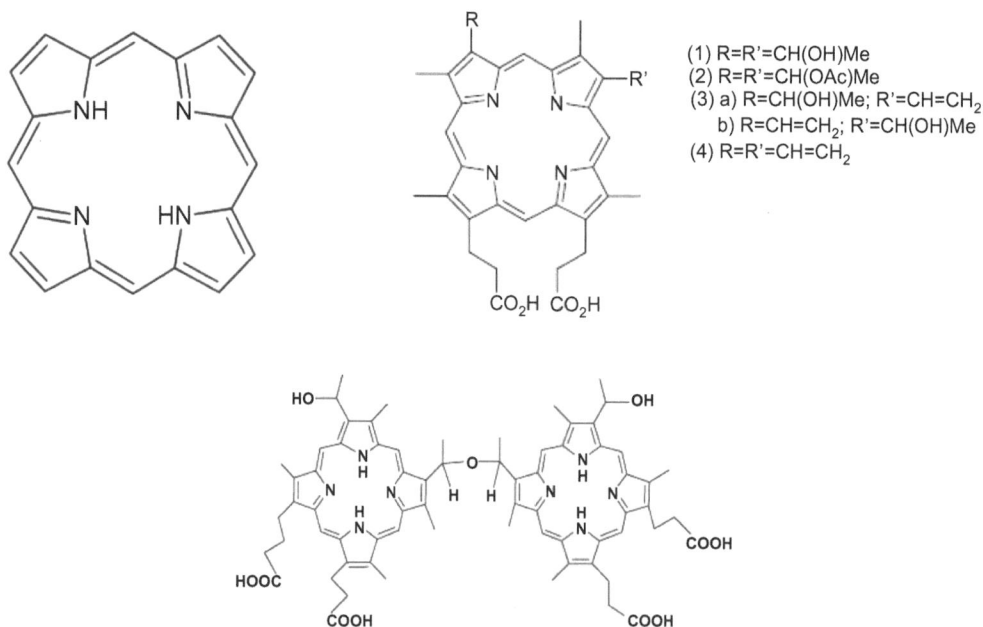

(1) R=R'=CH(OH)Me
(2) R=R'=CH(OAc)Me
(3) a) R=CH(OH)Me; R'=CH=CH$_2$
   b) R=CH=CH$_2$; R'=CH(OH)Me
(4) R=R'=CH=CH$_2$

Fig. (2). Structure of porphyrins.

Mikata *et al.* [61] investigated tetra-glycogenated and octa-glycogenated tetraphenyl chlorins and porphyrins which showed comparable singlet oxygen generation. It was found that octa-glycosated derivatives showed little phototoxicity, while porphyrin derivatives that were tetra-glycosated showed higher photosensitizing ability due to membrane permeability. Haylett *et al.* [62] prepared a series of asymmetric amide protoporphyrin derivatives with quantum yields in the range of 0.01-0.64 while Katona *et al.* [63] investigated tetrakismethoxyphenyl porphyrins and observed that appreciably high triplet quantum yields (0.63-0.84) were attained that were found to be dependent on the number and position of the MeO groups.

The structure of phthalocyanine can be differentiated from that of porphyrin by the presence of nitrogen atoms linked to individual pyrrole units. The presence of extended conjugation in these materials is provided by the peripheral benzene moieties (Fig. **3**).

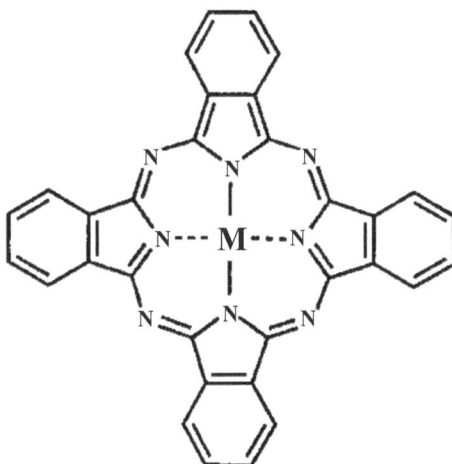

Fig. (3). Structure of phthalocyanine.

Metallo-phthalocyanines of diamagnetic metal ions such as Al^{3+} or Zn^{2+} are reported to reveal high quantum yields. Nesterova *et al.* [64] studied phthalocyanine dimerization for *in vivo* and *in vitro* DNA/RNA detection. Mantareva *et al.* [65] prepared zinc based phthalocyanines and utilized them as fluorescent contrast agents for pigmented melanoma tumors. However, one of the problems in using phthalocyanines is their intrinsic tendency towards aggregation. This effect reduces their photoactivity, shortens the singlet and triplet lifetimes and thereby significantly reduces the singlet oxygen production. The use of peripheral substituents and axial ligands has been shown to prevent this effect [66].

Research has also been carried out on the synthesis of fullerene–porphyrin hybrids, that were shown to exhibit excellent photo-absorption and solubility.

Fullerene (C60) is reported to be an excellent electron acceptor in the ground state exhibiting absorption around 1 μm, near the infrared region. A water soluble fullerene–porphyrin hybrid known as "porphylleren" has been patented [67], which is known to be a 'membranotropic fullerene'. This structure has been reported to possess good water solubility along with the capacity of being dispersed and retained in biological membranes lipid layers. Another interesting patent has been filed dealing with a C60 molecule covalently linked to a tetrapyrrole/chlorine-e6 series [68]. Studies on the synthesis of a complex of fullerene complex of poly(vinylpyrrolidone) (PVP)- tetraphenylporphyrin (TPP)–C60 have also been reported.

Polymer Nanoparticles and Dendrimers

Polymer-based nanoparticles are generally used for designing targeted delivery systems loaded with a photosensitizer in which the surface of photosensitizer particle can be modified by polymer encapsulation to improve its biocompatibility [69, 70]. It has been reported that encapsulation of photosensitizer by polymer nanoparticles improves the photodynamic properties. The commonly used techniques for designing such systems are emulsion polymerization or interfacial polymerization. Recently, poly(D,L-lactide-co-glycolide) and poly(D,L-lactide) nanoparticles entrapped within a hydroxyphenyl-porphyrin (p-THPP) have been reported [71]. Photosensitizer loaded polymeric nanoparticles have been developed by Moreno *et al.* [72] using polyacrylamide (PAA) and amine-functionalized polyacrylamide (AFPAA). The nanoparticles were entrapped with disulfonated 4,7-diphenyl-1,10-phenantroline ruthenium [Ru(dpp(SO$_3$)$_2$)$_3$], which exhibited high quantum yield and high photostability along with a hydrodynamic diameter of 20–25 nm. Poly (phthalocyanine-co-sebacic anhydride) (Pc-SA), has also been reported *via* polymerization of Zn(II) based phthalocyanine and sebacic anhydride units. The fluorescence quantum yield of the product was reported to be quite close to that of free phthalocyanine [73].

Cheng *et al.* [74] encapsulated a silicon based phthalocycanine photosensitizer (*Pc*4) using polyethyleneglycol (PEG) molecules grafted on the surface of gold nanoparticles. Citrate-stabilized gold nanoparticles encapsulated with a poly(allylamine hydrochloride) (PAH) layer have also been reported which were found to bear negative charge [75]. Zinc phthalocyanine (ZnPcF16), was encapsulated in poly(ethylene glycol)-coated poly(lactic acid) nanoparticles by 'salting-out' method [76] and the synthesized nanoparticles revealed an average diameter, of 1000 nm.

McCarthy *et al.* [77] also used poly(lactic-co-glycolic acid) (PLGA) nanoparticles to encapsulate meso tetraphenyl porpholactol (m-TPPPL) and meso-tetraphenyl porphyrin (m-TPP). A comparison of the photocytotoxity between internal and external m-TPPPL nanoparticles demonstrated cell specific killing of cancer cell lines. Konan *et al.* [78] prepared sub-150 nm nanoparticles (50:50 PLGA:poly(DL-lactide) (PLA), 75:25 PLGA:PLA and PLA) loaded with a second generation photosensitizer meso-tetra(p-hydroxy-phenyl) porphyrin (p-THPP) and the physical characteristics of the nanoparticles were not found to be influenced by the polymer composition.

Metallic and Metal Oxide Nanoparticles

Nanomaterials have proved to be promising photosensitizers as they can be rendered hydrophilic, and also their surface properties can be modified with a variety of functional groups. Their sub-cellular and sub-micron size, facilitates deeper penetration into tissues. Numerous strategies have been adopted for the preparation of nanomaterials and a variety of inorganic oxides, and metals have been reported to be used as photosensitizers [79 - 85]. Gold nanoparticles (particle size 2-4 nm) using phthalocyanine (Pc), were reported by *Russell et al.,* [86] and they prepared a three component system comprising of the photosensitizer, gold nanoparticles along with a phase transfer reagent) .The system was reported to generate singlet oxygen with enhanced quantum yields as compared to the free phthalocyanine. Reum *et al.* [87] functionalized the surface gold nanoparticles using a layer by layer technique (14.5 nm diameter) while Lopez *et al.* [88, 89] studied the synergistic interaction of zinc based phthalocyanines (ZnPc), with TiO$_2$ nanoparticles and investigated their photoactivity, on cancer cells as well as Leishmania parasites. The same authors compared the photosensitizing effect of zinc based phthalocyanines ZnPc, nano-TiO$_2$, and composites of ZnPc-TiO$_2$ [90]. Although the composite was reported to be phototoxic against *L. Chagasi* or *L. Panamensis* promastigotes, it was active against tumor and non-tumor mammalian cells due to lower cells internalization as compared to ZnPc. Yamaguchi *et al.* [91] demonstrated the photocatalytic anti-tumor activity of PEG modified water-dispersible TiO$_2$ nanoparticles on a glioma cell line. In a similar way, the effect of ultrasound and TiO$_2$ nanoparticle on melanoma cells and HepG2 cancer cells was studied by Harada *et al.* [92] and Ogino *et al.* [93]. The cancer-cell killing effect of platinum based Titania (Pt/TiO$_2$), TiO$_2$ and gold based Titania nanoparticles (Au/TiO$_2$) nanoparticles was investigated by Liu *et al.* and *found* high photodynamic efficiency of Pt/TiO$_2$ under a mild ultraviolet irradiation [94]. Drug loaded ZnO nanorods were designed by Zhang *et al.* [95] to kill human hepatocarcinoma cells (SMMC-7721 cells) while photodynamic response of conjugated magnetic nanoparticles (~20 nm diameter) was studied by Nie *et al.* [96] for gastric cancer imaging of MGC803. The magnetic nanoparticles offered guided drug delivery while covalently incorporated Ce6 molecules retained their functional properties in near infrared fluorescence imaging and PDT response.

Magnetic nanoparticles wrapped in a silica shell were formulated by Shi *et al.* [97] for fluorescence imaging, and tetra-substituted carboxyl-aluminium phthalocyanine (AlC4Pc) was covalently linked in the mesoporous silica framework. The photodynamic experiments revealed 70% cell death at a concentration of 200 µg ml^{-1}.

Quantum Dots

NIR fluorescence imaging of tumor is widely preferred in biomedical imaging because the absorbance spectra in the said region reaches minima for all the biomolecules present in tumor site, which provides a clear window for *in-vivo* optical imaging of the tumor. Quantum dots (QDs) of metal oxides such as cadmium oxide, cadmium selenide, zinc sulphide *etc.* are found to be highly active in the NIR region (700–900 nm) [98 - 100]. Bruchez *et al.* [100] used red nanocrystal probes to label F-actin filaments with biotin whereas a series of aptamers conjugated with different QDs were prepared by Kim *et al.* [101] to image the proteins in the cancer cell membranes. Compared with conventional dye molecules, the nanocrystal-labeled samples showed minimal photobleaching. QD 630-streptavidin (red) and QD 535-streptavidin (green) were utilized by Wu *et al.* [102] to stain microtubules and actin filaments and results revealed that QD-based probes were effective in labeling fine cellular structures. Bright QDs conjugated with anti-tubulin antibody were investigated by Higuchi *et al.* [103] which showed binding to microtubules in cancer cells. A non-cadmium QD coated with mercaptopropionic acid (QD800-MPA) showing λ_{em} of about 800 nm was reported by Gao *et al.* [104]. The material was seen to exhibit excellent contrast of tumor to surrounding tissues due to the enhanced permeability and retention effect. QDs of cadmium selenide and zinc sulphite (CdSe/ZnS) were labeled with proliferating cell nuclear antigens (PCNA) in breast cancer tissues by Tang *et al.* [105].

QDs *were also* developed by Gambhir *et al.* [106] for real-time imaging of luminal endothelium in mouse tumor neovasculature and it was found that the QDs specifically bound to tumor vessel and showed improved performance over organic dyes. Folic acid (FA) conjugated QDs were designed by Heyo *et al.* [107] *that* were used as photosensitizers for cancer treatment while Burda *et al.* [108] developed cadmium selenide based (CdSe) QDs to generate reactive $^1O_2$ species using the mechanism of fluorescence resonance energy transfer (FRET), (Fig. **4**). Results revealed that the QDs could be used to sensitize phthalocyanines or themselves act as photosensitizers to generate singlet oxygen ($^1O_2$). Multifunctional QDs based on doxorubicin were conjugated to QDs for cancer-targeted imaging [110] while Jon *et al.* [109], studied aptamer binding efficiency to prostate-specific membrane antigen (PSMA), (Fig. **5**).

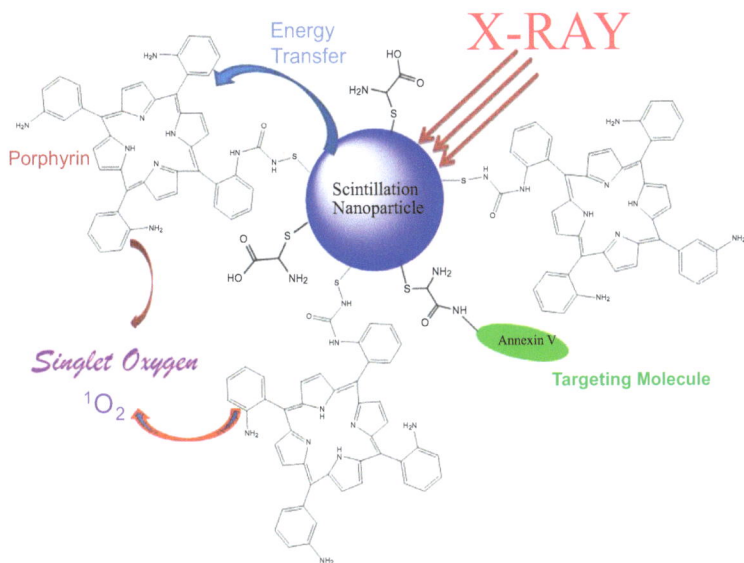

Fig. (4). Schematics of the singlet oxygen generation in QD-based PDT systems.
Reprinted with permission from American Chemical Society, A.C.S. Samia, X.B.Chen, C.Burda, Semiconductor quantum dots for photodynamic therapy. J. Am. Chem. Soc. 2003, 125, 15736–15737.

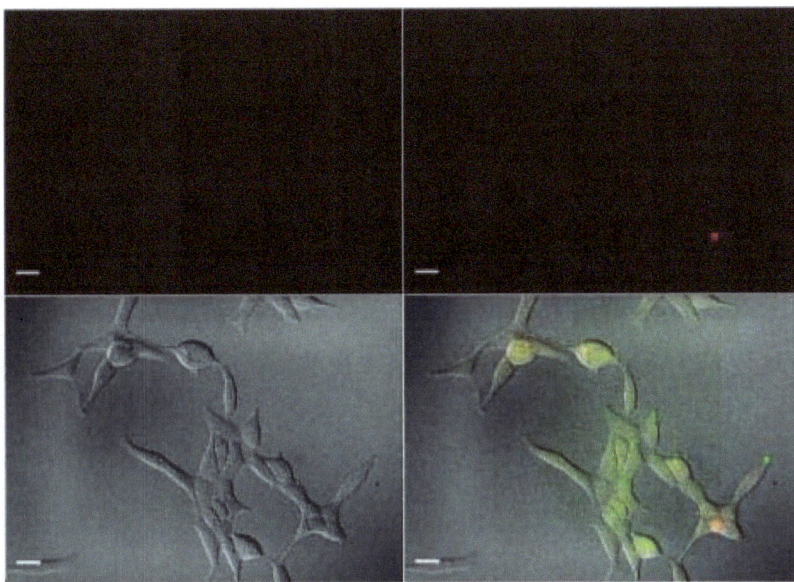

(a)

(Fig 5) contd.....

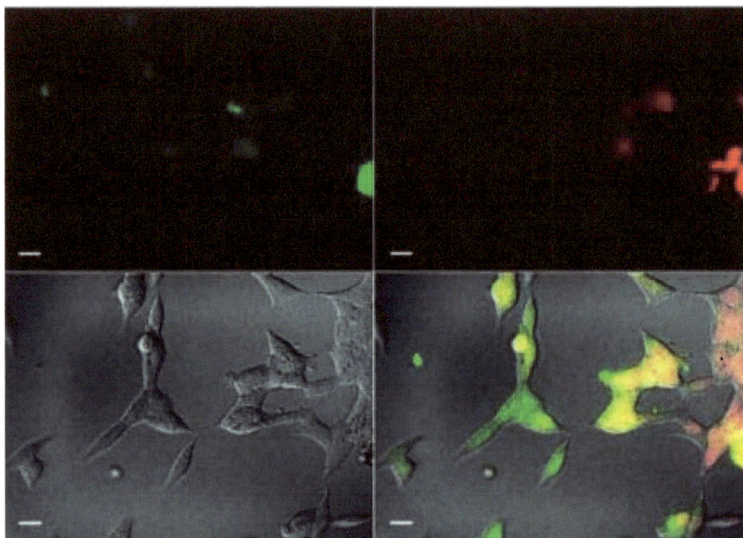

(b)

Fig. (5). Confocal laser scanning microscopy images of PSMA expressing LNCaP cells. after incubation with 100 nM QD-Apt-(Dox) conjugates for 0.5 h at 37 °C, washed two times with PBS buffer, and further incubated at 37 °C for (a) 0 h and (b) 1.5 h.
Reprinted with permission from American Chemical Society, V. Bagalkot, L.Zhang, E.Levy-Nissenbaum, S.Jon, P.W. Kantoff, R. Langer, O.C. Farokhzad, Quantum dot–aptamer conjugates for synchronous cancer imaging, therapy, and sensing of drug delivery based on bi-fluorescence resonance energy transfer. Nano Lett. 2007, 7, 3065–3070.

Silica Nanoparticles

Prasad *et al.* [111], developed a photosensitizer *via* controlled hydrolysis of triethylvinylsilane in micellar media for imaging esophageal cancer. The same authors also studied PEG-coated silica nanoparticles (50–350 nm) for their photodynamic activity [112]. Wieder *et al.* [113] synthesized silicon (Si) and gold (Au) based photosensitizers and showed that Au nanoparticles exhibited higher diffusion of singlet oxygen due to extremely small particle size, large surface area as compared to Si nanoparticles [114, 115]. Tada *et al.* [116] designed magnetic particles of around 11 nm in diameter encapsulated in silica spheres having diameters of 30 nm. Methylene blue was entrapped in the silica matrix and singlet oxygen was generated through the solution *via* diffusion.

Photosensitizers can simultaneously absorb two photons of lower energy, which facilitates deeper penetration of light into the tissue. Moreover, it involves photons of lesser energy to produce excitation as compared to the absorption of a single photon of higher energy and therefore, the concept of two-photon absorption (TPA) induced excitation of photosensitizers has also been utilized to enhance light penetration [117 - 120]. Zhang *et al.* [119] prepared photon upconverting nanoparticles (PUNPs) using $NaYF_4:Yb^{3+}$, Eu^{3+}, which are normally known as efficient photon up converting phosphors. The formation of singlet oxygen was comparable to that observed by Tang *et al.* [105].

CURRENT PROBLEMS, CHALLENGES AND FUTURE PROSPECTS

Despite the many advantages of PDT and the growing knowledge in recent decades, it is evident that clinical applications of PDT are still not very widespread [121]. The mechanisms involving PDT are still undergoing clinical and laboratory trials and to date it is still unclear which reactions take place or are involved in PDT and what their significance is for the process. The first step in considering the quality of a photosensitizer is based on its chemical and photophysical properties, such as singlet oxygen generation, light absorption at specific range of wavelengths, the occurrence of concurrent photoprocesses and reactions, molecular aggregation, hydrophobicity and photobleaching. However, these parameters are far from sufficient to conclude whether a substance can be successfully applied as photosensitizer for real applications. This is because of the very complex character of PDT itself. In order to understand the complexity of PDT, important targets of photooxidation reactions in the cells must be identified. Various factors affect the process of PTD, such as the interactions of photosensitizer molecules with all the types of substances present in biological tissues and cells. The action of some agents is limited or prevented due to the drug efflux by multi-drug resistance pumps from the cells. Incorporating a photosensitizer molecule into a certain kind of nanoparticle can prevent such elimination. Highly efficient photosensitizers in terms of high yields of photoactivity and singlet oxygen generation can completely fail in biological experiments or when used in real applications. Many interactions between the photosensitizer and the targets in cells result in the deactivation of the triplet excited state of agent molecules, without any subsequent chemical reaction or photophysical process. An example is the quenching followed by recovering of the ground state of the photosensitizer by the mechanisms of either electron or energy transfer processes, which eventually ends up in the transfer of light energy to heat. The direct activation of molecular oxygen to its singlet state is an ideal pathway of PDT, and so is the main objective of PDT. However, the lifetime of singlet oxygen can be significantly shortened upon contact with various substances present in biological tissues. These shortened lifetimes can significantly reduce overall efficiency. Another aspect is the migration of oxygen molecules, which is a necessary part of the process for PDT to reach a target. With short lifetimes, the activated oxygen inside cells cannot diffuse to a sufficient distance. Another route of PTD is the formation of radical species, which proceeds via electron transfer. The radical products undergo very complex changes, and their reactions are dependent on the properties of the substances they interact with. Therefore, the activity and significance of the radical mediated processes for the efficiency of PDT is not well understood yet. Moreover, radicals can easily react and thus deactivate photosensitizers. Photosensitized radical chain reactions involving proteins, nucleic acids, lipids, carbohydrates, as well as antioxidants are still yet to be investigated in detail, and their significance for PDT should be addressed. Therefore, nowadays the main target of PDT is still singlet oxygen formation. Unfavorable properties of some photosensitizers such as low stability or reactivity can be improved by incorporating them into nanoparticles, carriers, micelles, polymeric particles or hybrid materials. For example, methylene blue is a cheap, common and efficient photosensitizer, which exhibits high yields of triplet states, strong light absorption in the red region and good water solubility. However, it can be rapidly eliminated in biological systems by reduction to its leuco form [122]. Protecting this dye can significantly improve its PDT action, as has already been indicated by some experiments [123]. In a similar way, commonly used chlorin e6 exhibits advantageous properties when tested in hybrid materials with polyelectrolytes, human serum albumin, polymeric nanoparticles and inorganic oxides [124]. The major limitation of the use of curcumins in PDT is due to its low solubility in water. This obstacle can be dealt with by incorporating this compound in drug delivery systems based on polymeric particles. The hybrids based on curcumin incorporated in cellulose nanoparticles have been shown to exhibit improved properties compared to the free compound [125]. Nanoparticles could even help in PDT targeting, enhancing the permeability and retention time of the photosensitizer, but may also prevent the accumulation of photosensitizer molecules outside the targets, thus eliminating some adverse effects of photosensitization. The toxicity of some photosensitizers is the main obstacle to using them in PDT. Incorporating the molecules into the body of nanocarriers can eliminate the adverse effects of such agents. The doses of the agents confined in nanoparticles can be much larger than that of the free toxic molecules applied directly [126]. The limitations of the substances commonly used or considered to be promising photosensitizers can be overcome by seeking target-specific agents for PDT. The ideal solution to this problem would be identifying such formulations which can selectively kill tumor cells without affecting healthy ones. Certainly, achieving this goal requires a much deeper understanding of all the significant processes involved in PDT to specify a suitable photosensitizer, but also to build more complex systems where other, secondary components may play important roles. The advantage of nanomaterials is apparent here: Nanomaterials and hybrid materials are ideal types of substances which can play the role of nanocarriers in multicomponent systems. Immunoconjugates, antimicrobial agents or chemotherapeutics with other bases of action, antibodies and other pharmaceutical agents can be involved together with photosensitizers in such hybrids and ternary systems. The resulting nanomaterials can be complex systems exhibiting multifunctional properties. Nanotechnology and recent developments in research related to nanomaterials can provide valuable tools for designing new types of photosensitizers and formulations which would exhibit superior properties and high PDT efficiency. There are numerous parameters of nanomaterials which still have to be optimized, such as particle size, morphology, their interactions with photosensitizers, materials' intrinsic photoactivity, materials' toxicity vs. biocompatibility, etc. The properties of PDT formulations change significantly upon involving additional components in the systems. Even more complex behaviour is expected for very complex, multicomponent materials interacting with the substances in cells and biological tissues. Thus, the success of PDT based on nanomaterials will greatly depend on the focused research of cooperating experts from diverse research areas, such as chemistry, materials science, biology, medicine, physics and pharmacy.

CONCLUSIONS

The role of nanomaterials as photosensitizing agents is still in its infancy and much remains to be learned. The investigations pertaining to photodynamic therapy discussed in this chapter may open new avenues for nanotechnology in the near future. The concept of utilizing upconversion nanoparticles, for tumour imaging along with photodynamic therapy of malignancy in cells/tissues can prove to be a major breakthrough in cancer diagnosis. The rapid technological development of new photosensitizers of low-cost, low-toxicity, high singlet oxygen generating efficiency, and high photostability, would not only help in developing a secure market place for these materials but would also promote deeper understanding of multiplexed bioimaging processes.

CONFLICT OF INTEREST

The authors declare no conflict of interest, financial or otherwise.

ACKNOWLEDGEMENTS

The corresponding author Dr Ufana Riaz wishes to thank the Department of Science and technology (DST)-science and engineering research board DST-SERB, India vide sanction no. SB/S1/PC-070/2013 for providing a grant for this major research project. This work was also supported by the Slovak Research and Development Agency under contract No. APVV-15-0347.

REFERENCES

[1]　Hainfeld, J.F.; Slatkin, D.N.; Smilowitz, H.M. The use of gold nanoparticles to enhance radiotherapy in mice. *Phys. Med. Biol.,* **2004**, *49*(18), N309-N315.
[http://dx.doi.org/10.1088/0031-9155/49/18/N03] [PMID: 15509078]

[2]   Juzenas, P.; Chen, W.; Sun, Y.P.; Coelho, M.A.; Generalov, R.; Generalova, N.; Christensen, I.L. Quantum dots and nanoparticles for photodynamic and radiation therapies of cancer. *Adv. Drug Deliv. Rev.,* **2008**, *60*(15), 1600-1614.
        [http://dx.doi.org/10.1016/j.addr.2008.08.004] [PMID: 18840487]

[3]   Kennedy, J.E. High-intensity focused ultrasound in the treatment of solid tumours. *Nat. Rev. Cancer,* **2005**, *5*(4), 321-327.
        [http://dx.doi.org/10.1038/nrc1591] [PMID: 15776004]

[4]   Yu, T.; Wang, Z.; Mason, T.J. A review of research into the uses of low level ultrasound in cancer therapy. *Ultrason. Sonochem.,* **2004**, *11*(2), 95-103.
        [http://dx.doi.org/10.1016/S1350-4177(03)00157-3] [PMID: 15030786]

[5]   Huang, Z. A review of progress in clinical photodynamic therapy. *Technol. Cancer Res. Treat.,* **2005**, *4*(3), 283-293.
        [http://dx.doi.org/10.1177/153303460500400308] [PMID: 15896084]

[6]   Lucky, S.S.; Soo, K.C.; Zhang, Y. Nanoparticles in photodynamic therapy. *Chem. Rev.,* **2015**, *115*(4), 1990-2042.
        [http://dx.doi.org/10.1021/cr5004198] [PMID: 25602130]

[7]   Lu, K.; He, C.; Lin, W. A Chlorin-Based Nanoscale Metal-Organic Framework for Photodynamic Therapy of Colon Cancers. *J. Am. Chem. Soc.,* **2015**, *137*(24), 7600-7603.
        [http://dx.doi.org/10.1021/jacs.5b04069] [PMID: 26068094]

[8]   Samia, A.C.; Chen, X.; Burda, C. Semiconductor quantum dots for photodynamic therapy. *J. Am. Chem. Soc.,* **2003**, *125*(51), 15736-15737.
        [http://dx.doi.org/10.1021/ja0386905] [PMID: 14677951]

[9]   Louis, P.C.; Dubbelman, T.M. Fundamentals of photodynamic therapy: cellular and biochemical aspects. *Anticancer Drugs,* **1994**, *5*(2), 115-243.
        [PMID: 8049494]

[10]   Brown, S.B.; Brown, E.A.; Walker, I. The present and future role of photodynamic therapy in cancer treatment. *Lancet Oncol.,* **2004**, *5*(8), 497-508.
        [http://dx.doi.org/10.1016/S1470-2045(04)01529-3] [PMID: 15288239]

[11]   Ozlem, S.; Akkaya, E.U. Thinking outside the silicon box: molecular and logic as an additional layer of selectivity in singlet oxygen generation for photodynamic therapy. *J. Am. Chem. Soc.,* **2009**, *131*(1), 48-49.
        [http://dx.doi.org/10.1021/ja808389t] [PMID: 19086786]

[12]   Wang, S.; Gao, R.; Zhou, F.; Selke, M. Nanomaterials and singlet oxygen photosensitizers: potential applications in photodynamic therapy. *J. Mater. Chem.,* **2004**, *14*, 487-493.
        [http://dx.doi.org/10.1039/b311429e]

[13]   Kim, S.; Ohulchanskyy, T.Y.; Pudavar, H.E.; Pandey, R.K.; Prasad, P.N. Organically modified silica nanoparticles co-encapsulating photosensitizing drug and aggregation-enhanced two-photon absorbing fluorescent dye aggregates for two-photon photodynamic therapy. *J. Am. Chem. Soc.,* **2007**, *129*(9), 2669-2675.
        [http://dx.doi.org/10.1021/ja0680257] [PMID: 17288423]

[14]   Konan, Y.N.; Gurny, R.; Allémann, E. State of the art in the delivery of photosensitizers for photodynamic therapy. *J. Photochem. Photobiol. B,* **2002**, *66*(2), 89-106.
        [http://dx.doi.org/10.1016/S1011-1344(01)00267-6] [PMID: 11897509]

[15]   Castano, A.P.; Demidova, T.N.; Hamblin, M.R. Mechanisms in photodynamic therapy: part one-photosensitizers, photochemistry and cellular localization. *Photodiagn. Photodyn. Ther.,* **2004**, *1*(4), 279-293.
        [http://dx.doi.org/10.1016/S1572-1000(05)00007-4] [PMID: 25048432]

[16]   Daniell, M.D.; Hill, J.S. A history of photodynamic therapy. *Aust. N. Z. J. Surg.,* **1991**, *61*(5), 340-348.

[http://dx.doi.org/10.1111/j.1445-2197.1991.tb00230.x] [PMID: 2025186]

[17] Dougherty, T.J.; Gomer, C.J.; Henderson, B.W.; Jori, G.; Kessel, D.; Korbelik, M.; Moan, J.; Peng, Q. Photodynamic therapy. *J. Natl. Cancer Inst.,* **1998**, *90*(12), 889-905.
[http://dx.doi.org/10.1093/jnci/90.12.889] [PMID: 9637138]

[18] Ochsner, M. Light scattering of human skin: a comparison between zinc (II)-phthalocyanine and photofrin II. *J. Photochem. Photobiol. B,* **1996**, *32*(1-2), 3-9.
[http://dx.doi.org/10.1016/1011-1344(95)07209-8] [PMID: 8725049]

[19] Plaetzer, K.; Krammer, B.; Berlanda, J.; Berr, F.; Kiesslich, T. Photophysics andphotochemistry of photodynamic therapy: fundamental aspects. *Lasers Med. Sci.,* **2008**, *1*, 1-15.
[PMID: 18247081]

[20] Pazos, Md.; Nader, H.B. Effect of photodynamic therapy on the extracellular matrix and associated components. *Braz. J. Med. Biol. Res.,* **2007**, *40*(8), 1025-1035.
[http://dx.doi.org/10.1590/S0100-879X2006005000142] [PMID: 17665038]

[21] Saikali, S.; Avril, T.; Collet, B.; Hamlat, A.; Bansard, J.Y.; Drenou, B.; Guegan, Y.; Quillien, V. Expression of nine tumour antigens in a series of human glioblastoma multiforme: interest of EGFRvIII, IL-13Ralpha2, gp100 and TRP-2 for immunotherapy. *J. Neurooncol.,* **2007**, *81*(2), 139-148.
[http://dx.doi.org/10.1007/s11060-006-9220-3] [PMID: 17004103]

[22] De Rosa, F.S.; Bentley, M.V. Photodynamic therapy of skin cancers: sensitizers, clinical studies and future directives. *Pharm. Res.,* **2000**, *17*(12), 1447-1455.
[http://dx.doi.org/10.1023/A:1007612905378] [PMID: 11303952]

[23] Castano, A.P.; Demidova, T.N.; Hamblin, M.R. Mechanisms in photodynamic therapy: part one-photosensitizers, photochemistry and cellular localization. *Photodiagn. Photodyn. Ther.,* **2004**, *1*(4), 279-293.
[http://dx.doi.org/10.1016/S1572-1000(05)00007-4] [PMID: 25048432]

[24] Zou, H.; Li, Y.; Liu, X.; Wang, X. An APAF-1.cytochrome c multimeric complex is a functional apoptosome that activates procaspase-9. *J. Biol. Chem.,* **1999**, *274*(17), 11549-11556.
[http://dx.doi.org/10.1074/jbc.274.17.11549] [PMID: 10206961]

[25] Lauber, K.; Appel, H.A.; Schlosser, S.F.; Gregor, M.; Schulze- Osthoff, K.; Wesselborg, S. The adapter protein apoptotic proteaseactivating factor-1 (Apaf-1) is proteolytically processed during apoptosis. *J. Biol. Chem.,* **2001**, *276*, 29772-29781.
[http://dx.doi.org/10.1074/jbc.M101524200] [PMID: 11387322]

[26] Adams, J.M.; Cory, S. Apoptosomes: engines for caspase activation. *Curr. Opin. Cell Biol.,* **2002**, *14*(6), 715-720.
[http://dx.doi.org/10.1016/S0955-0674(02)00381-2] [PMID: 12473344]

[27] Krieser, R.J.; White, K. Engulfment mechanism of apoptotic cells. *Curr. Opin. Cell Biol.,* **2002**, *14*(6), 734-738.
[http://dx.doi.org/10.1016/S0955-0674(02)00390-3] [PMID: 12473347]

[28] Dahle, J.; Kaalhus, O.; Moan, J.; Steen, H.B. Cooperative effects of photodynamic treatment of cells in microcolonies. *Proc. Natl. Acad. Sci. USA,* **1997**, *94*(5), 1773-1778.
[http://dx.doi.org/10.1073/pnas.94.5.1773] [PMID: 9050854]

[29] Tan, K.H.; Hunziker, W. Compartmentalization of Fas and Fas ligand may prevent auto- or paracrine apoptosis in epithelial cells. *Exp. Cell Res.,* **2003**, *284*(2), 283-290.
[http://dx.doi.org/10.1016/S0014-4827(02)00056-3] [PMID: 12651160]

[30] Zhuang, S.; Demirs, J.T.; Kochevar, I.E. p38 mitogen-activated protein kinase mediates bid cleavage, mitochondrial dysfunction, and caspase-3 activation during apoptosis induced by singlet oxygen but not by hydrogen peroxide. *J. Biol. Chem.,* **2000**, *275*(34), 25939-25948.
[http://dx.doi.org/10.1074/jbc.M001185200] [PMID: 10837470]

[31]   Zhuang, S.; Kochevar, I.E. Ultraviolet A radiation induces rapid apoptosis of human leukemia cells by Fas ligand-independent activation of the Fas death pathways. *Photochem. Photobiol.,* **2003**, *78*(1), 61-67.
[PMID: 12929750]

[32]   Ormond, A.B.; Freeman, H.S. Dye sensitizers for photodynamic therapy. *Materials (Basel),* **2013**, *6*(3), 817-840.
[http://dx.doi.org/10.3390/ma6030817]

[33]   Hone, D.C.; Walker, P.I.; Gowing, R.E.; FitzGerald, S.; Beeby, A.; Chambrier, I.; Cook, M.J.; Russell, D.A. Generation of cytotoxic singlet oxygen *via* phthalocyanine-stabilized gold nanoparticles: a potential delivery vehicle for photodynamic therapy. *Langmuir,* **2002**, *18*(8), 2985-2987.
[http://dx.doi.org/10.1021/la0256230]

[34]   Gorman, S.A.; Brown, S.B.; Griffiths, J. An overview of synthetic approaches to porphyrin, phthalocyanine, and phenothiazine photosensitizers for photodynamic therapy. *J. Environ. Pathol. Toxicol. Oncol.,* **2006**, *25*(1-2), 79-108.
[http://dx.doi.org/10.1615/JEnvironPatholToxicolOncol.v25.i1-2.50] [PMID: 16566711]

[35]   Bonnett, R. Photosensitizers of the porphyrin and phthalocyanine series for photodynamic therapy. *Chem. Soc. Rev.,* **1995**, *24*, 19-33.
[http://dx.doi.org/10.1039/cs9952400019]

[36]   Wolf, P.; Rieger, E.; Kerl, H. Topical photodynamic therapy with endogenous porphyrins after application of 5-aminolevulinic acid. An alternative treatment modality for solar keratoses, superficial squamous cell carcinomas, and basal cell carcinomas? *J. Am. Acad. Dermatol.,* **1993**, *28*(1), 17-21.
[http://dx.doi.org/10.1016/0190-9622(93)70002-B] [PMID: 8318069]

[37]   Hamblin, M.R.; Hasan, T. Photodynamic therapy: a new antimicrobial approach to infectious disease? *Photochem. Photobiol. Sci.,* **2004**, *3*(5), 436-450.
[http://dx.doi.org/10.1039/b311900a] [PMID: 15122361]

[38]   Kennedy, J.C.; Pottier, R.H. Endogenous protoporphyrin IX, a clinically useful photosensitizer for photodynamic therapy. *J. Photochem. Photobiol. B,* **1992**, *14*(4), 275-292.
[http://dx.doi.org/10.1016/1011-1344(92)85108-7] [PMID: 1403373]

[39]   Ethirajan, M.; Chen, Y.; Joshi, P.; Pandey, R.K. The role of porphyrin chemistry in tumor imaging and photodynamic therapy. *Chem. Soc. Rev.,* **2011**, *40*(1), 340-362.
[http://dx.doi.org/10.1039/B915149B] [PMID: 20694259]

[40]   O'Connor, A.E.; Gallagher, W.M.; Byrne, A.T. Porphyrin and nonporphyrin photosensitizers in oncology: preclinical and clinical advances in photodynamic therapy. *Photochem. Photobiol.,* **2009**, *85*(5), 1053-1074.
[http://dx.doi.org/10.1111/j.1751-1097.2009.00585.x] [PMID: 19682322]

[41]   Kessel, D. Relocalization of cationic porphyrins during photodynamic therapy. *Photochem. Photobiol. Sci.,* **2002**, *1*(11), 837-840.
[http://dx.doi.org/10.1039/b206046a] [PMID: 12659521]

[42]   Mang, T.S.; Dougherty, T.J.; Potter, W.R.; Boyle, D.G.; Somer, S.; Moan, J. Photobleaching of porphyrins used in photodynamic therapy and implications for therapy. *Photochem. Photobiol.,* **1987**, *45*(4), 501-506.
[http://dx.doi.org/10.1111/j.1751-1097.1987.tb05409.x] [PMID: 3575444]

[43]   Bown, S.G.; Tralau, C.J.; Smith, P.D.; Akdemir, D.; Wieman, T.J. Photodynamic therapy with porphyrin and phthalocyanine sensitisation: quantitative studies in normal rat liver. *Br. J. Cancer,* **1986**, *54*(1), 43-52.
[http://dx.doi.org/10.1038/bjc.1986.150] [PMID: 2942166]

[44]   Moan, J. Effect of bleaching of porphyrin sensitizers during photodynamic therapy. *Cancer Lett.,* **1986**, *33*(1), 45-53.

[http://dx.doi.org/10.1016/0304-3835(86)90100-X] [PMID: 2945634]

[45]   Spikes, J.D.; Jori, G. Photodynamic therapy of tumours and other diseases using porphyrins. *Lasers Med. Sci.,* **1987**, *2*(1), 3-15.
[http://dx.doi.org/10.1007/BF02594124]

[46]   Chakraborty, A.; Held, K.D.; Prise, K.M.; Liber, H.L.; Redmond, R.W. Bystander effects induced by diffusing mediators after photodynamic stress. *Radiat. Res.,* **2009**, *172*(1), 74-81.
[http://dx.doi.org/10.1667/RR1669.1] [PMID: 19580509]

[47]   Batlle, A.M. Porphyrins, porphyrias, cancer and photodynamic therapy--a model for carcinogenesis. *J. Photochem. Photobiol. B,* **1993**, *20*(1), 5-22.
[http://dx.doi.org/10.1016/1011-1344(93)80127-U] [PMID: 8229469]

[48]   Peng, Q.; Berg, K.; Moan, J.; Kongshaug, M.; Nesland, J.M. 5-Aminolevulinic acid-based photodynamic therapy: principles and experimental research. *Photochem. Photobiol.,* **1997**, *65*(2), 235-251.
[http://dx.doi.org/10.1111/j.1751-1097.1997.tb08549.x] [PMID: 9066303]

[49]   Arnbjerg, J.; Jiménez-Banzo, A.; Paterson, M.J.; Nonell, S.; Borrell, J.I.; Christiansen, O.; Ogilby, P.R. Two-photon absorption in tetraphenylporphycenes: are porphycenes better candidates than porphyrins for providing optimal optical properties for two-photon photodynamic therapy? *J. Am. Chem. Soc.,* **2007**, *129*(16), 5188-5199.
[http://dx.doi.org/10.1021/ja0688777] [PMID: 17397157]

[50]   Rück, A.; Köllner, T.; Dietrich, A.; Strauss, W.; Schneckenburger, H. Fluorescence formation during photodynamic therapy in the nucleus of cells incubated with cationic and anionic water-soluble photosensitizers. *J. Photochem. Photobiol. B,* **1992**, *12*(4), 403-412.
[http://dx.doi.org/10.1016/1011-1344(92)85044-U] [PMID: 1578298]

[51]   Kuimova, M.K.; Collins, H.A.; Balaz, M.; Dahlstedt, E.; Levitt, J.A.; Sergent, N.; Suhling, K.; Drobizhev, M.; Makarov, N.S.; Rebane, A.; Anderson, H.L.; Phillips, D. Photophysical properties and intracellular imaging of water-soluble porphyrin dimers for two-photon excited photodynamic therapy. *Org. Biomol. Chem.,* **2009**, *7*(5), 889-896.
[http://dx.doi.org/10.1039/b814791d] [PMID: 19225671]

[52]   Stilts, C.E.; Nelen, M.I.; Hilmey, D.G.; Davies, S.R.; Gollnick, S.O.; Oseroff, A.R.; Gibson, S.L.; Hilf, R.; Detty, M.R. Water-soluble, core-modified porphyrins as novel, longer-wavelength-absorbing sensitizers for photodynamic therapy. *J. Med. Chem.,* **2000**, *43*(12), 2403-2410.
[http://dx.doi.org/10.1021/jm000044i] [PMID: 10882367]

[53]   Ogawa, K.; Hasegawa, H.; Inaba, Y.; Kobuke, Y.; Inouye, H.; Kanemitsu, Y.; Kohno, E.; Hirano, T.; Ogura, S.; Okura, I. Water-soluble bis(imidazolylporphyrin) self-assemblies with large two-photon absorption cross sections as potential agents for photodynamic therapy. *J. Med. Chem.,* **2006**, *49*(7), 2276-2283.
[http://dx.doi.org/10.1021/jm051072+] [PMID: 16570924]

[54]   Hilmey, D.G.; Abe, M.; Nelen, M.I.; Stilts, C.E.; Baker, G.A.; Baker, S.N.; Bright, F.V.; Davies, S.R.; Gollnick, S.O.; Oseroff, A.R.; Gibson, S.L.; Hilf, R.; Detty, M.R. Water-soluble, core-modified porphyrins as novel, longer-wavelength-absorbing sensitizers for photodynamic therapy. II. Effects of core heteroatoms and *meso*-substituents on biological activity. *J. Med. Chem.,* **2002**, *45*(2), 449-461.
[http://dx.doi.org/10.1021/jm0103662] [PMID: 11784149]

[55]   Atilgan, S.; Ekmekci, Z.; Dogan, A.L.; Guc, D.; Akkaya, E.U. Water soluble distyryl-boradiazaindacenes as efficient photosensitizers for photodynamic therapy. *Chem. Commun. (Camb.),* **2006**, (42), 4398-4400.
[http://dx.doi.org/10.1039/b612347c] [PMID: 17057856]

[56]   Sharman, W.M.; Allen, C.M.; van Lier, J.E. Photodynamic therapeutics: basic principles and clinical applications. *Drug Discov. Today,* **1999**, *4*(11), 507-517.
[http://dx.doi.org/10.1016/S1359-6446(99)01412-9] [PMID: 10529768]

[57] Kudinova, N.V.; Berezov, T.T. Photodynamic therapy of cancer: Search for ideal photosensitizer, Biochem.(Moscow). *Supplement Series B: Biomed.Chem.,* **2010**, *4*(1), 95-103.

[58] Kozyrev, A.N.; Efimov, A.V.; Efremova, O.A.; Perepyolkin, P.Y.; Mironov, A.F. New chlorin and bacteriochlorine-type photosensitizers for photodynamic therapy *Proc. SPIE, Photodynamic Therapy of Cancer II,* **1995**.
[http://dx.doi.org/10.1117/12.199158]

[59] Sessler, J.L.; Miller, R.A. Texaphyrins: new drugs with diverse clinical applications in radiation and photodynamic therapy. *Biochem. Pharmacol.,* **2000**, *59*(7), 733-739.
[http://dx.doi.org/10.1016/S0006-2952(99)00314-7] [PMID: 10718331]

[60] Blumenkranz, M.S.; Woodburn, K.W.; Qing, F.; Verdooner, S.; Kessel, D.; Miller, R. Lutetium texaphyrin (Lu-Tex): a potential new agent for ocular fundus angiography and photodynamic therapy. *Am. J. Ophthalmol.,* **2000**, *129*(3), 353-362.
[http://dx.doi.org/10.1016/S0002-9394(99)00462-6] [PMID: 10704552]

[61] Mikata, Y.; Onchi, Y.; Shibata, M.; Kakuchi, T.; Ono, H.; Ogura, S.; Okura, I.; Yano, S. Synthesis and phototoxic property of tetra- and octa-glycoconjugated tetraphenylchlorins. *Bioorg. Med. Chem. Lett.,* **1998**, *8*(24), 3543-3548.
[http://dx.doi.org/10.1016/S0960-894X(98)00645-3] [PMID: 9934468]

[62] Haylett, A.K.; McNair, F.I.; McGarvey, D.; Dodd, N.J.; Forbes, E.; Truscott, T.G.; Moore, J.V. Singlet oxygen and superoxide characteristics of a series of novel asymmetric photosensitizers. *Cancer Lett.,* **1997**, *112*(2), 233-238.
[http://dx.doi.org/10.1016/S0304-3835(96)04577-6] [PMID: 9066733]

[63] Katona, Z.; Grofcsik, A.; Baranyai, P.; Bitter, I.; Grabner, G.; Kubinyi, M.; Vidóczy, T. Triplet state spectroscopic studies on some 5,10,15,20-tetrakis(methoxyphenyl)porphyrins. *J. Mol. Struct.,* **1998**, *450*, 41-45.
[http://dx.doi.org/10.1016/S0022-2860(98)00411-6]

[64] Nesterova, I.V.; Erdem, S.S.; Pakhomov, S.; Hammer, R.P.; Soper, S.A. Phthalocyanine dimerization-based molecular beacons using near-IR fluorescence. *J. Am. Chem. Soc.,* **2009**, *131*(7), 2432-2433.
[http://dx.doi.org/10.1021/ja8088247] [PMID: 19191492]

[65] Mantareva, V.; Kussovski, V.; Angelov, I.; Borisova, E.; Avramov, L.; Schnurpfeil, G.; Wöhrle, D. Photodynamic activity of water-soluble phthalocyanine zinc(II) complexes against pathogenic microorganisms. *Bioorg. Med. Chem.,* **2007**, *15*(14), 4829-4835.
[http://dx.doi.org/10.1016/j.bmc.2007.04.069] [PMID: 17517508]

[66] Angelov, I.; Mantareva, V.; Kussovski, V.; Woehrle, D.; Borisova, E.; Avramov, L. Improved antimicrobial therapy with cationic tetra- and octa-substituted phthalocyanines/ *Proc. SPIE 7027, 15th International School on Quantum Electronics: Laser Physics and Applications,* , p. 702717.

[67] Pharmaceutical Composition for Photodynamic Therapy and a Method for Treating Oncological Diseases by Using Said Composition. *LADAS & PARRY LLP; IPC8Class: AA61K3805FI; USPC Class: 514 19,*

[68] New Water Soluble Porphylleren Compounds. *KENYON & KENYON LLP; IPC8 Class: AC07D48722FI; USPC Class: 540145 ,*

[69] Kim, H.; Csaky, K.G. Nanoparticle-integrin antagonist C16Y peptide treatment of choroidal neovascularization in rats. *J. Control. Release,* **2010**, *142*(2), 286-293.
[http://dx.doi.org/10.1016/j.jconrel.2009.10.031] [PMID: 19895863]

[70] Lee, Y.E.; Kopelman, R. Polymeric nanoparticles for photodynamic therapy. *Methods Mol. Biol.,* **2011**, *726*, 151-178.
[http://dx.doi.org/10.1007/978-1-61779-052-2_11] [PMID: 21424449]

[71] Chatterjee, D.K.; Fong, L.S.; Zhang, Y. Nanoparticles in photodynamic therapy: an emerging paradigm. *Adv. Drug Deliv. Rev.,* **2008**, *60*(15), 1627-1637.

[http://dx.doi.org/10.1016/j.addr.2008.08.003] [PMID: 18930086]

[72]   Moreno, M.J.; Monson, E.; Reddy, R.G.; Rehemtulla, A.; Ross, B.D.; Philbert, M.; Schneider, R.J.; Kopelman, R. Production of singlet oxygen by Ru(dpp(SO$_3$)$_2$)$_3$ incorporated in polyacrylamide PEBBLES. *Sens. Act. B: Chem.,* **2003**, *90*(1-3), 82-89.

[73]   Fu, J.; Li, X-Y.; Nag, D.K.; Wu, C. Encapsulation of phthalocyanines in biodegradable poly(sebacic anhydride) nanoparticles. *Langmuir,* **2002**, *18*(10), 3843-3847.
       [http://dx.doi.org/10.1021/la011764a]

[74]   Cheng, Y.; C Samia, A.; Meyers, J.D.; Panagopoulos, I.; Fei, B.; Burda, C.; Gonzales, M.A.; Mascharak, P.K. Highly efficient drug delivery with gold nanoparticle vectors for *in vivo* photodynamic therapy of cancer. *J. Am. Chem. Soc.,* **2008**, *130*(32), 10643-10647.
       [http://dx.doi.org/10.1021/ja801631c] [PMID: 18642918]

[75]   Reum, N.; Fink-Straube, C.; Klein, T.; Hartmann, R.W.; Lehr, C.M.; Schneider, M. Multilayer coating of gold nanoparticles with drug-polymer coadsorbates. *Langmuir,* **2010**, *26*(22), 16901-16908.
       [http://dx.doi.org/10.1021/la103109b] [PMID: 20964349]

[76]   Allémann, E.; Brasseur, N.; Benrezzak, O.; Rousseau, J.; Kudrevich, S.V.; Boyle, R.W.; Leroux, J.C.; Gurny, R.; Van Lier, J.E. PEG-coated poly(lactic acid) nanoparticles for the delivery of hexadecafluoro zinc phthalocyanine to EMT-6 mouse mammary tumours. *J. Pharm. Pharmacol.,* **1995**, *47*(5), 382-387.
       [http://dx.doi.org/10.1111/j.2042-7158.1995.tb05815.x] [PMID: 7494187]

[77]   McCarthy, J.R.; Perez, J.M.; Brückner, C.; Weissleder, R. Polymeric nanoparticle preparation that eradicates tumors. *Nano Lett.,* **2005**, *5*(12), 2552-2556.
       [http://dx.doi.org/10.1021/nl0519229] [PMID: 16351214]

[78]   Konan, Y.N.; Berton, M.; Gurny, R.; Allémann, E. Enhanced photodynamic activity of meso-tetra(--hydroxyphenyl)porphyrin by incorporation into sub-200 nm nanoparticles. *Eur. J. Pharm. Sci.,* **2003**, *18*(3-4), 241-249.
       [http://dx.doi.org/10.1016/S0928-0987(03)00017-4] [PMID: 12659935]

[79]   Allison, R.R.; Mota, H.C.; Bagnato, V.S.; Sibata, C.H. Bio-nanotechnology and photodynamic therapy--state of the art review. *Photodiagn. Photodyn. Ther.,* **2008**, *5*(1), 19-28.
       [http://dx.doi.org/10.1016/j.pdpdt.2008.02.001] [PMID: 19356632]

[80]   Chen, W. *Nanoparticle Based Photodynamic Therapy for Cancer Treatment*; American Scientific Publishers, **2007**.

[81]   Panyam, J.; Zhou, W-Z.; Prabha, S.; Sahoo, S.K.; Labhasetwar, V. Rapid endo-lysosomal escape of poly(DL-lactide-co-glycolide) nanoparticles: implications for drug and gene delivery. *FASEB J.,* **2002**, *16*(10), 1217-1226.
       [http://dx.doi.org/10.1096/fj.02-0088com] [PMID: 12153989]

[82]   Ricci-Júnior, E.; Marchetti, J.M. Preparation, characterization, photocytotoxicity assay of PLGA nanoparticles containing zinc (II) phthalocyanine for photodynamic therapy use. *J. Microencapsul.,* **2006**, *23*(5), 523-538.
       [http://dx.doi.org/10.1080/02652040600775525] [PMID: 16980274]

[83]   Zeisser-Labouèbe, M.; Lange, N.; Gurny, R.; Delie, F. Hypericin-loaded nanoparticles for the photodynamic treatment of ovarian cancer. *Int. J. Pharm.,* **2006**, *326*(1-2), 174-181.
       [http://dx.doi.org/10.1016/j.ijpharm.2006.07.012] [PMID: 16930882]

[84]   Pegaz, B.; Debefve, E.; Borle, F.; Ballini, J-P.; van den Bergh, H.; Kouakou-Konan, Y.N. Encapsulation of porphyrins and chlorins in biodegradable nanoparticles: the effect of dye lipophilicity on the extravasation and the photothrombic activity. A comparative study. *J. Photochem. Photobiol. B,* **2005**, *80*(1), 19-27.
       [http://dx.doi.org/10.1016/j.jphotobiol.2005.02.003] [PMID: 15963434]

[85]   Vargas, A.; Eid, M.; Fanchaouy, M.; Gurny, R.; Delie, F. *In vivo* photodynamic activity of

photosensitizer-loaded nanoparticles: formulation properties, administration parameters and biological issues involved in PDT outcome. *Eur. J. Pharm. Biopharm.,* **2008**, *69*(1), 43-53.
[http://dx.doi.org/10.1016/j.ejpb.2007.09.021] [PMID: 18023564]

[86] Hone, D.C.; Walker, P.I.; Evans-Gowing, R.; FitzGerald, S.; Beeby, A.; Chambrier, I.; Cook, M.J.; Russell, D.A. Generation of Cytotoxic Singlet Oxygen *via* Phthalocyanine-Stabilized Gold Nanoparticles: A Potential Delivery Vehicle for Photodynamic Therapy. *Langmuir,* **2002**, *8*, 2985-2990.
[http://dx.doi.org/10.1021/la0256230]

[87] Reum, N.; Fink-Straube, C.; Klein, T.; Hartmann, R.W.; Lehr, C.M.; Schneider, M. Erratum: Multilayer coating of gold nanoparticles with drug- polymer coadsorbates (Langmuir (2010) 26 (16901)). *Langmuir,* **2011**, *27*(2), 861-865.
[http://dx.doi.org/10.1021/la104746c]

[88] López, T.; Figueras, F.; Manjarrez, J.; Bustos, J.; Alvarez, M.; Silvestre-Albero, J.; Rodríguez-Reinoso, F.; Martínez-Ferre, A.; Martínez, E. Catalytic nanomedicine: a new field in antitumor treatment using supported platinum nanoparticles. In vitro DNA degradation and in vivo tests with C6 animal model on Wistar rats. *Eur. J. Med. Chem.,* **2010**, *45*(5), 1982-1990.
[http://dx.doi.org/10.1016/j.ejmech.2010.01.043] [PMID: 20153564]

[89] López, T.; Ortiz, E.; Alvarez, M.; Navarrete, J.; Odriozola, J.A.; Martinez-Ortega, F.; Páez-Mozo, E.A.; Escobar, P.; Espinoza, K.A.; Rivero, I.A. Study of the stabilization of zinc phthalocyanine in sol-gel TiO2 for photodynamic therapy applications. *Nanomedicine (Lond.),* **2010**, *6*(6), 777-785.
[http://dx.doi.org/10.1016/j.nano.2010.04.007] [PMID: 20493967]

[90] Lopez, T.; Ortiz, E.; Alvarez, M.; Navarrete, J.; Odriozola, J.A.; Martinez-Ortega, F.; Páez-Mozo, E.A.; Escobar, P.; Espinoza, K.A.; Rivero, I.A. Study of the stabilization of zinc phthalocyanine in sol-gel TiO_2 for photodynamic therapy applications. *Nanomedicine (Lond.),* **2010**, *6*(6), 777-785.
[http://dx.doi.org/10.1016/j.nano.2010.04.007] [PMID: 20493967]

[91] Yamaguchi, S.; Kobayashi, H.; Narita, T.; Kanehira, K.; Sonezaki, S.; Kubota, Y.; Terasaka, S.; Iwasaki, Y. Novel photodynamic therapy using water-dispersed tio_2–polyethylene glycol compound: evaluation of antitumor effect on glioma cells and spheroids *in vitro* photochemistry and photobiology. *Special Issue: Symposium in Print: "Phototoxicity of the Skin and Eye" in honor of Dr Colin Chignell,* **2010**, *86*(4), pp. 964-971.

[92] Harada, Y.; Ogawa, K.; Irie, Y.; Endo, H.; Feril, L.B., Jr; Uemura, T.; Tachibana, K. Ultrasound activation of TiO_2 in melanoma tumors. *J. Control. Release,* **2011**, *149*(2), 190-195.
[http://dx.doi.org/10.1016/j.jconrel.2010.10.012] [PMID: 20951750]

[93] Ninomiya, K.; Noda, K.; Ogino, C.; Kuroda, S.; Shimizu, N. Enhanced OH radical generation by dual-frequency ultrasound with TiO_2 nanoparticles: its application to targeted sonodynamic therapy. *Ultrason. Sonochem.,* **2014**, *21*(1), 289-294.
[http://dx.doi.org/10.1016/j.ultsonch.2013.05.005] [PMID: 23746399]

[94] Liu, L.; Miao, P.; Xu, Y.; Tian, Z.; Zou, Z.; Li, G. Study of Pt/TiO_2 nanocomposite for cancer-cell treatment. *J. Photochem. Photobiol. B,* **2010**, *98*(3), 207-210.
[http://dx.doi.org/10.1016/j.jphotobiol.2010.01.005] [PMID: 20149675]

[95] Zhang, H.; Shan, Y.; Dong, L. A comparison of TiO_2 and ZnO nanoparticles as photosensitizers in photodynamic therapy for cancer. *J. Biomed. Nanotechnol.,* **2014**, *10*(8), 1450-1457.
[http://dx.doi.org/10.1166/jbn.2014.1961] [PMID: 25016645]

[96] Lin, J.; Wang, S.; Huang, P.; Wang, Z.; Chen, S.; Niu, G.; Li, W.; He, J.; Cui, D.; Lu, G.; Chen, X.; Nie, Z. Photosensitizer-loaded gold vesicles with strong plasmonic coupling effect for imaging-guided photothermal/photodynamic therapy. *ACS Nano,* **2013**, *7*(6), 5320-5329.
[http://dx.doi.org/10.1021/nn4011686] [PMID: 23721576]

[97] Chen, Y.; Chen, H.; Zeng, D.; Tian, Y.; Chen, F.; Feng, J.; Shi, J. Core/shell structured hollow mesoporous nanocapsules: a potential platform for simultaneous cell imaging and anticancer drug

delivery. *ACS Nano,* **2010**, *4*(10), 6001-6013.
[http://dx.doi.org/10.1021/nn1015117] [PMID: 20815402]

[98] Ghasemi, Y.; Peymani, P.; Afifi, S. Quantum dot: magic nanoparticle for imaging, detection and targeting. *Acta Biomed.,* **2009**, *80*(2), 156-165.
[PMID: 19848055]

[99] Cai, W.; Shin, D.W.; Chen, K.; Gheysens, O.; Cao, Q.; Wang, S.X.; Gambhir, S.S.; Chen, X. Peptide-labeled near-infrared quantum dots for imaging tumor vasculature in living subjects. *Nano Lett.,* **2006**, *6*(4), 669-676.
[http://dx.doi.org/10.1021/nl052405t] [PMID: 16608262]

[100] Bruchez, M., Jr; Moronne, M.; Gin, P.; Weiss, S.; Alivisatos, A.P. Semiconductor nanocrystals as fluorescent biological labels. *Science,* **1998**, *281*(5385), 2013-2016.
[http://dx.doi.org/10.1126/science.281.5385.2013] [PMID: 9748157]

[101] Kang, W.J.; Chae, J.R.; Cho, Y.L.; Lee, J.D.; Kim, S. Multiplex imaging of single tumor cells using quantum-dot-conjugated aptamers. *Small,* **2009**, *5*(22), 2519-2522.
[http://dx.doi.org/10.1002/smll.200900848] [PMID: 19714733]

[102] Wu, X.; Liu, H.; Liu, J.; Haley, K.N.; Treadway, J.A.; Larson, J.P.; Ge, N.; Peale, F.; Bruchez, M.P. Immunofluorescent labeling of cancer marker Her2 and other cellular targets with semiconductor quantum dots. *Nat. Biotechnol.,* **2003**, *21*(1), 41-46.
[http://dx.doi.org/10.1038/nbt764] [PMID: 12459735]

[103] Gonda, K.; Watanabe, T.M.; Ohuchi, N.; Higuchi, H. *In vivo* nano-imaging of membrane dynamics in metastatic tumor cells using quantum dots. *J. Biol. Chem.,* **2010**, *285*(4), 2750-2757.
[http://dx.doi.org/10.1074/jbc.M109.075374] [PMID: 19917603]

[104] Gao, J.; Chen, K.; Xie, R.; Xie, J.; Lee, S.; Cheng, Z.; Peng, X.; Chen, X. Ultrasmall near-infrared non-cadmium quantum dots for *in vivo* tumor imaging. *Small,* **2010**, *6*(2), 256-261.
[http://dx.doi.org/10.1002/smll.200901672] [PMID: 19911392]

[105] Xu, H.; Chen, C.; Peng, J.; Tang, H.W.; Liu, C.M.; He, Y.; Chen, Z.Z.; Li, Y.; Zhang, Z.L.; Pang, D.W. Evaluation of the bioconjugation efficiency of different quantum dots as probes for immunostaining tumor-marker proteins. *Appl. Spectrosc.,* **2010**, *64*(8), 847-852.
[http://dx.doi.org/10.1366/000370210792081154] [PMID: 20719046]

[106] Smith, B.R.; Cheng, Z.; De, A.; Koh, A.L.; Sinclair, R.; Gambhir, S.S. Real-time intravital imaging of RGD-quantum dot binding to luminal endothelium in mouse tumor neovasculature. *Nano Lett.,* **2008**, *8*(9), 2599-2606.
[http://dx.doi.org/10.1021/nl080141f] [PMID: 18386933]

[107] Morosini, V.; Bastogne, T.; Frochot, C.; Schneider, R.; François, A.; Guillemin, F.; Barberi-Heyob, M. Quantum dot-folic acid conjugates as potential photosensitizers in photodynamic therapy of cancer. *Photochem. Photobiol. Sci.,* **2011**, *10*(5), 842-851.
[http://dx.doi.org/10.1039/c0pp00380h] [PMID: 21479314]

[108] Samia, A.C.; Chen, X.; Burda, C. Semiconductor quantum dots for photodynamic therapy. *J. Am. Chem. Soc.,* **2003**, *125*(51), 15736-15737.
[http://dx.doi.org/10.1021/ja0386905] [PMID: 14677951]

[109] Bagalkot, V.; Zhang, L.; Levy-Nissenbaum, E.; Jon, S.; Kantoff, P.W.; Langer, R.; Farokhzad, O.C. Quantum dot-aptamer conjugates for synchronous cancer imaging, therapy, and sensing of drug delivery based on bi-fluorescence resonance energy transfer. *Nano Lett.,* **2007**, *7*(10), 3065-3070.
[http://dx.doi.org/10.1021/nl071546n] [PMID: 17854227]

[110] Gillies, E.R.; Fréchet, J.M. pH-Responsive copolymer assemblies for controlled release of doxorubicin. *Bioconjug. Chem.,* **2005**, *16*(2), 361-368.
[http://dx.doi.org/10.1021/bc049851c] [PMID: 15769090]

[111] Roy, I.; Ohulchanskyy, T.Y.; Pudavar, H.E.; Bergey, E.J.; Oseroff, A.R.; Morgan, J.; Dougherty, T.J.;

Prasad, P.N. Ceramic-based nanoparticles entrapping water-insoluble photosensitizing anticancer drugs: a novel drug-carrier system for photodynamic therapy. *J. Am. Chem. Soc.,* **2003**, *125*(26), 7860-7865.
[http://dx.doi.org/10.1021/ja0343095] [PMID: 12823004]

[112]   Yan, F.; Kopelman, R. The embedding of meta-tetra(hydroxyphenyl)-chlorin into silica nanoparticle platforms for photodynamic therapy and their singlet oxygen production and pH-dependent optical properties. *Photochem. Photobiol.,* **2003**, *78*(6), 587-591.
[http://dx.doi.org/10.1562/0031-8655(2003)078<0587:TEOMIS>2.0.CO;2] [PMID: 14743867]

[113]   Wieder, M.E.; Hone, D.C.; Cook, M.J.; Handsley, M.M.; Gavrilovic, J.; Russell, D.A. Intracellular photodynamic therapy with photosensitizer-nanoparticle conjugates: cancer therapy using a 'Trojan horse'. *Photochem. Photobiol. Sci.,* **2006**, *5*(8), 727-734.
[http://dx.doi.org/10.1039/B602830F] [PMID: 16886087]

[114]   Wang, S.; Gao, R.; Zhou, F.; Selke, M. Nanomaterials and singlet oxygen photosensitizers: potential applications in photodynamic therapy. *J. Mater. Chem.,* **2004**, *14*, 487-493.
[http://dx.doi.org/10.1039/b311429e]

[115]   Hone, D.C.; Walker, P.I.; Gowing, R.E.; Gerald, S.F.; Beeby, A.; Chambrier, I.; Cook, M.J.; Russell, D.A. Generation of cytotoxic singlet oxygen *via* phthalocyanine-stabilized gold particles: a potential delivery vehicle for photodynamic therapy. *Langmuir,* **2002**, *18*, 2985-2987.
[http://dx.doi.org/10.1021/la0256230]

[116]   Tada, D.B.; Vono, L.L.; Duarte, E.L.; Itri, R.; Kiyohara, P.K.; Baptista, M.S.; Rossi, L.M. Methylene blue-containing silica-coated magnetic particles: a potential magnetic carrier for photodynamic therapy. *Langmuir,* **2007**, *23*(15), 8194-8199.
[http://dx.doi.org/10.1021/la700883y] [PMID: 17590032]

[117]   Gao, D.; Agayan, R.R.; Xu, H.; Philbert, M.A.; Kopelman, R. Nanoparticles for two-photon photodynamic therapy in living cells. *Nano Lett.,* **2006**, *6*(11), 2383-2386.
[http://dx.doi.org/10.1021/nl0617179] [PMID: 17090062]

[118]   Kim, S.; Ohulchanskyy, T.Y.; Pudavar, H.E.; Pandey, R.K.; Prasad, P.N. Organically modified silica nanoparticles co-encapsulating photosensitizing drug and aggregation-enhanced two-photon absorbing fluorescent dye aggregates for two-photon photodynamic therapy. *J. Am. Chem. Soc.,* **2007**, *129*(9), 2669-2675.
[http://dx.doi.org/10.1021/ja0680257] [PMID: 17288423]

[119]   Zhang, P.; Steelant, W.; Kumar, M.; Scholfield, M. Versatile photosensitizers for photodynamic therapy at infrared excitation. *J. Am. Chem. Soc.,* **2007**, *129*(15), 4526-4527.
[http://dx.doi.org/10.1021/ja0700707] [PMID: 17385866]

[120]   Heer, S.; Kömpe, K.; Güdel, H.U.; Haase, M. Highly efficient multicolour upconversion emission in transparent colloids of lanthanide-doped $NaYF_4$ nanocrystals. *Adv. Mater.,* **2004**, *16*, 2102-2105.
[http://dx.doi.org/10.1002/adma.200400772]

[121]   Banerjee, S.M.; MacRobert, A.J.; Mosse, C.A.; Periera, B.; Bown, S.G.; Keshtgar, M.R. Photodynamic therapy: Inception to application in breast cancer. *Breast,* **2017**, *31*, 105-113.
[http://dx.doi.org/10.1016/j.breast.2016.09.016] [PMID: 27833041]

[122]   DiSanto, A.R.; Wagner, J.G. Pharmacokinetics of highly ionized drugs. II. Methylene blue--absorption, metabolism, and excretion in man and dog after oral administration. *J. Pharm. Sci.,* **1972**, *61*(7), 1086-1090.
[http://dx.doi.org/10.1002/jps.2600610710] [PMID: 5044807]

[123]   Bujdák, J.; Jurečeková, J.; Bujdáková, H.; Lang, K.; Šeršeň, F. Clay mineral particles as effficient carriers of methylene blue used for antimicrobial treatment. *Environ. Sci. Technol.,* **2009**, *43*(16), 6202-6207.
[http://dx.doi.org/10.1021/es900967g] [PMID: 19746714]

[124]   Master, A.; Livingston, M.; Sen Gupta, A. Photodynamic nanomedicine in the treatment of solid

tumors: perspectives and challenges. *J. Control. Release,* **2013**, *168*(1), 88-102.
[http://dx.doi.org/10.1016/j.jconrel.2013.02.020] [PMID: 23474028]

[125] Yallapu, M.M.; Dobberpuhl, M.R.; Maher, D.M.; Jaggi, M.; Chauhan, S.C. Design of curcumin loaded cellulose nanoparticles for prostate cancer. *Curr. Drug Metab.,* **2012**, *13*(1), 120-128.
[http://dx.doi.org/10.2174/138920012798356952] [PMID: 21892919]

[126] Bujdáková, H. Management of Candida biofilms: state of knowledge and new options for prevention and eradication. *Future Microbiol.,* **2016**, *11*(2), 235-251.
[http://dx.doi.org/10.2217/fmb.15.139] [PMID: 26849383]

SUBJECT INDEX

www.ingramcontent.com/pod-product-compliance
Lightning Source LLC
Chambersburg PA
CBHW041726210326
41598CB00008B/792